Springer Series in Advanced Manufacturing

Series Editor

Duc Truong Pham, University of Birmingham, Birmingham, UK

The **Springer Series in Advanced Manufacturing** includes advanced textbooks, research monographs, edited works and conference proceedings covering all major subjects in the field of advanced manufacturing.
The following is a non-exclusive list of subjects relevant to the series:

1. Manufacturing processes and operations (material processing; assembly; test and inspection; packaging and shipping).
2. Manufacturing product and process design (product design; product data management; product development; manufacturing system planning).
3. Enterprise management (product life cycle management; production planning and control; quality management).

Emphasis will be placed on novel material of topical interest (for example, books on nanomanufacturing) as well as new treatments of more traditional areas.

As advanced manufacturing usually involves extensive use of information and communication technology (ICT), books dealing with advanced ICT tools for advanced manufacturing are also of interest to the Series.

Springer and Professor Pham welcome book ideas from authors. Potential authors who wish to submit a book proposal should contact Anthony Doyle, Executive Editor, Springer, e-mail: anthony.doyle@springer.com.

More information about this series at http://www.springer.com/series/7113

Sotiris Makris

Cooperating Robots for Flexible Manufacturing

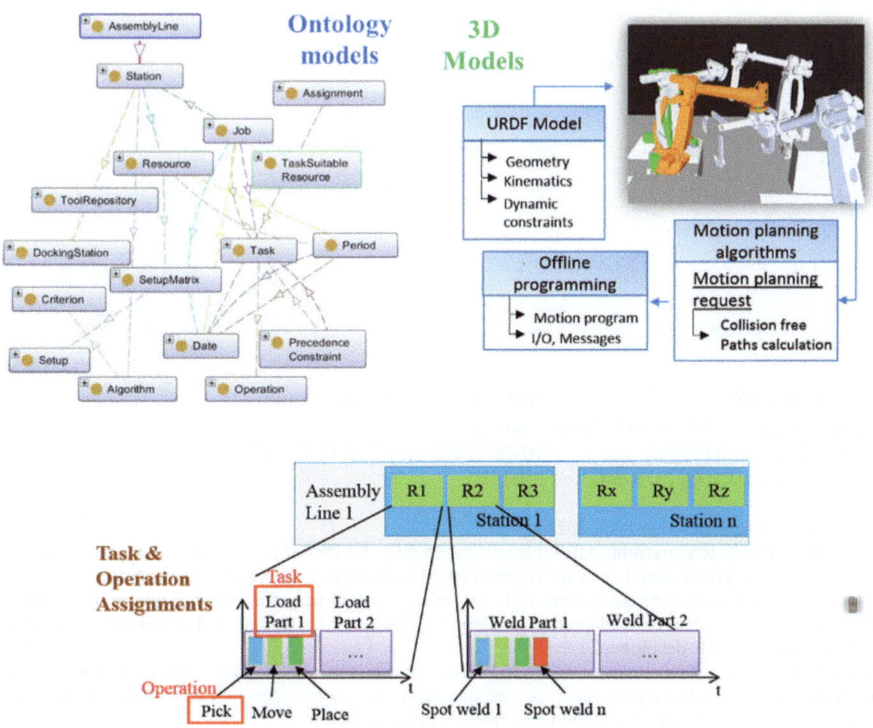

Springer

Sotiris Makris
Laboratory for Manufacturing Systems
and Automation, Department for Mechanical
Engineering and Aeronautics
University of Patras
Patras, Greece

ISSN 1860-5168 ISSN 2196-1735 (electronic)
Springer Series in Advanced Manufacturing
ISBN 978-3-030-51593-5 ISBN 978-3-030-51591-1 (eBook)
https://doi.org/10.1007/978-3-030-51591-1

This Springer imprint is published by the registered company Springer Nature Switzerland AG
The registered company address is: Gewerbestrasse 11, 6330 Cham, Switzerland

Ει γάρ ηδύνατο έκαστον των οργάνων (κελευσθέν ή προαισθανόμενον) αποτελείν το αυτού έργον, ουδέν αν έδει ούτε τοις αρχιτέκτοσιν υπηρετών, ούτε τοις δεσπόταις δούλων

Aristotele, Politics

καὶ τὸν ἐπιχειροῦντα λύειν τε καὶ ἀνάγειν, εἴ πως ἐν ταῖς χερσὶ δύναιντο λαβεῖν καὶ ἀποκτείνειν

Plato, Republic

To Sophia, John and Dimitra

Preface

This book is the derivation of my research work at the Laboratory for Manufacturing Systems and Automation, University of Patras in Greece. It is based on close interaction with both industry and research colleagues. There have been many great ideas conceived over these years of research on the topic of cooperating robots for flexible manufacturing. This book writing is an effort to share these experiences with a broader audience in a format that is concise enough to summarize many different concepts but also provides adequate detail and references to allow those interested to follow similar a direction.

This book aims to consolidate the content of discussions and experiences with experts, practitioners and engineers in manufacturing systems. With the great variety of manufacturing systems, we had the opportunity to study a noteworthy set of methods and tools have been produced. The aim of the book is sharing this experience with academia and industry practitioners hoping to contribute to improving manufacturing practice. While there is a plethora of books detailed enough to teach principles of robotics, this book offers a unique opportunity to dive into the practical aspects of implementing real-world complex robotic applications.

The term "Cooperating robots" in this book refers to robots that either cooperate between themselves or cooperate with people. The book investigates aspects of cooperation towards implementing flexible manufacturing systems. Therefore, manufacturing systems are the main driver behind the discussion on implementing such robotic systems.

Numerous methods have been proposed to design and to operate manufacturing systems. This book aims to introduce a novel set of methods for designing and operating manufacturing systems consisting of cooperating robots. Many methods are available in the literature on designing or operating robotized manufacturing systems when the main priority is efficiency and robustness; the essential element of these robotized manufacturing systems is the need for flexibility.

Initially, the concept of a manufacturing system will be briefly introduced, followed by aspects of flexibility. Following the key aspects of robotic systems will be introduced while the discussion will be streamlined towards implementing systems of cooperating robots towards flexible manufacturing systems. Aspects of designing such systems, such as considering material flow, logistics, processing times, shop floor footprint and design of flexible handling systems, are going to be discussed. Additionally, key issues in operating such systems involve decision making, autonomy, cooperation, communication, task scheduling, motion generation and distribution of control at the control level of different devices among others.

The book consolidates knowledge published in papers with co-authors; however, it introduces several novel concepts that have not been published before. It presents a number of chapters in the form of technical papers discussing industrial challenges and approaches taken. These chapters are organized in four major parts. Part I introduces the topics of cooperating robots in two dimensions. On the one hand, there are the topics of robots cooperating among themselves, and on the other hand, there are the aspects of robots cooperating with humans. Part II includes aspects of robot to robot cooperation. Part III elaborated aspects of collaborative robotics, namely humans cooperating with robots. Part IV summarizes with an outlook for the future.

I would like to thank the European Commission for the financial support in my research over the years. Thanks to a number of funding programs, namely the Factories of the Future program and the Robotics initiative it has been possible to facilitate this work in cooperation with the European industry and realize the journey towards implementing the vision discussed throughout this book. Moreover, I am grateful for a number of leading European manufacturing companies for the great cooperation, namely Daimler, PSA, FCA, Volvo, Ford, Bic Violex, Electrolux, Comau, Prima Industrie, Siemens, Festo and Pilz. I have enjoyed extensive and fruitful conversations over the years.

My very good friends and associates in our Laboratory for Manufacturing Systems and Automation, in the University of Patras, have been invaluable in both performing research activities and also achieving this first edition of the manuscript. Special thanks to the colleague and friend, Dr. George Michalos for his persistence and hard work in many fronts and for sharing the vision for researching the issues of cooperating robots in industry.

In addition, I am grateful for the support and good cooperation with Professor Dimitris Mourtzis from the manufacturing systems group in LMS, Dr. Kosmas Alexopoulos from the software development group in LMS and Professor Panagiotis Stavropoulos from the manufacturing processes group in LMS.

I would like to thank the great team I had the opportunity to cooperate with over the years, in the Laboratory for Manufacturing Systems and Automation, in the University of Patras. They have been a great help in developing these topics, and I am grateful for their passion and commitment; Niki Kousi, Panagiotis Karagiannis, Panagiotis Aivaliotis, Apostolis Papavasileiou, Dionisis Andronas, Christos Gkournelos, Stereos Matthaiakis, Andreas Sardelis, Konstantinos Dimoulas, Nikolaos Nikolakis, Charalampos Kouros, Spyros Koukas, Plato Sipsas, Evangelos Xanthakis, Dr. Loukas Renztos, Dr. Konstantinos Efthymiou and Dr. Panagiota Tsarouchi have greatly contributed to the work behind the actual manuscript over the years. There are many others not being mentioned, and I would like to thank them all.

Above all, I would like to thank my teacher and mentor, Prof. George Chryssolouris, for his encouragement and enlightenment over the years.

Finally, I would like to greatly and warmly thank my wife, Sophia, for the tolerance and support over the years as well as my children, John and Dimitra, wishing them a bright future.

Patras, Greece
June 2020

Sotiris Makris

Contents

Part I

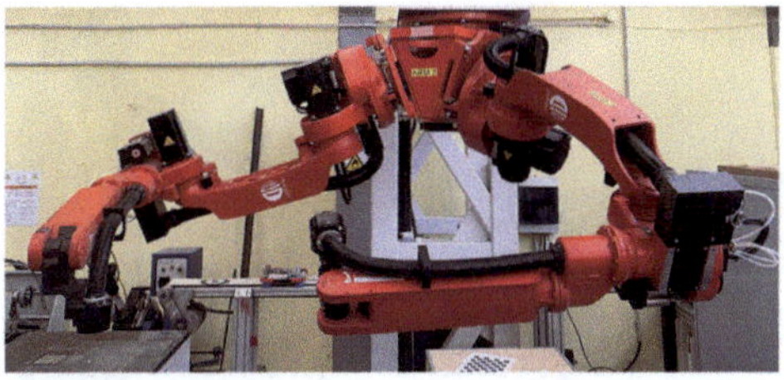

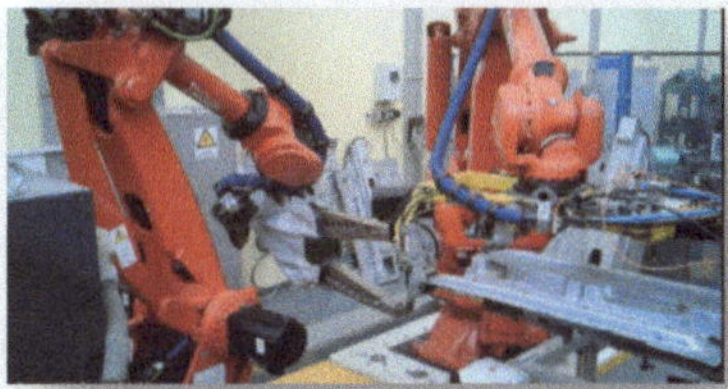

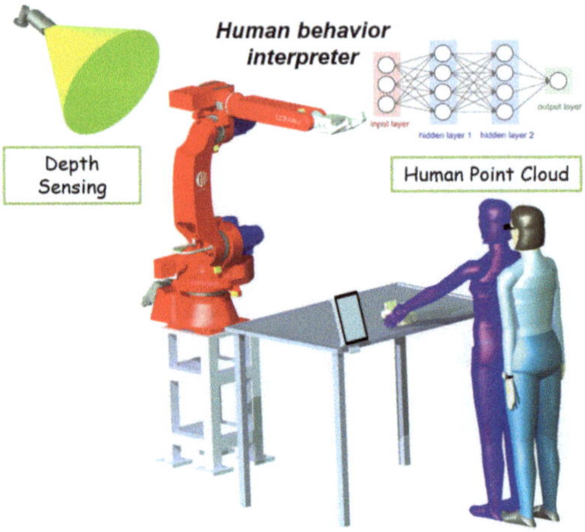
Human behavior interpreter
Depth Sensing
Human Point Cloud
input layer
hidden layer 1
hidden layer 2
output layer

Chapter 1
Introduction to Cooperating Robots and Flexible Manufacturing

1.1 Overview

Automation is a central competitive factor for large manufacturers but is also becoming increasingly important for small and medium-sized enterprises around the world. Growth in robots installation in advanced manufacturing regions of the World, namely Europe, Japan, China, USA is closely linked with increase of industrial production and therefore number of jobs. There is a stronger economic driver for adopting robots in higher-wage economies than there is in lower-wage economies. Atkison investigated where nations stand in robot adoption if we consider wage levels. He found that Southeast Asian nations significantly outperform the rest of the world, and Europe and the United States lag significantly behind [1]. Therefore, the driver for robotics is not always the one of cost reduction but there are also other factors. The need for robots is constantly increasing due to labor shortage, low birthrate, the aging population including developed countries, the demand for reducing quality errors in production, and the removing heavy and repetitive work, from humans aiming to enhancing the quality of the working environment [2].

Considering the global population, one may identify plenty of opportunities to reach new markets and introduce new products. Increasing from a 3 Billion global population in 1960, to 3.6 Billion in 1970, 4.4 Billion in 1980, 5.2 in 1990, 6.1 in 2000, 6.9 Billion in 2010 reaching to 7.5 Billion in 2018, the growth is steady and continuously increasing at a high rate [3]. However, current production methods need to be seriously advanced in order to address such a global market. Increasing production rate, lowering the cost of ensuring product quality and achieving competitive delivery times can only be achieved by increasing automation. In this perspective, employing robots is a key enabler of a new manufacturing paradigm.

This book aims to consolidate the content of the research, investigation and experiences with experts, practitioners and engineers in applying cooperating robots technology in manufacturing. With the great variety of manufacturing systems that we had the opportunity to study a noteworthy set of methods and tools have been produced.

S. Makris, *Cooperating Robots for Flexible Manufacturing*,
Springer Series in Advanced Manufacturing,
https://doi.org/10.1007/978-3-030-51591-1_1

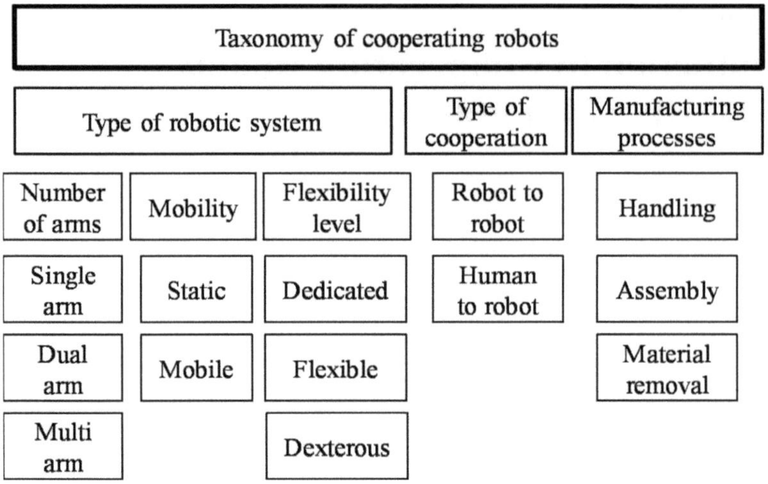

Fig. 1.1 Taxonomy of cooperating robots' systems addressed in the book

The main objective is sharing this experience with academia and industry practitioners hoping to contribute to improving manufacturing practice. While there a plethora of manuscripts detailed enough to teach principles of robotics, this text offers a unique opportunity to dive into the practical aspects of implementing real world complex robotic applications.

The term "Cooperating robots" in this book, refers to robots that either cooperate between themselves or cooperate with people. The attempt made herein is to investigate aspects of cooperation towards implementing flexible manufacturing systems. The level of flexibility to achieve by each configuration is a matter of research and several these aspects are discussed with the help of real world industrial examples. The taxonomy of systems covered in the book is summarized in Fig. 1.1, which aims to help he reader navigate through the possibilities offered in in our days.

The level of cooperation can be considered through a number of combinations of robotic systems as shown in Fig. 1.2. Increasing the level of cooperation between robots can help achieve higher autonomy and more intuitive task sharing. On the other hand, introducing humans cooperating with robots allows for greater flexibility due to involving the problem solving capabilities of the human in cooperation with the strength of robots. Finally, combining the ability of robots to move, cooperate between themselves and cooperate with hums offers the maximum levels of flexibility possible today.

Chapters in part 2 of this book presets a number of approaches on the topic of robot to robot cooperation. Part 3 discusses a number of approaches on human–robot collaboration. Finally Part 4 aims to summarize the discussion and discuss the outlook for the future.

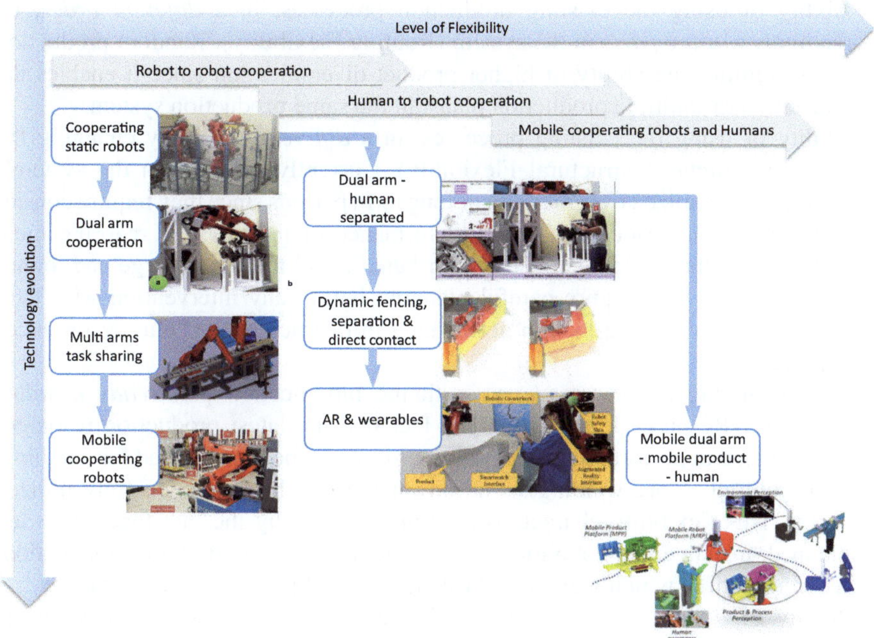

Fig. 1.2 Flexibility achieved with increasing the level of interaction

1.2 Cooperating Robots for Flexible Manufacturing

The introduction of the serial production has been a breakthrough for manufacturing since it allowed to increase production volumes and reduce costs of production [4, 5]. After several decades of applying this paradigm, the world prosperity has grown to unprecedented levels. Modern industrial practice today is to apply robots wherever there are repetitive tasks in the manufacturing process. Robots offered a number of advantages, namely Quality improvement, Workers' health by undertaking unhealthy manufacturing operations and Reduction of throughput time [6].

However, in order for these benefits to be achieved, production systems make use of rigid flow line structures employing model-dedicated handling/transportation with fixed control logic signal—based tasks sequencing that requires high manual effort for changes. These assembly systems make the line of introducing/modifying a product cumbersome and this comes into conflict with the diversification sought after by production firms.

Introducing or varying a product in the production line means that the process plan of the specific product can be accommodated by the current line setup. Four main directions can be followed for the successful adaptation of the new production requirements [7]:

- Ability of using the existing production processes, in a *different order*, by randomly *routing parts* in the system (Routing Flexibility). Random production flows signifies the ability of higher product diversification since it enables the realization of multiple production plans, in the same production system.
- Ability of using the existing processes, in a *different order*, by changing the *system's structure* (Structural Flexibility). Currently, changes in the system's structure are realized over medium or long term periods since they require considerable time and resources for the performance of the physical rearrangement and setup of the equipment. This is attributed to (a) the use of large and immobile resources that require careful planning before any intervention with their installation and (b) the lack in networking infrastructure for "plug & produce" approach.
- Ability of *adding new processes* with the modification of *already installed resources* (Resource Flexibility). The fact that the latest production resources are designed to perform multiple processes (e.g. a robot can be used for handling, spot welding or arc welding) is an advantage that has not been fully utilized so far. This flexibility characteristic is undermined by the fact that robots are planned, installed and not allowed to change their roles until the next shop floor reconfiguration, which may take place after several months or even years.
- Ability of *adding new processes* with the *addition of resources* (Expansion Flexibility). Should a system require the adoption of new production processes not having been implemented before, it will have to introduce new resources. Although simple it may sound, there are many complications to be handled and are related to (a) ensuring the required installation space, (b) minimizing the installation time, (c) handling the complexity of the integration with the existing control systems and (d) maintaining cost efficiency. All these signify a wide time frame that is not satisfactory.

Figure 1.3, depicts a qualitative comparison between a conventional line and a line with mobile cooperating robots. The visualization in this figure is mainly based on the following advantages that mobile cooperating robots can offer:

- Reduction in reconfiguration time—The system's structure can be changed with the relocation of the mobile units, thus obtaining the structure that better suits the production at each period. The same resources are used in different areas to perform a range of processes. The reconfiguration should take place in minutes or hours rather than days or weeks.
- Enhanced system reliability and reduction in breakdown times. Considering the fact that in serial assembly lines such as those in the automotive industry, there are very small or no buffers at all between the stations, any resource breakdown can result in a stoppage. Since the time required for the repair is based on the type of malfunction, the mobile robots can promptly replace the problematic resources.
- Reduction in commissioning time. The advantage of mobile manipulators is that they include standardized mechanical and electrical interfaces allowing them to plug into the system, setup their parameters and start operating. The installation of fixed robots requires a line stoppage.

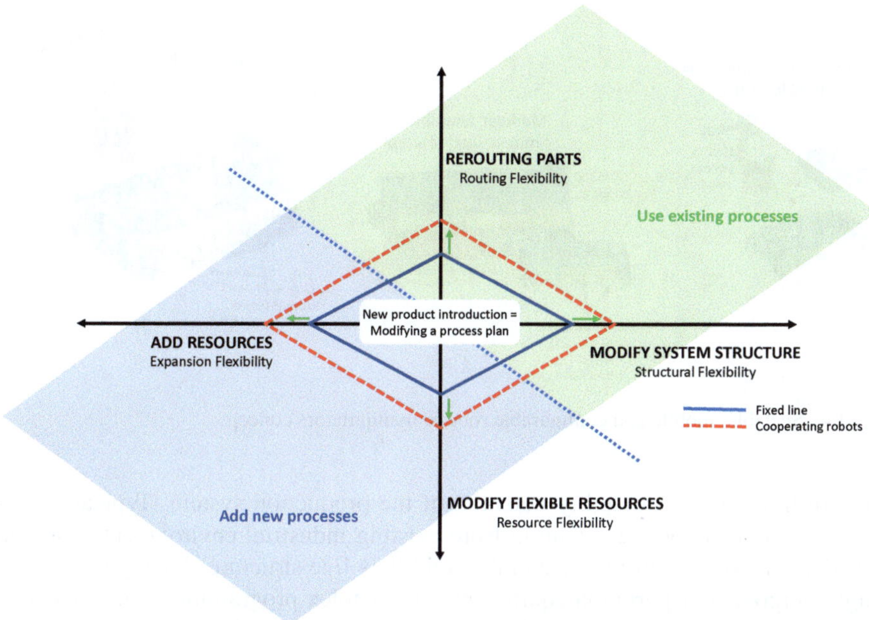

Fig. 1.3 A model for flexibility assessment using cooperating robots

- Reduction in the cycle time through the minimization of picking/placing operations. The use of flexible and reusable tooling can eliminate the existing stationary tooling. In this sense, the products will be continuously handled by the robots thus reducing any extra handling operations.
- Enabling higher product variability through robot to robot handling. The aforementioned ability for parts to be transferred between the robots, overcomes the limitation of fixed, on-ground, tooling with respect to product routing. This can be translated into higher plant and product variability.
- Reducing planning and control efforts by automated task allocation and resource integration. This means that the system will have to decide on the reaction steps to be followed by evaluating its state and capabilities. The autonomous resources will be able to decide about the kind of task to be undertaken and then automatically navigate them to the specific area, plug into the system and carry out the task.

1.2.1 Flexible Production Systems with Cooperating Robots

Cooperating robots offer a radically new paradigm for upligting the capability of manufacturing systems to deal with the flexibility requirements discussed so far. Such a new paradigm aims to radically advance the way that today's production lines are conceived, designed and built allowing the elimination or reduction of technologies

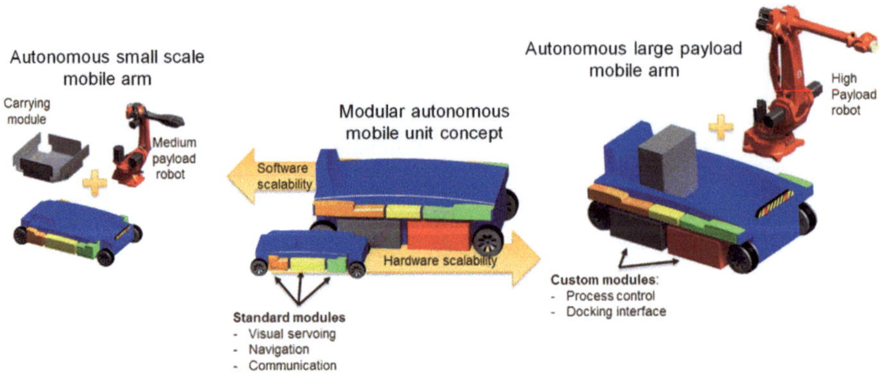

Fig. 1.4 Scalable, mobile and configurable robotic manipulators concept

that are limiting the flexibility potential of the production system. Typical examples of these technologies, coming from existing industrial environments, include the adopted fixed control logic and the rigid flow line structures having expensive, single-purpose transportation equipment and complex programming methods. The thesis made and demonstrated in Part 2 of this book is towards introducing a radical change in the paradigm: in the same plant the sequence can be changed by introducing autonomous production/handling units which can change task (from joining to handling and vice versa) and position (around the shop floor), eventually cooperating among themselves, based on current process sequences, auto reconfiguring themselves, the tools and the line to answer quickly to internal/external demands. Based on the above, the aim is to radically advance the way that today's production lines are conceived, designed and built allowing the elimination or reduction of technologies that are limiting the flexibility potential of the production system.

The underlying concepts for realizing this overall vision are the following:

- Mobile and Reconfigurable robotic devices. By employing mobile manipulators that are able to perform different manufacturing processes and navigate across the shopfloor, the system shall be able to change its structure in an automated way aiming at the optimization of the changeover times and costs. Scalability and reconfigurability are the most important characteristics of the mobile units in order to be deployable to a wide application range. This requires the generic design of mobile platforms using standard software and hardware components as shown in Fig. 1.4.

Given the diversity of the application of this technology, the robotic components will be driven by considering the optimization of:

- Mechanics design
  - Docking subsystems for the mobile units for enabling the execution of precise operations by the robot arm, such as welding or handling, ensuring accuracy and repeatability.

- Structure: allowing the integration of required process equipment on the mobile unit. This involves the design of the mobile unit main structure so that it can support the high payload robot together with the controller and peripheral/auxiliary equipment for executing the assembly/handling process.
- Dimensions: ensuring a trouble free navigation and manoeuvrability in tight industrial environments.

- Control architecture for autonomous operation of mobile robotic devices [8]:
  - Exploration and mapping: The mobile units will need to get knowledge of their surroundings and need to be able to maintain maps of dynamic environments over longer periods of time. These maps will identify key environmental features that will allow the platform to determine its whereabouts. Occupancy is a 2D representation of the tessellated space and laser scanners can be used to create such grids.
  - Localization in dynamic environments: The effort will be towards increasing the accuracy of localisation on maps constructed from sensed data and to be able to merge and combine multi-scale maps derived from different data sets. Odometric pose estimation combined with visual odometry and Inertial Measurement Unit (IMU) can help get a rather precise odometry which nonetheless has only relative localization information and accumulates drift.
  - Navigation: Software modules for navigation, obstacle avoidance and path planning will be developed in an architecture specific for the task. The moving robot needs to be able to navigate its way to the station, avoiding static and mobile obstacles and place itself with precision in the working area. This involves the development of global and local planners based on occupancy grids.

- Robot arms optimized for mobile applications as shown in Fig. 1.5.

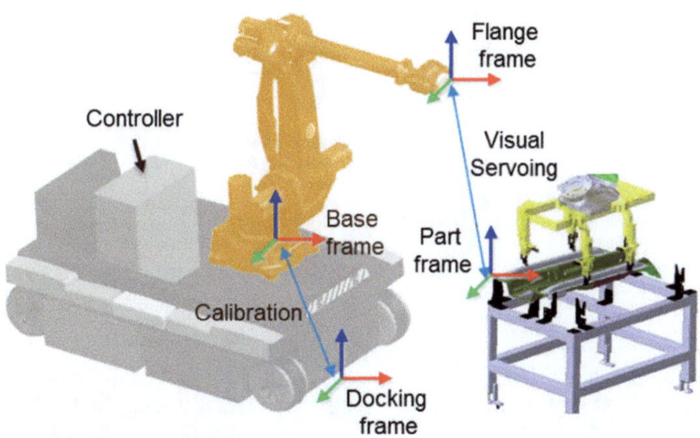

Fig. 1.5 Robots for mobile applications

- Achieving open and compact controllers, fitting on a mobile platform and not interfering with robot workspace.
- Higher accuracy—precisely bringing the robot end effector to a certain position, compensating for errors due to the mobile nature of the problem (robot base calibration). Additionally, vision guided control can help achieve precise positioning even in uncontrolled environments based on data from a 2D camera.
- Advanced stiffness/compliance control of robot arms, making the robot suitable/efficient in handling contact tasks. For this purpose the sensing and control functionalities need to be directly integrated into mechanical structures. Moreover fault-tolerant and resilient control methods are needed to minimize the required human intervention.

• Distributed re-planning and control. In terms of planning, it needs to be able to optimize achieving both vertical (from production control down to resource level) and horizontal (resource to resource) integration. Using the abstracted Line, Station and Robot level the production system needs to be modeled, allowing the deployment in highly different environments. Figure 1.6 provides a graphical representation of these levels along with the functionalities required by each one of them and the tools required to implement these functionalities. More specifically the hierarchical control levels include:

  - Line level re-configuration: This control level, undertakes the determination of the required reorganization—action planning of the work in the line. Tools for automatically identifying the proper station where a mobile robot should be deployed will be realized.
  - Station level organization: The focus of this level is towards distributing the work in the resources within the station. At the station level, decision making

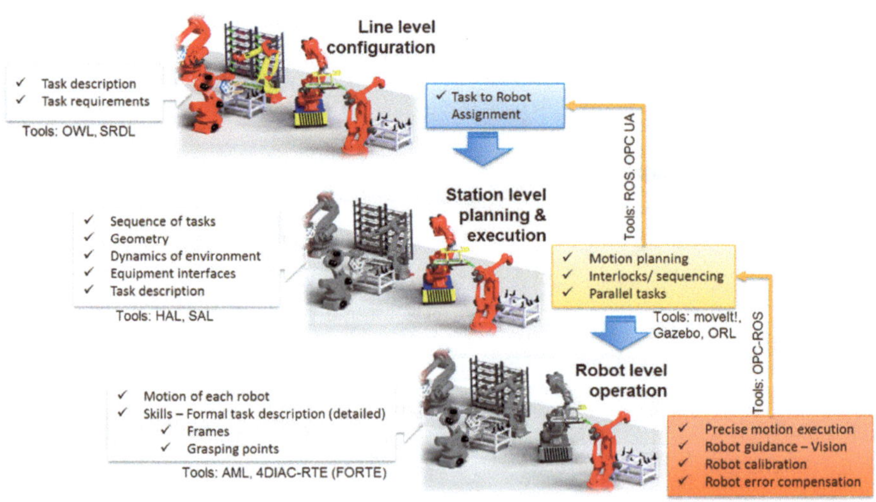

Fig. 1.6 Multi level distributed planning and control

logic enables the integration of planning (action and motion) in the execution loop to address changing, complex and unpredictable application domains, especially those where multiple types of robots operate.
- Unit level control: The lowest control level provides the tools for adjusting the robot behavior in order to execute the station level control derived tasks (avoid obstacles, perform contact tasks, perform accurate motion, use data from sensors, exchange tools between robots). Additionally, tools for efficiently modelling resources and processes will be of primary importance in terms of.

• Integration & communication architecture to allow easier integration and networking of the control systems utilizing agent-based, web-services and ontology technologies. The architecture involves all the mechanisms for communication and message exchange that allows to:
  - plug new components/robot allowing their automatic set-up and operation,
  - enable robot to robot co-operation, allowing robot-based product parts transfer without the use of traditional transfer lines or fixed on-ground tooling.

Figure 1.7 describes in a higher level of detail the structure of such an open architecture as well as the decentralized nature of the local resource controls schemes that usch a paradigm enables and these asects are thoroughly discussed throughout the book.

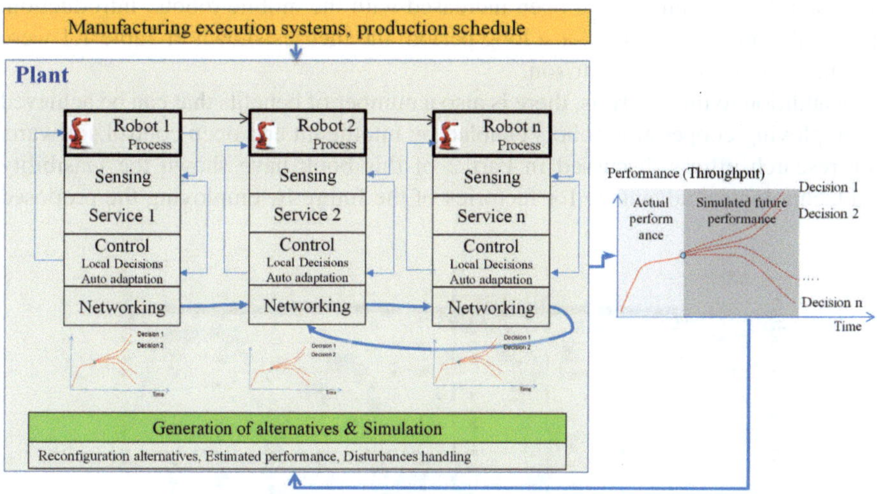

Fig. 1.7 Service oriented control of an assembly line with cooperating robot

1.2.2 *Efficiency Aspects of Cooperating Robots Technology*

In a number of case studies, the serial assembly line was compared to setups comprising cooperating robots. In the example of Fig. 1.8, there is a presentation of the daily demand profile [7]. The study modeled with the help of discrete event simulation tecnhiques the performance of a serial assemby line and compared it with a system of cooperating robots. The production capacity of both systems exceeds the maximum volume of the demand profile. This is a common practice used to accommodating fluctuations in the demand. Everyday failures of the resources result in parts being gathered into the buffers thus, increasing the future daily demand. As a result, the systems use their extra capacity to absorb this extra demand. The fact that the serial line presents more frequent breakdowns results in a lower availability and production volume. Thanks to the mobile cooperating robots, the new system can approximately produce more than a hundred parts per day.

The enclosed area in Fig. 1.8 highlights the daily production of both systems during facelift period, where changes are introduced to the production system in order to accommodate a product facelift. The area is enlarged in Fig. 1.9. The production of the new paradigm exhibits a faster recovery. This more efficient performance is attributed to the introduction of the mobile cooperating robots that reduce the time required for modifications.

As it can be observed, the production volume and the utilization as well as the system's availability have been increased with the mobile robots' introduction. Having the models running for a long period, the metrics shown in Table 1.1 were obtained and used for comparison.

In addition to this analysis, there is also a number of benefits that can be achieved by employing cooperating robots enabled by intelligent and open control software. The research efforts discussed in Part 2 of this book have shown the feasibility and technological advantage for factories of the future by employing the proposed

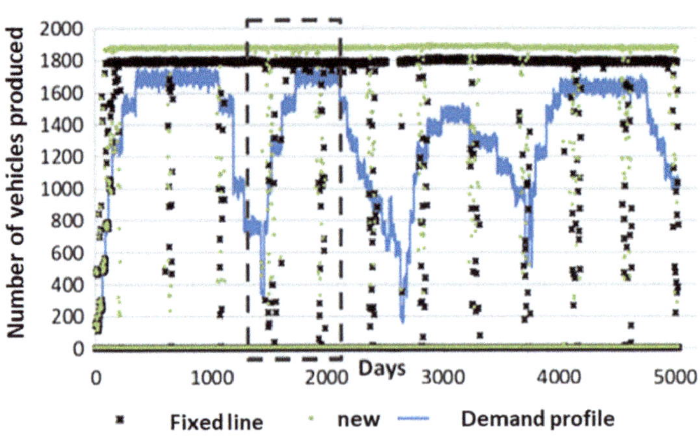

Fig. 1.8 Daily production of **a** the conventional and **b** the new paradigm

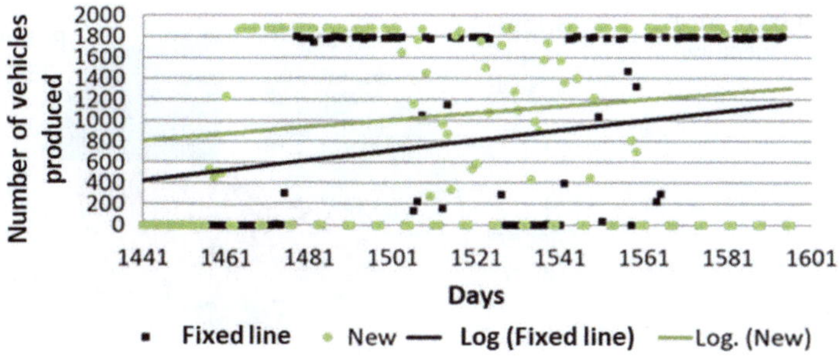

Fig. 1.9 The daily production of the two systems for the facelift period

Table 1.1 Comparison of the two types of production systems

KPI	Fixed line	Cooperating robots	Increase (%)
Number of vehicles	5.636.925	6.339.134	10.7
Resource utilization (%)	68.76	76.518	7.758
Availability (%)	90.59	96.328	5.738

methods. Applying cooperating robots in a variety of industrial sectors, has been challenged due to the highly dynamic nature of the production environment, which is imposed by the need for multi-variant production in a random production flow in industrial sectors like the automotive. In addition, considering more industrial sectors with large number of subcomponents that need to be handled at high speed, high quality requirements that impose high precision in handling and control has helped to explore how the technology can become more mature for industry wide adoption.

Today's main challenges in adopting new technologies involve the lack of open and robust software and the lack of standardized interfaces towards sophisticated sensors [9, 10]. The highly reconfigurable solutions discussed in this book integrated under the open architecture provide the means to reduce effort and complexity that they offer during the introduction of new equipment. Openness is achieved in two dimensions, hardware and software.

- In the hardware level, interoperable and exchangeable end effectors are able to be exchanged among production units/robots according to standardized interfaces. The mobile nature of the production units, offers the potential for and standardized hardware architecture, for enabling a highly cooperative nature of production equipment.
- In the software level, the control technology utilizes the service oriented architecture that helps standardize the architecture of control software. In addition, this control architecture enables a standardized mechanism of communication among

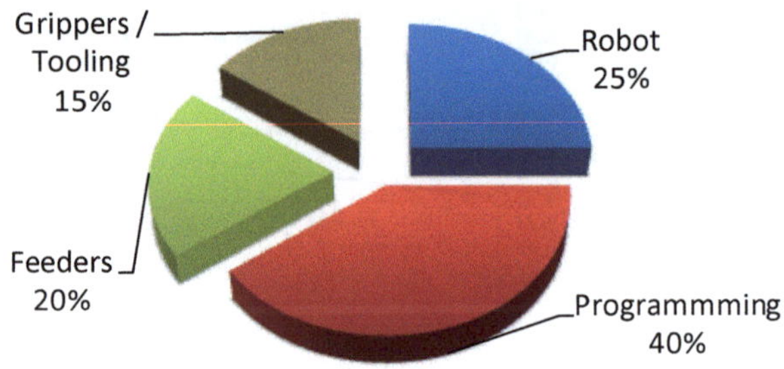

Fig. 1.10 Cost breakdown for an assembly operation

production units, and results in a production environment of extreme cooperative potential, thus the manufacturers will be able to operate more openly (Fig. 1.10).

Openness to adopt new technologies is heavily affected by the cost of the technologies to be introduced. For instance, in industrial robotic assembly application, the cost breakdown is shown in Fig. 1.10. The minimization of programming effort and time pursued by the methods discussed in this book, affects the biggest part of the cost pie (programming—40%) thus making the introduction of such technologies more appealing [9].

Employing this technology, results in a number of benefits, for example producing customized products is becoming affordable since faster and cost-effectively smaller lot sizes of diverse products is becoming possible. Production of goods of higher quality is also feasible since the robotic technology helps to improve the quality of products on the one hand by enabling producers to make more and cost-effective optimizing steps and on the other hand, to react faster on errors or uncertainty.

Moreover, intelligent control of machinery can help to achieve better energy management. New tooling technology can result in significant energy savings. For example eco-efficient clamping saves up to 45% compressed air [11]. It can be calculated that for 5.000 clamps at an assembly line level a total of 0.5 t less CO_2 can be expected in a day, which corresponds to 125 t less CO_2 in a year. Additionally, scrap material and parts can be reduced, following the collaborative quality control and monitoring that the advanced autonomous systems can offer.

1.3 Human–Robot Collaboration

While the concept of robot to robot cooperation has been studied and it is expected to offer breakthrough advances, the manufacturing industry has been considering an intermediate step where several human operations are assigned to robots while others are still managed by people. There are numerous industrial applications where the assembly process is mainly performed by human operators due to the fact that (a) operations require a human like sensitivity, (b) handled materials are different and often show a compliant, unpredictable behavior and (c) often more than one operators are active in each station to perform cooperative or parallel operations. Nevertheless, the automation of operations in manual assembly stations and lines is highly demanded so that quality levels are increased mainly in terms of precision and repeatability, throughput time is decreased in assembly stations, traceability of the performed operations is allowed and in the meanwhile the ergonomic workload for the operators is reduced.

Industrial automation systems for assembly operations have to integrate the needed human capabilities with the characteristics of robotic automation such as strength, velocity, predictability, repeatability, precision and so on. Additionally the introduction of robots to support assembly operators reduces the need for physical strength especially in the cases of large part assembly such as in the capital goods industry. Therefore it is possible that people of greater age can continue to work inside the production facility, undertaking mostly the cognitive tasks (coordination, troubleshooting etc.) which benefit by their large experience and intuition while letting the robot handle the physical requirements.

Nevertheless the advantages of industrial robots are not exploited in their full potential within the production plants. Rather than automating the relevant assembly processes, the concept of Human–robot collaboration promotes a hybrid solution involving the safe cooperation of human operators with autonomous and self-learning/adapting robotic systems through a user-friendly interaction. The synergy effect of the robot's precision, repeatability and strength with the human's intelligence and flexibility will be much greater especially for the case of small scale production where reconfigurability and adaptability are of great importance. In this direction the concept of human–robot collaboration works towards achieving the following technology directions as shown in Fig. 1.11.

- Intuitive interfaces for safe human–robot cooperation during assembly and heavy part manipulation by using sensors, visual servoing, speech recognition, advanced control algorithms (such as force/impedance control) to regulate the manipulation of the parts by the robots and close the gap between the human and the robot in the assembly line.
- Introduction of advanced safety strategies and equipment allowing fenceless human robot assembly cells. Different levels of interaction will be supported: common workspace sharing, small scale cooperation outside the task and joint human robot assembly task execution.
- Robust methods and software tools for determining the optimal planning of assembly/disassembly operations using a multi-criteria, simulation enabled approach. Ergonomics, resource utilization and safety will be the prevailing criteria for designing the hybrid production process.
- Simplified and user-friendly robot programming by means of: (a) Programming by Demonstration (PbD) and (b) Robot instructions libraries which allow the robot program to be incrementally and automatically created.
- Introduction of mobile robots acting as assistants to the human operators. The scope involves autonomously coordinated mobile platforms for supplying parts

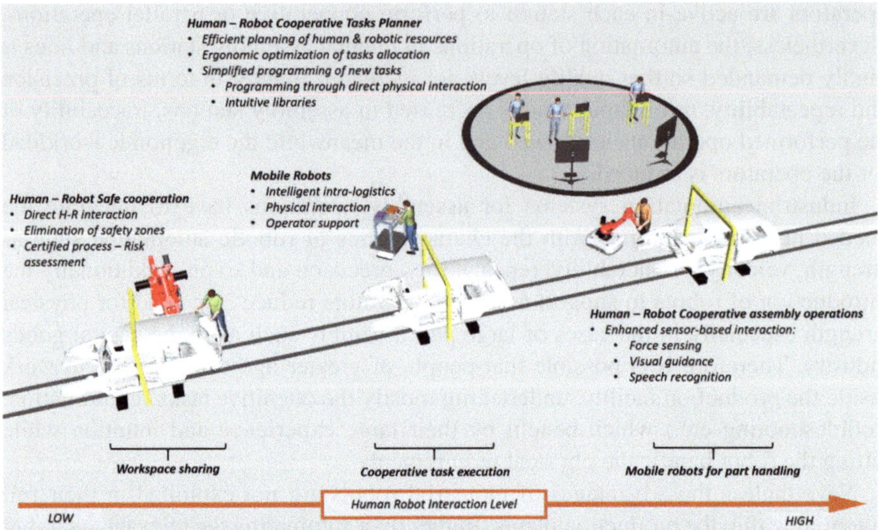

Fig. 1.11 Concept of human–robot collaboration

to the assembly line and high payload and overhead, general purpose handling robots with advanced navigation and interaction capabilities.

- Introduction of more flexible integration and communication system for the shared data (both control and sensor) by utilizing a distributed computing model and ontology services, in order to properly distribute the acquired data to every relevant resource, networking all possible resources and link them all for higher level coordination by the Task planner.

These aspects are discussed in detail in Part 3 of this book with the help of a number of real world applications. Facing the highly demanding industrial challenges is possible by pursuing the following scientific and technical technology streams.

1.3.1 Safe Human–Robot Cooperative Assembly Operations

1.3.1.1 Human Robot Interaction (HRI)

Methods for HRI intend to enable the cooperation of humans and robots during the execution of the assembly task at different cooperation levels. Figure 1.11, shows three indicative cooperation cases.

- The first one involves the concurrent execution of different assembly tasks by the robot and the human, while sharing a common workspace. No fences or other physical safety devices need to be present while the robot is always aware of

Fig. 1.12 Human robot interaction concepts—multi modal interfaces

human presence by utilizing a plethora of force/vision/presence sensors. This allows it to implement a safety- first behavior,

- In the second level the cooperation is carried out mainly in the cognitive level since the mobile robot can provide the operator with the correct parts for the assembly thus reducing the time to identify and retrieve them from areas far from the assembled product,
- The final level of cooperation is the execution of the same assembly task by the robot and the human which are in direct physical interaction. The approach allows to combine human skills like perception and dexterity with robot strength, accuracy and repeatability into efficiently performing the same task. The involvement of the robot also allows the automated quality check through the robot sensors.

In order to realize the direct human–robot cooperation concept a number of advanced control algorithms and multi modal interfaces are required in order to regulate the movement of the part by both operator and robot [12]. For example the operator is capable of moving the robot TCP (tool center point) with bare hands, exploiting force sensors and standardized voice commands or gestures to perform any additional functionality. At the same time the robot carries the part's payload and ensures the collision free path through virtual windows. State of the art voice recognition engines allow to transform the most common operator commands (e.g. point definition, gestures, voice instructions etc.) to robot programming code (e.g. PDL2). The HRI concepts are graphically shown in Fig. 1.12.

1.3.1.2 Human Safety

Given the safety issues that arise from the coexistence of robots and humans, several means for detecting/monitoring human presence and adjusting the behavior of the robots are required. This is possible by utilizing a combination of industrial sensors available in the market and robot control technology [13].

Since industrial robots are normally large, move fast and carry heavy or hazardous parts, a collision with a human being could cause severe injuries, even the death. Current manufacturing practices require complete physical separation between people and active industrial Robots (typically achieved using fences or similar physical barriers) as precautions to ensure safety. Overcoming the inefficiency in terms of time and resources precautions on the types of tasks that can be performed by introducing new sensors, robust path planning etc., for ensuring the safety of people in close proximity to robots in an industrial work cell, or even in contact with it. In this context, the system is expected to be able to i.e. adjust the speed of the robots upon the detection of humans in the work cell. Ways of adjusting the trajectory in real time in order to avoid collisions so that the process is not interrupted will also be explored, utilizing the latest advances in robust robot control systems and algorithms.

1.3.1.3 Human Robot Cooperative Task Planning

Under this technology stream, the focus is on the derivation of robust methods for determining an efficient planning of assembly/disassembly operations, utilizing to the highest possible extent the capabilities of both the human and the robot. Towards this direction and referring to Fig. 1.13, the Task Planner implements the following functionalities:

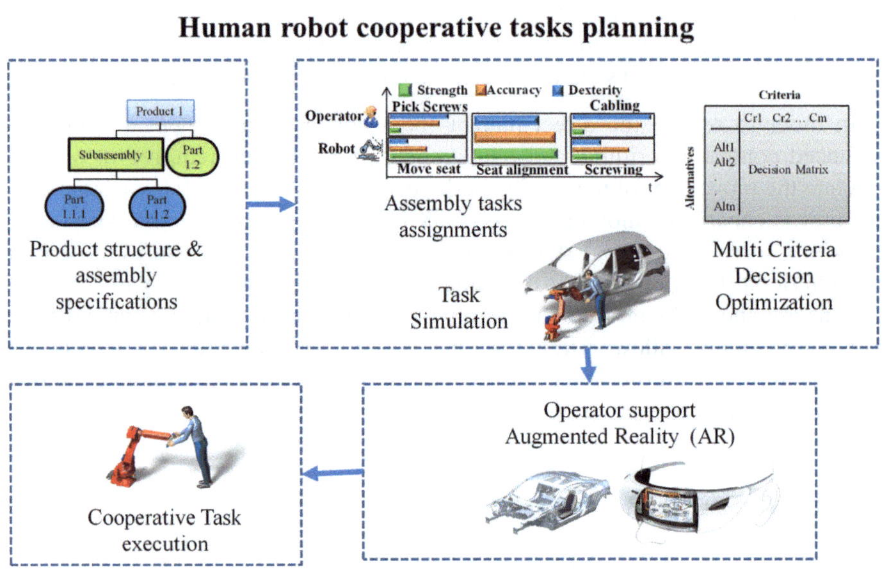

Fig. 1.13 Human robot cooperative task planner concept

- Based on the product structure and the assembly specifications, the extraction of assembly tasks and the related requirements (physical strength, accuracy etc.) will take place.
- Following these requirements, the planning of the assembly processes and the assignment of tasks to the most suitable human/robot entities takes place.
- Human and robot simulation tools help to evaluate the ergonomics and feasibility of the assignments in a structured and semi/fully automated way. The emphasis in this area is to automatically generate and evaluate the numerous possible combinations of human and robot collaborative poses and eventually to select one good compromise in terms of process efficiency, ergonomics and other criteria.
- The derived task assignments are evaluated against user criteria (e.g. operator and resource utilization, matching of operators' skills and task requirements etc.) by the Task Planner using proven multiple criteria decision-making methods. This ensures that the process is executed in an efficient time while at the same time ensuring that the skills of each entity are efficiently exploited.
- Additionally, the outcome of the planning/simulation activities is further used to support the operators through the integration of the latest technologies, such as Augmented Reality. An example would involve the operator helping a robot to move a part in the 3D space between obstacles. The 3D models from the simulation can be used at this time to superimpose the final position of the part on the assembly so that the operator can visualize and confirm the correct position where he will guide the robot.

1.3.1.4 Innovative Robot Programming Techniques

The focus of this technology stream is in reducing the time and simplifying the robot programming by developing user friendly intuitive robot instruction libraries, capable of simplifying the programming effort. These libraries include common robot programming instructions (pick, move, place, copy etc.) and intelligent algorithms to combine the routines from the library in order to achieve the designated task, if this is required according to the Task Planner as shown in Fig. 1.14.

1.3.1.5 Mobile Units for Smart Logistics and Operator Support

This technology stream aims to the introduction of mobile units and/or mobile robotic manipulators that act as assistants to the human operators during the assembly operations. Therefore, the aim is to enable autonomous mobile units able to provide the operator with the correct parts/tools for the assembly punctually, at the right place. The design of the unit allows minimizing the operator's:

- physical strain by providing the parts at a comfortable posture and
- cognitive load by providing the correct parts thus relieving the operator from the task of consulting the product documentation, identifying and retrieving the correct part from the storage area (Fig. 1.15).

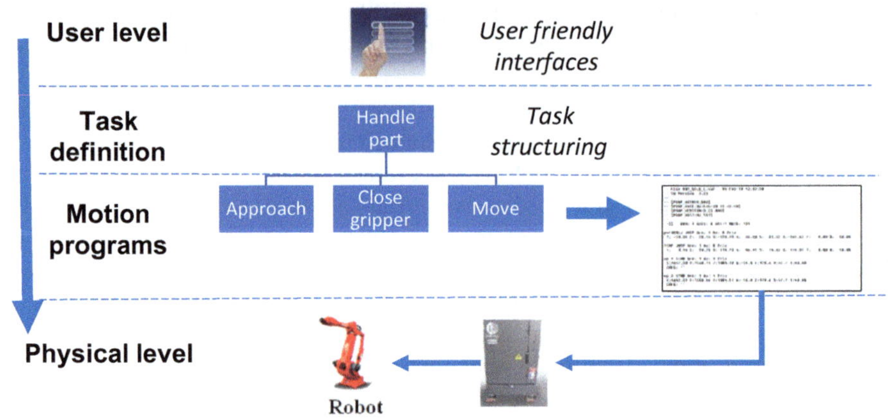

Fig. 1.14 Intuitive robot instruction libraries and multimodal interfaces

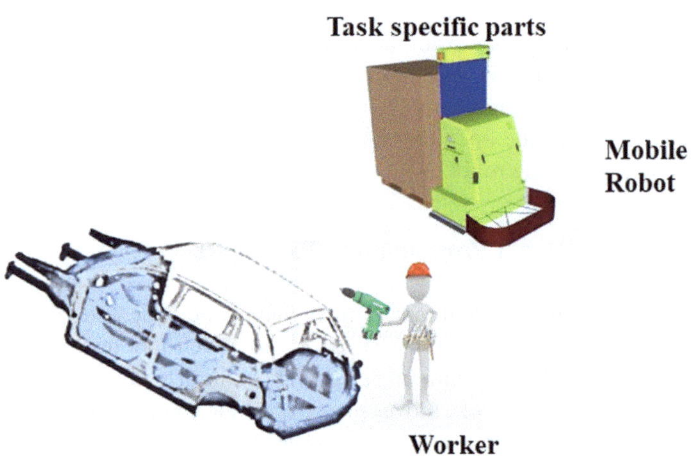

Fig. 1.15 Human operator assistance by mobile units

This type of support is aimed at minimizing or even eliminating the human errors that are related to the variability and complexity of the parts that are being assembled. Moreover the use of mobile units with onboard sensors assists in implementing smart logistics processes by allowing the automated part tracking and inventory reporting at the shop floor.

1.3.1.6 Integration and Communication Architecture

It is essential for robots to be connected to the rest of the resources on the shop floor, knowing their status and autonomously schedule automatically their operations in combination with the operations of the human workers. The scope of this technology stream is to introduce the easier integration and networking of the control and sensor data utilizing agent based, web and ontology services. The expected result in terms of communication will be the distribution of these acquired data to every relevant resource (e.g. robot, machine, human workforce via multi-modal interfaces etc.). The advantages and challenges simultaneously concern the robustness, flexibility, autonomous behavior and openness of this architecture in case of failure. The minimizing of programming effort for the automated and distributed way of realizing the suitable data from the appropriate resource between cooperative production entities is the motivation for an open architecture development. This architecture undertakes all the local controls in the production facilities and helps to:

- minimize the use of existing complex programming methods for integrating and networking purposes (e.g. PLCs)
- eliminate or drastically reduce existing centralized decision making with Fixed control logic
- reduce the time for supporting the additional of resources to the networking and integration system, without requiring a lot of changes and expertise workforce
- provide with generic control models that aim to replace large monolithic software packages which are developed and adapted case by case
- avoid the large configuration costs, by not requiring specific devices and software packages that accompany them
- overcome the high cost implication when implementing, maintaining or reconfiguring the control application
- support efficiently the current requirements in terms of flexibility, expansibility, agility and re-configurability.

Several open source and commercial tools help in implementing this technology. by promoting the current initiatives- mainly open source- solutions. Indicative examples are the ROS-Robot Operating System. ROS includes among others, specific libraries allowing to build architectures for communication between robots giving the possibility to extend these libraries by adding new functionalities. Furthermore the integration architecture aspires to address the lack of international open standards for interfacing with sophisticated sensors [9]. The sensing and control logics are used to identify the status of each machine and also to generate a set of local alternative reaction plans accordingly, by using the Task Planner development. Nevertheless it is expected that the Task planner's alternatives may have an impact on the operation of other machines/robots/ human which might need to change their working plan (e.g. take up some tasks from another robot or the human take tasks from the robot). In this case the network interface and responsible services are triggered to communicate the selected actions to the affected machines so that their control services can respond positively/negatively according to their status and capabilities.

An example of such am open architecture is shown in Fig. 1.16 and supports functionalities such as data exchanges for communication and messages that allow to easily plug new operators/robots that are able to cooperate automatically in the same fenceless workspace.

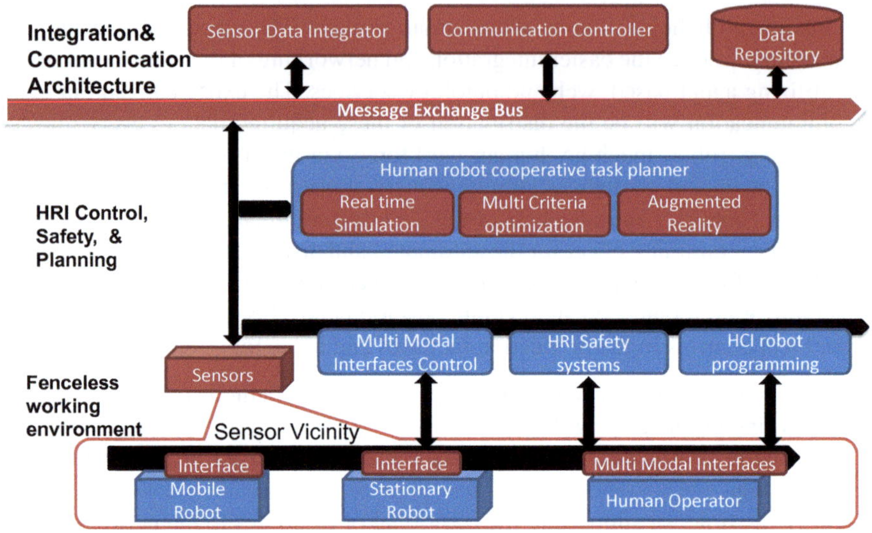

Fig. 1.16 Concept of communication and integration for Human robot interaction

The concepts discussed in this section are explained in further detail with the help of real world example and industrial cases in Part 3 of this book.

1.3.2 Efficiency Aspects in Human–Robot Collaboration

Through the enablement of the direct human robot cooperation under a safe environment, Human robot collaboration aspires to promote of advanced robotic solutions, giving special emphasis to robotic co-workers for operator support, within manufacturing and assembly sectors that are traditionally dominated by human resources and low degree of automation.

Major manufacturers are concerned about the wellness of workers aiming towards reduction of injuries at work. The potential for intuitive robotic cooperating systems employment would eliminate difficult and repetitive assembly tasks is thus tremendous. For example, automotive companies estimate that advanced cooperating robotics technology can lead to large scale deployment of robots in the final assembly and engine assembly workshops. The main target is the partial automation of final assembly tasks which are currently mainly manual, with the goal of relieving workers from repetitive tasks as shown in Fig. 1.17.

The impact of robotics is expected to be substantial considering the robot acquisition and operating costs. Assuming that robots will perform at least one-quarter of the manufacturing tasks that can be automated, the average savings in total labour costs until 2025 could be 16% lower. By installing advanced collaborative robots,

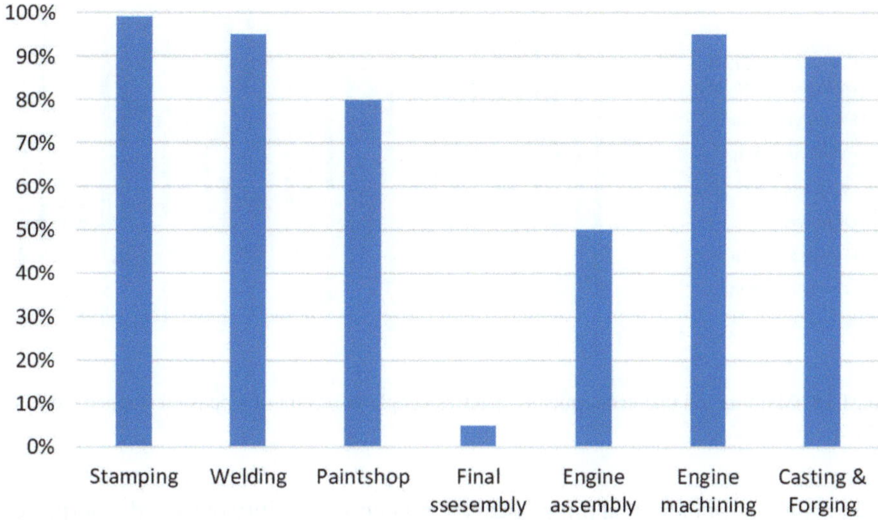

Fig. 1.17 Percentage of tasks automated in automotive factories

and depending on the location, output per worker in manufacturing industries can be 10 to 30% higher than the case of using only humans in the shop floor.

The new generation of robots has "come out of the cage" for 24-h shifts, working alongside human counterparts. Increasing returns on investment, demand and advances in HRC increase can lead to their adoption to 25–45% of production tasks by 2030, beyond their use in the automotive and electronics industries. Adopting advanced robotics and AI could boost productivity in many industries by 30%, while cutting labor costs by 18–33%, yielding a positive economic impact of between $600 billion and $1.2 trillion by 2025 [14]. Therefore, adoption of cooperating robots technology is directly aligned with the objective as it promotes the automation of existing task by flexible and cognitive HRC robotics. Autonomous robots provide the potential for frequent reallocation of resources reducing the costs for physical changes. The sustainability is further promoted by the integration of intuitive Human and robot interfaces through smart wearable devices and perception that allow the robot to be easily operated and programmed by non-expert users.

Another point is that a significant number of workers has to give up their work for health/disability reasons and also are unable to find another job as shown in Fig. 1.18. Cooperating robots technology contributes to the reduction of these percentages by the introduction of technologies that can supplement the abilities of these people and support them in continuing to work under favorable conditions.

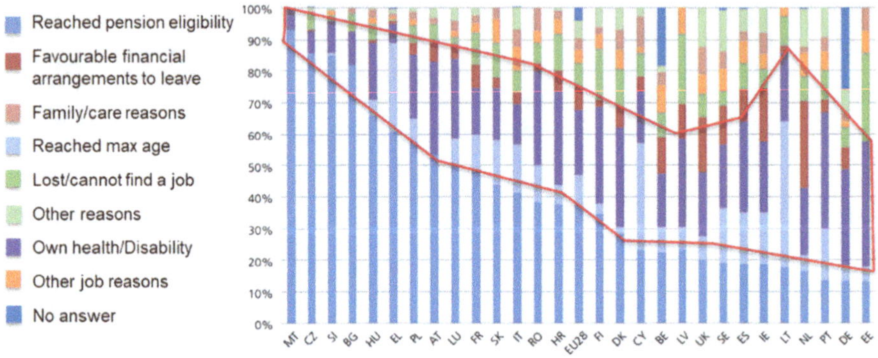

Fig. 1.18 Main reason that economically inactive people in receipt of a pension stopped working [15]

In a number of industrial pilot cases, the aspects of viability of hybrid workplaces has been studied. The following analysis concerns the introduction of one Dual arm robots in cooperation with a worker in order to undertake the work that normally a single worker performs. The analysis summarized hereafter is rather indicative of the way to assess the potential robotization of production processes and it is dependent on specific conditions in each factory, country, company.

In this case, the dual arm robot has undertaken the ergonomically difficult and heavy tasks, when human performs the flexible parts assembly such as the cable pack installation. The manual line in automotive industry is totally manual, including a number of fixtures and tools that help user to perform the assembly tasks. The data of the manual assembly line are summarized in the Table 1.2.

The cost for the manual assembly line, concerning the setup is totally 42.000 € including 35.000 € for equipment, 2.000 € for commissioning labor cost and 5.000 € for energy lines. Two different scenarios have been proposed for the hybrid assembly line. These scenarios are: (a) the optimistic case and (b) the pessimistic case. The following table summarizes the manual assembly line Investment cost and the hybrid line Investment cost for the two cases. In the optimistic case, the cost of the hybrid line is approximately 3.5 times more than the manual one, when in the pessimistic case it is 6 times more (Table 1.3).

Table 1.2 Manual assembly line data

Dashboard psre-assembly- manual assembly line	
Cycle time (min)	2
Annual production rate	140.000
Number of shifts (shifts/day)	2
Man hours (hours/day)	20
Working days (days/year)	233

Table 1.3 Investment cost of manual and hybrid assembly line- Dashboard pre-assembly case

	Manual line	Hybrid line (Optimistic case)	Hybrid line (Pessimistic case)
Equipment cost (robot cost, tooling, safety etc.)	35.000	135.000	210.000
Commissioning labor cost	2.000	5.000	25.000
Energy lines (electric, pressurized air,civil works)	5.000	9.000	20.000
Total initial investment cost (€)	42.000 €	149.000 €	255.000 €

The running costs of the above cases are summarized in the following table. The total running costs for the hybrid line both cases is the same and approximately 3 times less that the manual.

	Manual line	Hybrid line (Optimistic case)	Hybrid line (Pesimistic case)
Engineering cost (/year)	2.000	2.000	2.000
Maintenance cost (/year)	3.000	7.000	7.000
Operation cost (e.g. Electricity, pressurized air)	1.000	2.000	2.000
Labor cost (1 person, 2 shifts, 233 days)	64.800	13.000	13.000
Rework cost	3.000	100	100
Ergonomy cost	3240	650	650
Total running cost (per year) (€)	77.040 €	24.750 €	24.750 €

An analysis for calculating the Payback HTHP (Head To Head Point) has been done for both the hybrid lines cases. In the optimistic case the HTHP is calculated in 24 months and is illustrated in the following figure (Fig. 1.19).

Similarly, the estimation of HTHP is calculated 49 months for the pessimistic case and is visualized in Fig. 1.20.

While cost figure may vary per country and installation, the method presented shows the way to quantify such applications and that in many cases it is feasible to operate a hybrid line at a cost that is competitive to the one of using only workers, while at the same time achieving all the benefits of removing high physical load tasks to robots rather than humans.

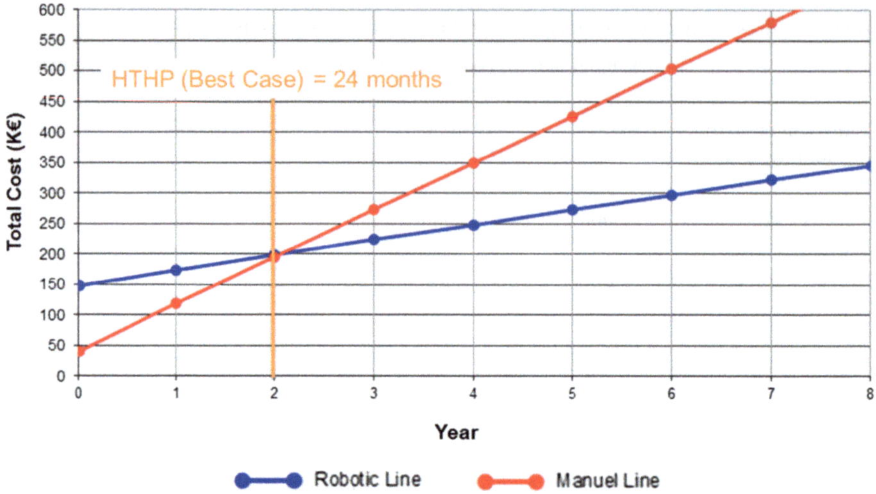

Fig. 1.19 HTHP calculation-Optimistic case

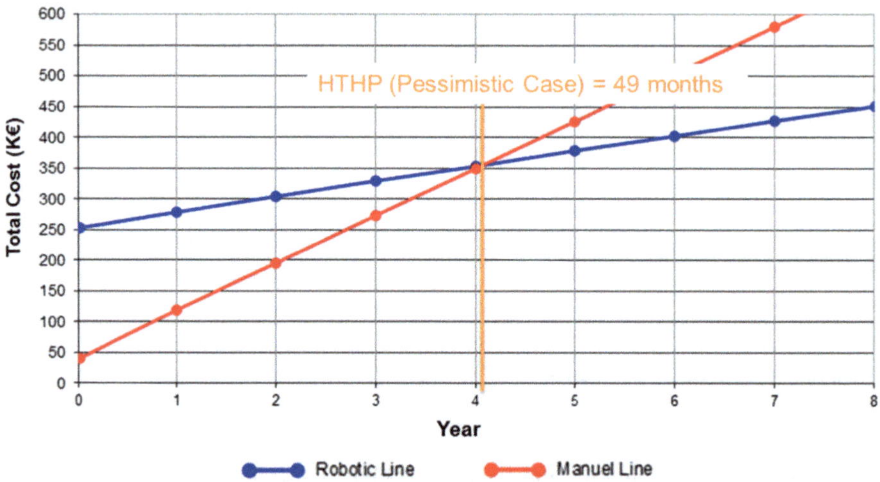

Fig. 1.20 HTHP calculation-Pessimistic case

1.4 Technology Perspective

Investigating the enablers to realize flexible factories with the help of cooperating robots, does involve a broad range of technology. The remaining chapters discuss in detail how this technology has been combined in numerous ways for achieving a broad range of production systems for a number of industrial sectors. The following sections briefly summarize the relevant technology, aiming to provide a first glance on what is following.

1.4.1 Robotic Perception of Shop Floor, Process and Human

Perception involves the use of several sensors in order to sense a number of parameters in the execution of a manufacturing process. Industrial applications require the recognition, handling and assembly of parts of different shapes by both stationary and moving robots. Significant capabilities of robot actuation and of sensor data processing and interpretation are required to reflect the environment and meet certain task specifications.

The handled/assembled products are in general either stationary or moving depending on the end user shop floor specification. As a result, it is necessary to bring together multiple sensing and perception technologies to enable dexterous and variable sensitivity operations.

A great deal of individual perception technologies and equipment are already available on the market but the customization for achieving robust application under each pilot fall within the scope of the following chapters in the book. Similarly, the systems for navigation of mobile robots are also handled by customizing 'off the shelf' solutions. For the navigation of a mobile platform the robot should be able to determine its position and orientation at least in a two-dimensional map of the environment. By contrast, the flexible manipulation of objects requires sensing and localization skills in six degrees of freedom. In addition, much higher accuracies in localization and actuation are necessary to successfully perform typical assembly tasks while the robot is on moving platform. The concept of integrating a variety of sensing technologies to control individual cooperating robots is shown in Fig. 1.21.

Such existing technology involves 2D and 3D vision systems by implementing algorithms that can enhance the accuracy and speed of the sensing process. Existing sensors provides accurate depth information along with standard images. The data includes a confidence measure on the 3D position of physical objects, which allows the evaluation of the robustness of the information perceived by the sensor. Additionally, vision sensors computes its ego-motion from images and the data of an integrated inertial measurement unit (IMU) [16]. Thus, such a sensor always knows its current pose relative to previous poses, which permits the registration of depth

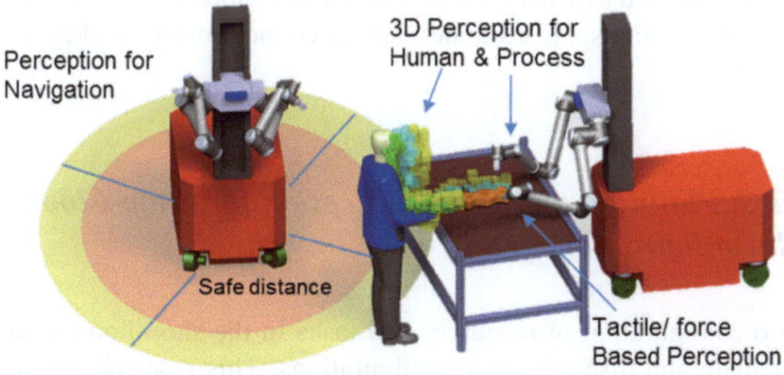

Fig. 1.21 Environment and process perception

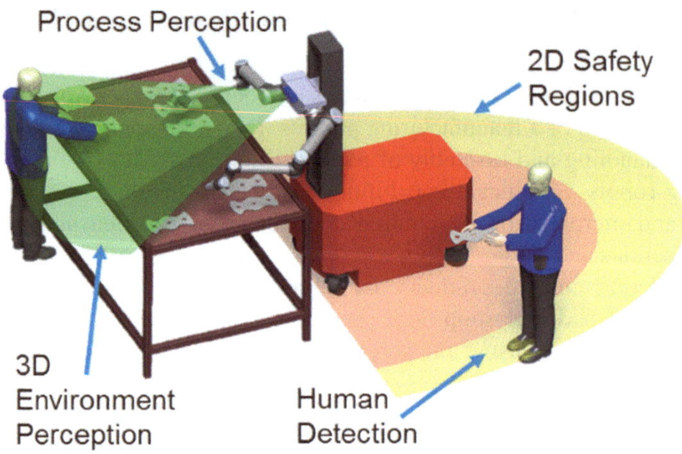

Fig. 1.22 Human sensing concepts

data to a larger 3D model. In other cases, industrial cameras and texture projectors are combined in order to provide high accuracy 3D sensors, based on stereovision or laser projection or structured light techniques. These customized solutions are used in its robotics applications for part features detection, online quality control, cell occupancy analysis, as well as autonomous navigation.

Besides process perception, and given the safety issues that arise from the coexistence of robots and humans, perception capabilities are required for detecting/monitoring human presence and adjusting the behavior of the robots. Current practices require physical separation between people and active robots as precautions to ensure safety. In a number of cases this book discusses methods to overcome this inefficiency through hybrid approaches using safety and perception sensors, robust path planning etc. in order to ensure the safety of people in proximity or even in contact with robots (Fig. 1.22).

The combination of 2D safety laser scanners with 3D sensors (e.g. 3D Time of Flight camera) helps to achieve workers' safety [17]. In a number of applications the 3D sensor is not needed to satisfy safety regulations, while being much more expensive. In other applications, 3D accurate workspace monitoring is a key requirement to introduce robots.

1.4.2 Task Planning and Communication for Shop Floor Reconfiguration

Achieving reconfiguration of resources and tasks in the shop floor, requires to be able to calculate and dispatch such configurations. This research area consists in achieving two major functionalities that will the main flexibility enablers:

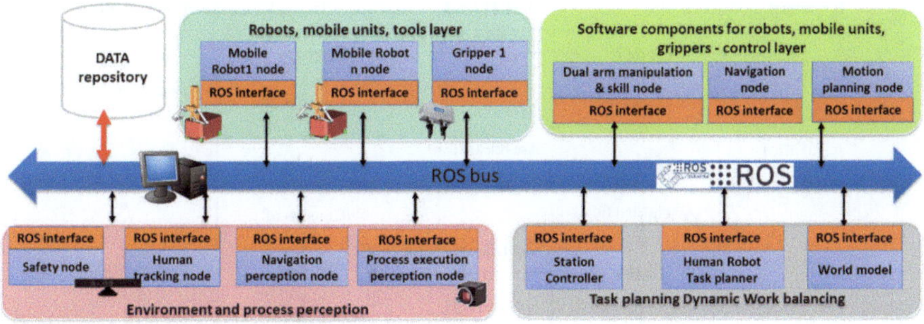

Fig. 1.23 Synchronization and control framework

- Simplifying integration and networking of the control and sensor data utilizing web based, and ontology services [18]. The expected result in terms of communication is the distribution of the acquired data to all relevant resources (e.g. mobile robots, humans via HMIs etc.). The advantages and challenges at the same time concern the robustness, flexibility, autonomy and openness of the architecture.
- Using the common integration architecture to monitor the execution of the production plan and dynamically redistribute the workload to adapt the production in run time. A digital world model in created for this purpose, that is continuously updated through the service network by all the perception components. Using this information and the status reporting of the autonomous robots, the workload balancing system is able to generate alternative allocations for human and robots. The Station Controller is a module capable of dispatching the assignments and monitoring the progress of the execution (Fig. 1.23).

1.4.3 Facility and Workload Modeling

In order for the manufacturing tasks to be undertaken by robots and humans, a hierarchical structure is adopted and is explained further in the following chapters [19]. Figure 1.24, represents this generic model representing the robotic system. The Factory corresponds to the entire production system and includes a number of Lines. Each Assembly Line consists of a number of Stations, which in turn, consist of a number of Resources. The Orders consist of Jobs, which in turn, consist of Tasks. The Orders correspond to the Factory and they are divided into Jobs released to Assembly Lines. A Job, based on its specification, can be processed only by one Assembly Line and is thus released to the proper Assembly Line. The Tasks that are included in a Job should be processed again by the designated Station and are therefore released to the corresponding Station. Finally, the Tasks can be processed by more than one of the Resources and the assignment of a Task to a Resource is made with the help either of complex decision making logic or a simple dispatching rule.

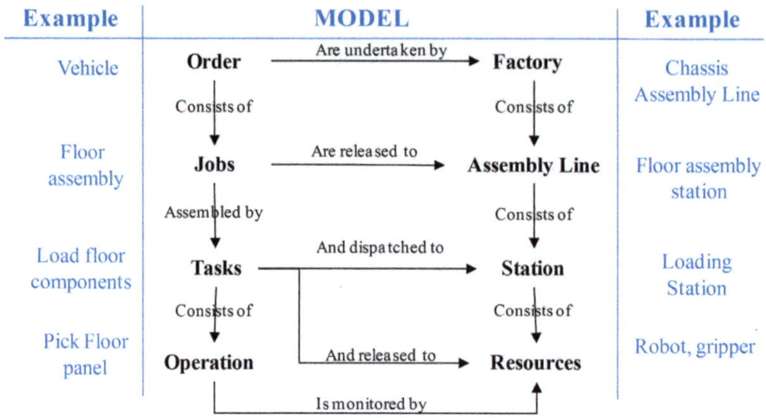

Fig. 1.24 Hierarchic model for the decision making

For example, the car tunnel assembly operation is one Job that is assigned to the floor Assembly Line. The Job consists of a number of handling and welding Tasks, such as moving or performing a weld. The moving Tasks are assigned to the loading Station, whereas the welding Tasks are assigned to the respective welding station. In each station, there are Resources which could perform the Tasks. Therefore, the Tasks have to be dispatched to the Resources according to the assignment logic. An important constraint in releasing and dispatching Jobs and Tasks is the precedence relationships among them. A number of examples in this book help to see the implementation of this generic model in the case of employing robots cooperating which each other as well as cooperating with humans.

References

1. Atkinson RD (2018) Which nations really lead in industrial robot adoption? Inf Technol Innov Found
2. Okuma T (2019) Editorial world robotics report 2019. Int Fed Rob
3. Population, total | Data. (2020) https://data.worldbank.org/indicator/SP.POP.TOTL?end=2018&start=1960&view=chart. Accessed 20 Apr 2020
4. Alexopoulos K, Makris S, Chryssolouris G (2018) Production. CIRP encyclopedia of production engineering. Springer, Berlin, Heidelberg, pp 1–5
5. Chryssolouris G (2006) Manufacturing systems: theory and practice. Springer, Berlin
6. Lien TK (2014) Robot. In: The international academy for production engineering. In: Laperrière L, Reinhart G (eds) CIRP encyclopedia of production engineering. Springer, Berlin Heidelberg, pp 1068–1076
7. Michalos G, Kousi N, Makris S, Chryssolouris G (2016) Performance assessment of production systems with mobile robots. In: Procedia CIRP. Naples, pp 195–200
8. Yoonseok Pyo DL, Cho H, Jung L (2017) ROS Robot programming (English). ROBOTIS
9. Forge S, Blackman C, Bogdanowicz M (2010) Helping Hand for Europe: the competitive outlook for the EU robotics industry

10. Makris S, Michalos G, Eytan A, Chryssolouris G (2012) Cooperating robots for reconfigurable assembly operations: review and challenges. Procedia CIRP 3:346–351. https://doi.org/10.1016/j.procir.2012.07.060
11. TÜNKERS Maschinenbau GmbH (2020). In: TÜNKERS Maschinenbau GmbH. https://www.tuenkers.com/d3/d3_product_detail.cfm?productID=P0017094. Accessed 5 Jan 2020
12. Krüger J, Lien TK, Verl A (2009) Cooperation of human and machines in assembly lines. CIRP Ann 58:628–646. https://doi.org/10.1016/j.cirp.2009.09.009
13. Michalos G, Makris S, Tsarouchi P, Guasch T, Kontovrakis D, Chryssolouris G (2015) Design considerations for safe human-robot collaborative workplaces. Procedia CIRP 37:248–253. https://doi.org/10.1016/j.procir.2015.08.014
14. World economic forum (2017) Technology and innovation for the future of production: accelerating value creation. World Economic Forum, Cologny/Geneva Switzerland
15. Eurofound (2016) Sustainable work throughout the life course: national policies and strategies. Publications Office of the EU, Luxembourg
16. rc_visard is a powerful and high performance 3D sensor for robotic systems. In: Roboception. https://roboception.com/en/rc_visard-en/. Accessed 23 Feb 2020

17. Safety laser scanners SICK (2020). https://www.sick.com/ag/en/opto-electronic-protective-devices/safety-laser-scanners/c/g187225. Accessed 23 Feb 2020
18. URL OWL Web Ontology Language OWL/W3C Semantic Web Activity (2012). https://www.w3.org/2004/OWL/. Accessed 13 Jan 2012
19. Chryssolouris G (2006) Manufacturing systems: theory and practice, 2nd edn. Springer, New York

Part II
Cooperating Robots: Robot–Robot Cooperation

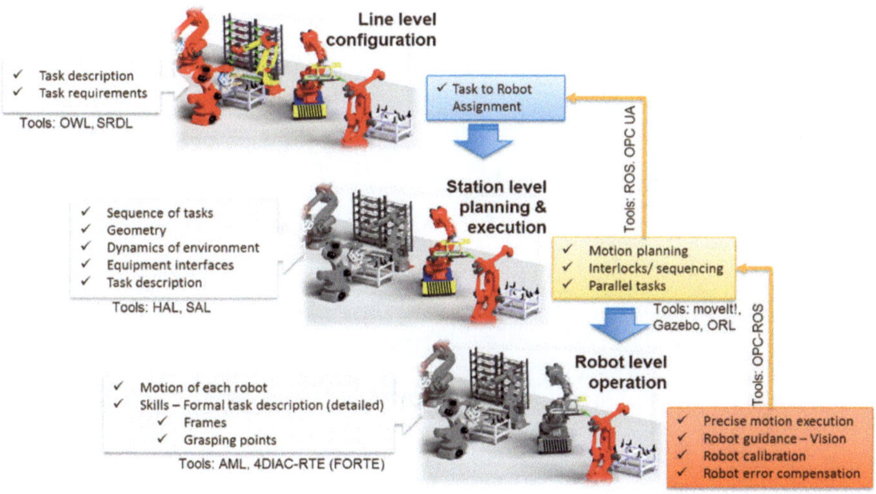

Chapter 2
Flexible Cooperating Robots for Reconfigurable Shop Floor

2.1 Introduction

Industrial production is typically organized following the fixed assembly line paradigm. This paradigm employs a combination of fixed linear sequences of operations. Manual and automated tasks are repeated in the same way within the cycle time of each production station. Processes are designed in the most suitable and optimized way in design time and then these are executed repeatedly in execution time. This paradigm is rather efficient when the production is set to the maximum capacity and considering no halt situation due to technical problems.

The optimized production capacity that is achieved by the fixed sequence of operations can no longer guarantee the sustainability inside a turbulent market that requests new models more frequently than ever. The robustness and efficiency of the serial production model is highly compromised by the need to perform changes in the production equipment that does not have the cognitive capabilities to support multiple operations in a dynamic environment.

On the other hand, the trend for producing more customized products and offering more products variants to the market, calls for production systems that are capable of producing in a more flexible manner. The industry is becoming more customer centric in an attempt to meet the varying customers' demand and minimize the costs of large inventories [1]. The paradigm shift is evident in multiple industrial sectors such as the automotive and aeronautics that have relied in the serial production line paradigm for decades.

The scientific community has highlighted the necessary elements that can lead to increasing the autonomy in production. Systems that are described by the term autonomous, should be considered as having, an independent decision making, without external instruction and the ability to perform actions without external stimuli. The autonomous systems possess self-x characteristics such as self-adaptability, self-optimization and are more flexible, robust and fault tolerant. In [2] three main technology and scientific areas have been highlighted as

S. Makris, *Cooperating Robots for Flexible Manufacturing*,
Springer Series in Advanced Manufacturing,
https://doi.org/10.1007/978-3-030-51591-1_2

enablers for advancing autonomy in assembly systems: (a) Reconfigurable Manufacturing/Assembly Systems (RMS/RAS) [3, 4], (b) intelligent control methods and architectures and (c) intelligent computing methods utilizing heterogeneous 'smart' objects.

Flexible manufacturing aims at introducing a radical change in production paradigm in the paradigm: in the same plant the sequence can be changed by introducing autonomous production and handling robotic resources, which can change task, for example from joining to handling and vice versa and change position on the shop floor. These robotic resources are eventually cooperating among themselves, based on running process sequences, nevertheless having always the possibility to recover eventual failures to any robot or tool by switching position and assigning a new job, auto reconfiguring themselves, the tools and the line to answer quickly to the stop of production and reducing losses as much as possible.

This chapter discusses the approach and implementation of a flexible assembly system, using robots working in cooperation as a major enabler for implementing this paradigm [5]. Cooperating robots, i.e. robots communicating with each other for carrying out common tasks, may expand their capabilities greatly. They can be used for reducing the number of required fixtures as well as for shortening the process cycle time, whilst addressing the accessibility constraints introduced by the use of fixtures [6].

2.1.1 Fixed Assembly Line

The typical assembly line comprises rigid flow line structures by employing model-dedicated handling and transportation equipment of raw materials and components, as shown in Fig. 2.1 [4, 7, 8]. The main problems encountered are the lack of flexi-

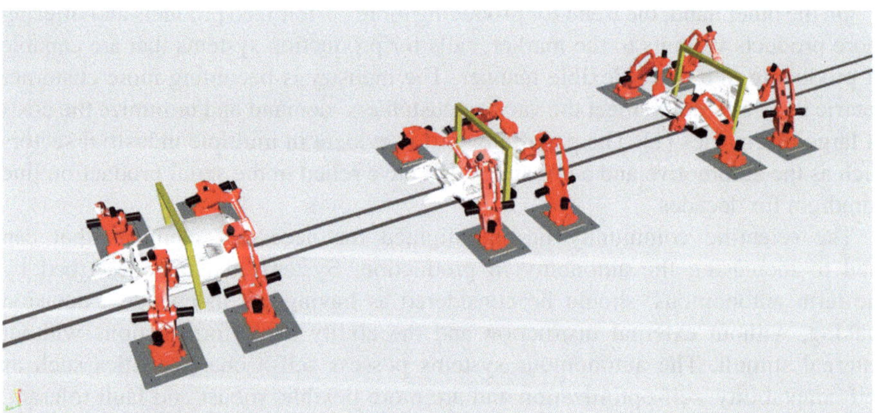

Fig. 2.1 Conventional arrangement of an assembly line

bility, since the system is hard to reconfigure, while its up-time is prone to resource failures, given that a single resource failure, as an effect, has the entire line remaining down until he resource has been repaired. In order to face these challenges, alterations of the production and logistics processes are required to enable the system's fast reconfiguration with minimal human intervention.

In terms of control, the traditional assembly line employs fixed control logic with signal-based tasks sequencing that requires significant effort for the implementation of changes in the production plan [5]. The current practices involve the use of Programmable Logical Controller-PLC signals to denote the start/stop of the operations, requiring a hard-coded approach that signifies high complexity and downtime in case of changes. These systems cannot follow the market needs, for fast introduction of new products or frequent improvement of the existing ones, because it requires manual reprogramming each elements of the production line.

New production systems should exhibit attributes such as flexibility, reusability, scalability and reconfigurability, therefore, a different control scheme has to be employed [1, 2, 9, 10]. Flexible Manufacturing System is a new manufacturing systems paradigm that aims at achieving cost-effective and rapid system changes, as needed and when needed, by incorporating principles of modularity, integrability, flexibility, scalability, convertibility, and ability to diagnose errors [11]. The flexible manufacturing system, will allow flexibility not only in producing a variety of parts, but also in changing the system itself [12].

While the term Flexible manufacturing systems has been introduced a few decades back, the term has been widely used for machining systems while robotics coordinated dynamically in assembly setups have not been thought as a place to widely apply this concept. For this reason, attempts to apply robots in the industry have resulted in employing large numbers of robots, exhibiting the problems identified in this section [13]. The following sections discuss the approach that would help overcome these limitations and would then turn the idea of flexible cooperating robots to a reality that would be achievable in a cost efficient manner.

2.1.2 Flexible Assembly Systems

Cooperating robots, i.e. robots communicating and physically interacting with each other for carrying out common tasks, may expand their capabilities greatly. They can be used for reducing the number of required fixtures as well as for shortening the process cycle time, whilst addressing the accessibility constraints introduced by the use of fixtures [5, 6, 14]. The conceptual structure of this type of system is shown in Fig. 2.2.

The main enabling elements of this approach are the following:

- A robot to robot cooperation approach, enabling task reallocation and collision free motion plans generation, and switching of robots control according to the needs of the assembly tasks.

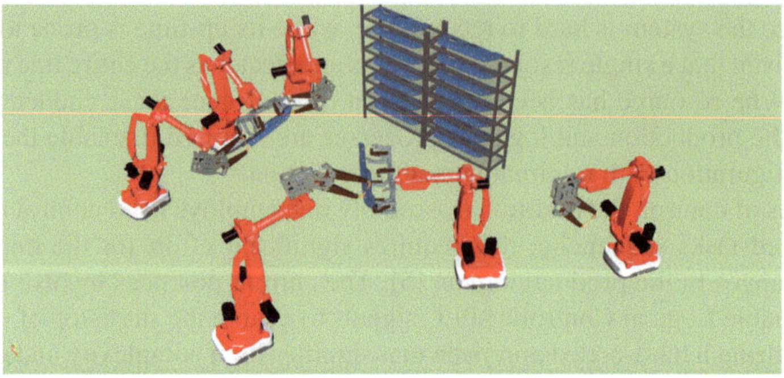

Fig. 2.2 Assembly concept with mobile robots

- A decision-making logic for flexible assembly lines reconfiguration applied in such a system.
- A shop floor control logic, capable of sensing disturbances in the shop floor and making decisions on the reconfiguration of the system.
- A service-oriented architecture for integrating a variety of robotic resources, namely robot arms, mobile platforms, flexible grippers.

Employing mobile robotic arms is adding a lot more flexibility to the manufacturing system. In addition, using flexible grippers and tooling, which are capable of handling parts of different geometries, it is possible for the system's flexibility to be increased even further. Although the discussed technologies have been under individual development over the last years, they have been oriented toward solving very focused, low-level technical problems and have not been considered under the potential of a complete production concept. Based on the analysis performed in this work, the advantages offered by the new paradigm are the following:

- Relocation of robots on the shop floor, by taking advantage of the mobile robots' ability to move.
- Implementation of dynamic parts' routings, by taking advantage of the robot to robot cooperation.
- Random material flow—Flexibility in re-routing parts: Unlimited by using exchangeable—reconfigurable tools and mobile robots.
- Handling of a variety of parts by reconfiguring the structure of the gripper.
- Flexibility of changing physical system structure: Very high—Units are able to relocate themselves in short time.
- Easiness of adding/removing resources: Very high—resources are automatically integrated and configured for operation.
- Flexibility of adding new processes: Very high—Robotic equipment can be configured to perform different processes.
- Flexibility of adding new products: High—Tools are already reconfigurable, programming required.
- Required space: Small to high depending on configuration.

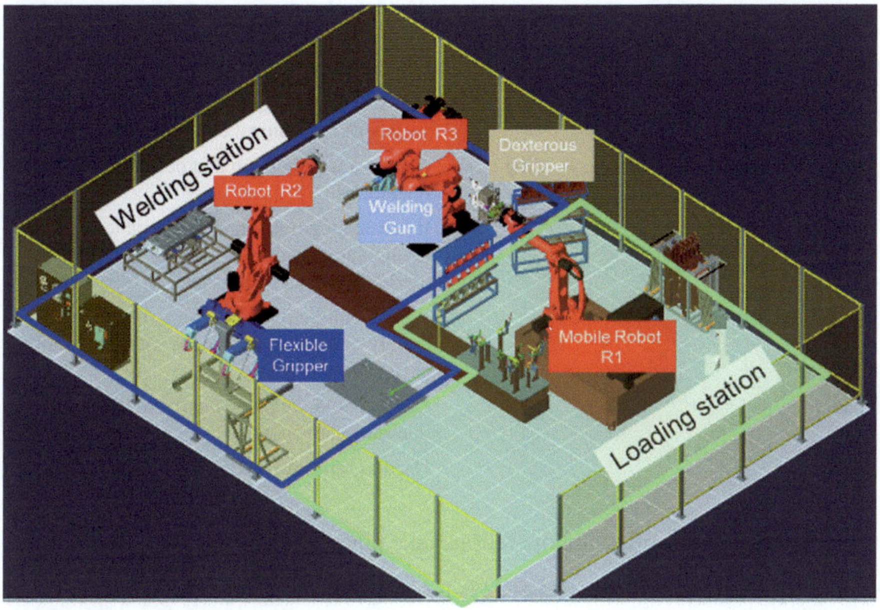

Fig. 2.3 Layout of the flexible manufacturing cell

- Production rate: Small to High.
- Cost: High to acquire and setup. Dramatic cost reduction in the long run.

2.1.3 Illustrative Industrial Example

In order to exemplify the discussion along the chapter, a specific industrial example of a reconfigurable system that was built for this purpose is going to be employed. This has been realised in a real-world assembly cell, the following layout has been implemented in the course of this investigation. This combines fixed robots, mobile robots, welding gun, flexible and exchangeable grippers, racks and vehicle parts to be welded as shown in Fig. 2.3.

This cell consists of the following resources, as shown in Fig. 2.3:

Resource	Functionality
One robot arm (Robot 1), mounted on a mobile platform	Able to move around the shop floor. This robot is used for two main types of tasks as follows: In the loading station, as shown in Fig. 2.3, the robot is used for loading the parts on the fixture In the welding station, this robot is cooperating with robots R2 and R3 for performing the welding the parts

(continued)

(continued)

Resource	Functionality
Fixed robot arm (Robot 2) in the welding station	Carries a flexible gripper capable of handling a variety of mechanical components
Fixed robot arm (Robot 3) in the welding station	Carries a welding gun and performs welding operations
Dexterous gripper	It is Reconfigurable and it is used to manipulate a variety of smaller parts and is used for loading operations
Flexible gripper	It is reconfigurable, it is used to handle a variety of bigger parts and is used for handling parts during welding
Fixture	Used to hold the parts and move them form one station to another
Racks	Used as repositories of parts

In this layout, the following operations take place:

- Initially the mobile robot (Robot 1) is located in the loading station and it is used for loading parts on the fixture using the dexterous gripper.
- The fixture is moved to the welding station for welding operations.
- The welding robot (Robot 3) performs geometrical welding operations.
- The second fixed robot (Robot 2), having the flexible gripper on its flange, lifts the part form the fixture.
- The welding robot (Robot 3) performs welding operations in cooperation with Robot 2.
- During this welding operation, a breakdown occurs. The breakdown occurs in a random pose of the robot R2, and this requires a number of actions as follows. The robot that is breaking down communicates this situation to the other robots in order for further action to be taken.
- At this stage, the system needs to be reconfigured. The mobile robot R1 is then commanded to navigate through the welding station.
- The recently arriving mobile robot (Robot 1) has to exchange information with the fixed robot in order to calculate its motion for picking up the gripper from the fixed robot (Robot 2). The gripper pickup should take place in a random pose of the broken robot.
- The mobile robot (Robot 1) should bring the part in proximity of the welding robot (Robot 3) enabling the welding robot (Robot 3) to finish the welding.

A vehicle floor tunnel is considered as the product being assembled, as shown in Fig. 2.4. The tunnel consists of nine components of variable geometry, requiring certain flexibility for handling them.

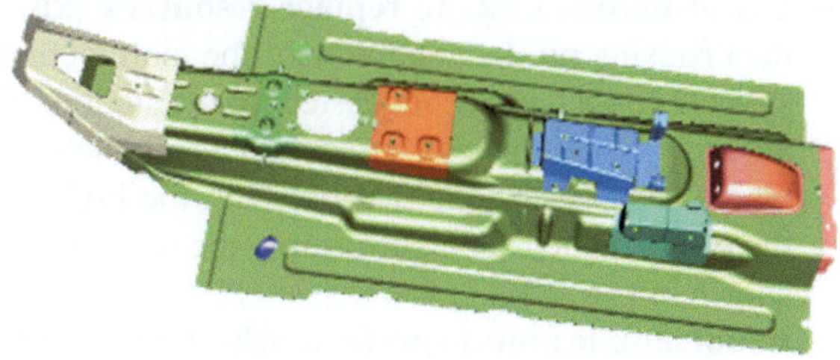

Fig. 2.4 Vehicle tunnel being assembled in the case study

2.2 Approach for Controlling Flexible Assembly Systems with Cooperating Robots

Controlling a flexible assembly system consisting of multiple mobile and cooperating robots, requires implementing intelligent control and monitoring systems essentially in two levels, namely unit level control and line level control.

- Unit level control algorithms for planning and executing assembly operations. Special emphasis is given on tasks such as the exchange of end effectors between robots and the tool pickup from stationary repositories.
- Line reconfiguration logic to automate the decision making logic and derive reconfiguration plans exploiting the mobile units and flexible equipment.

As it has already been stated, by introducing the concept of mobile robots and exchangeable/dexterous grippers several degrees of freedom are added to the manufacturing system. Given the fact that today's assembly environments do not employ such flexible production paradigms, it is essential to describe in higher detail the nature of the problem that this work is addressing.

In its general form, the challenge to be met involves the reconfiguration of the production/system in the case of unexpected events such as resources breakdown. Such events would otherwise bring production to a stop for an unpredictably long period of time until the malfunctioning unit is replaced. However, in the examined paradigm it is possible to resume operation by either employing a mobile unit to substitute the broken resource or by allocating the resource's tasks to adjacent equipment/stations.

From a more practical point of view, reconfiguration scenarios such as the following can be realized:

- **Decoupling of assembly tasks and stations/ability to transfer tasks along the line**: Currently tasks are strictly performed in specific stations according to the pre-programmed routines of the assembly equipment. The use of exchangeable grippers to transfer the part and gripper to an adjacent station allows for some of the tasks (or parts of them) to be performed in different stations. Of course this presumes that all required and suitable tools are present in the adjacent station (e.g. a welding gun that can access the welding areas on the part). An example is schematically shown in Fig. 2.5b.

- **Use of mobile units to replace resources experiencing malfunctions**. Rather than pausing production to repair the malfunctioning robot, a mobile unit with a suitable robot can be "requested by the station. If a unit is available it will either move to a tool repository to obtain a suitable tool or move directly to the station where it can exchange the tool with the broken down robot. The unit replaces the problematic robot, performing the tasks that were originally assigned to it as shown in Fig. 2.5c.
- **Use of mobile units to perform additional (not foreseen) tasks within a station**. Combining the two previous functionalities it is possible that a mobile unit can be assigned to a station further along the line where a docking station is available (Fig. 2.5d). In this case a new task has to be created and automatically tuned (robot program generation etc.) to the specific unit. The task will involve all the activities that were not carried out in the previous station due to the technical problems. Of course this approach also requires that the product's Bill of Process allows for the transfer of the task in a next station.

To physically realize such reconfiguration scenarios the functionalities presented in Table 2.1 need to be implemented by each resource respectively. These are the functionalities that are used as the building blocks for deriving reconfiguration plans as will be shown in the following paragraphs.

Summarizing the above, it is evident that in order to control and efficiently utilize these new degrees of freedom, a decision-making logic is required with capabilities that extend beyond traditional scheduling systems. Therefore the devised logic will need to:

- Evaluate the current status of the task's execution at the shop floor
- Identify possible breakdowns and determine the task execution state
- Create recovery plans by considering the product assembly specifications:
  - Task needs to be finalized in the specific station
  - Task can be transferred to next station
- Automatically reallocate resources (e.g. by combining the available robots and end effectors) that can potentially undertake the halted tasks
- For each of the plans in the previous steps, new tasks may need to be generated and inserted in the production plan. Indicatively:
  - Tasks representing the movement of the mobile unit to the station
  - Tasks representing the gripper exchange between robots etc.
- Further generate recovery plans based on the capabilities of the resources and the new tasks—assignment of resources
- Evaluate the recovery plans against user defined criteria and select the most efficient one.

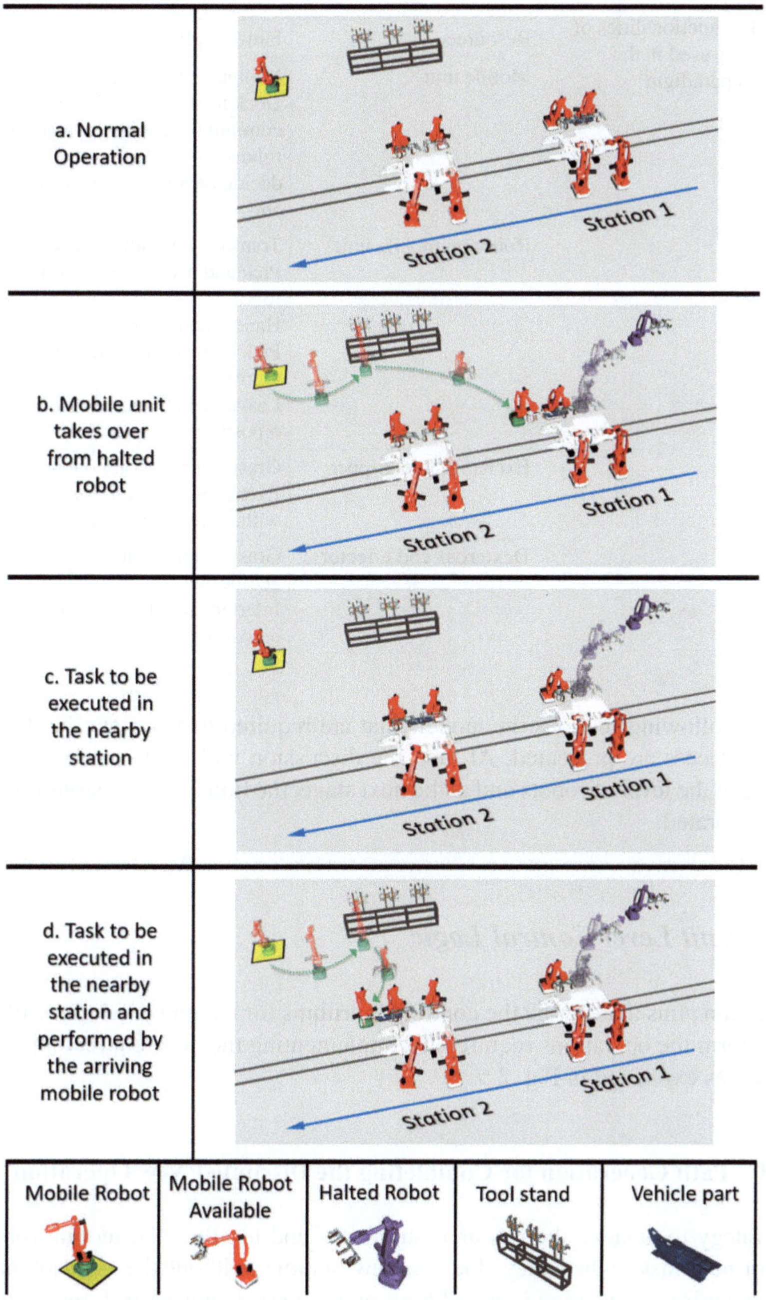

Fig. 2.5 Reconfiguration scenarios

Table 2.1 Functionalities of the resources used in the production paradigm

Resource	Functionality
Mobile unit	Navigate to area/station Dock to station communicate with the adjacent robots decide on a task to be carried out, and
Robot on mobile unit	Transact with other robot Pick and use end effector from other robot Handle end effector to other robot Pick and use end effector from tool repository Leave end effector on tool repository
Exchangeable gripper	Grasp specific components Exchangeable from robot to robot without releasing part
Dexterous end effector	Grasp components of variable geometry Interface for storing on tool repository

In the following sections the models that are required to represent the decision-making process are presented. At first, the discussion will focus on decisions and planning at the level of robots and at the next stage, the lion level decisions are going to be elaborated.

2.2.1 *Unit Level Control Logic*

This section aims to describe the control algorithms for the unit level that will allow it to perform the operations required for implementing the different reconfiguration scenarios as explained in Fig. 2.5.

2.2.1.1 Path Generation for Completing the Disrupted/New Operation

The strategy to achieve the resource autonomy and to allow the mobile robots to perform new tasks when they dock to new stations without the need of manual reprogramming or teaching is based both on the manufacturing task modeling and description within the overall data model, utilizing frames of reference and on the logic level of the robot resources as shown in Fig. 2.6. More information on the data model will be provided in the following section of this chapter, focusing on implementation. At this stage, it would suffice to refer to this data model as the

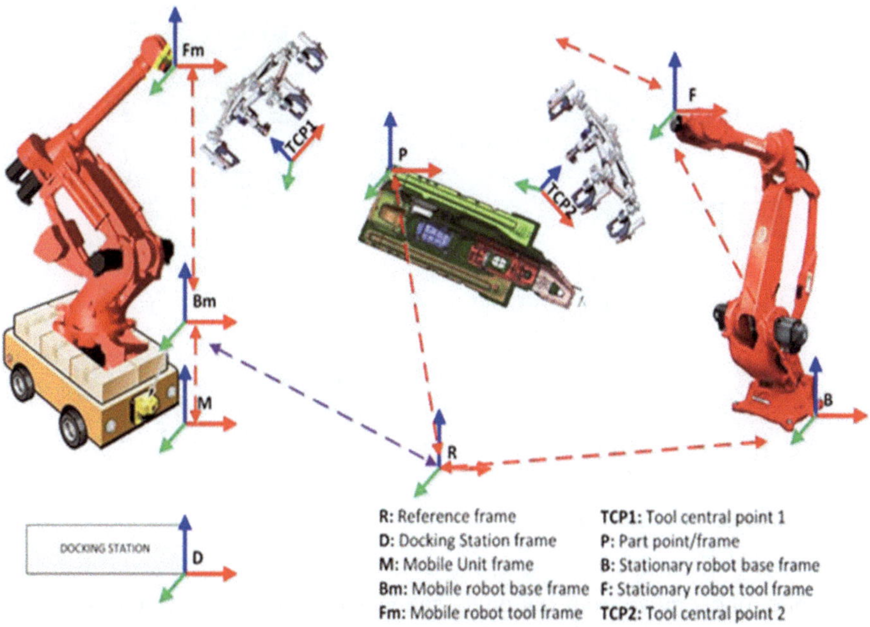

Fig. 2.6 Coordinate frames configuration for robot path generation

system ontology, since an ontology mechanism has been used for providing access to the relevant data.

On the one hand, the manufacturing task data is instantiated in the ontology and described as a set of operations, as shown at Fig. 2.7. Each operation describes robotic arm mechanical moves in a parametric way and information which allows the logic level of the robot resources to calculate the parameters and coordinate the mechanical moves. On the other hand, the robot resources run a local logic level which retrieves the robot resource schedule (assigned tasks and task's operation) when necessary, for example when a mobile robot docks to a new station. The logic level processes the operations of the resource assigned tasks, calculates the operation parameters consuming other services when needed (i.e. vision service) and finally monitors and coordinates the mechanical moves of the robot. In order to enhance the parameter calculation, the initial construction of each task's operations utilizes a number of coordinate frames of reference. In particular, as shown in Fig. 2.6, the initial definition of the operations will be based in the utilization of the reference frame (R), which is a fixed point within the station. Each robot, either stationary or mobile is locally calibrated in order to be able to translate its tool center point (TCP) moves according to this reference frame (R). In the case of the stationary robots, the Robot Controller is able to perform the necessary calculations. In the case of the mobile robots, an image processing system is used in order to define to exact position of the mobile unit and thus the mobile robots base. This is necessary since

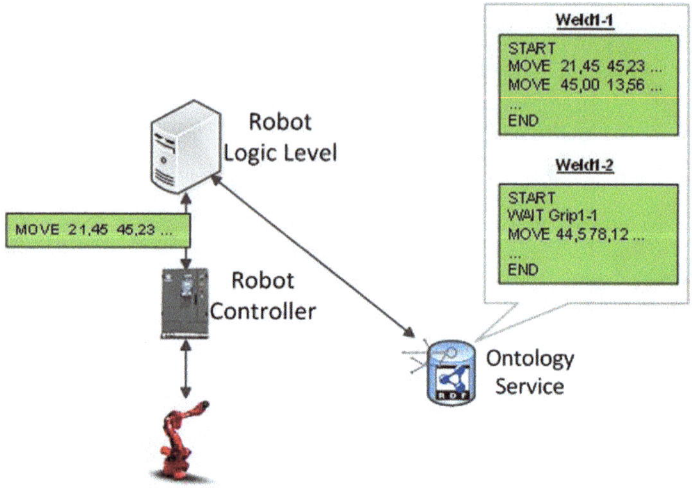

Fig. 2.7 Manufacturing task description

the mobile robots move from station to station and after their mechanical docking it is necessary to obtain a reliable referencing of their base frame (frame Bm) with respect to a common frame for the entire work station (frame R). This is done with the help of image processing to perform the necessary calculations.

The communication between the resources and the services utilizes the communication architecture and the defined resource or service interfaces as described in the following sections.

2.2.1.2 End Effector Exchange

This section is dedicated to providing the control logic design for implementing the functionality of a gripper exchange between two robots. This is a rather advanced behavior, which increases dramatically the shop floor flexibility potential. The logic includes all the steps to be followed by the involved specific robots actions namely robot picking the gripper and robot releasing the gripper. The coordinate frames along with the relations between all these frames are shown in Fig. 2.8.

In order for the two robots to share a common coordinate system, all the coordinates which are exchanged among the robot on the mobile unit and the stationary robot, are based on the reference frame (R) which will be a fix point within the station. For the stationary robot, calibration is needed for the definition of the relationships among the robot base frame, the reference frame and the tool center point. For the mobile robot the use of a vision system is needed in order to define the relationship between the mobile robot base frame and the reference frame. This is essential, because the exact position of the mobile unit after it has docked to a docking station cannot be determined with high accuracy without the use of a sensing system.

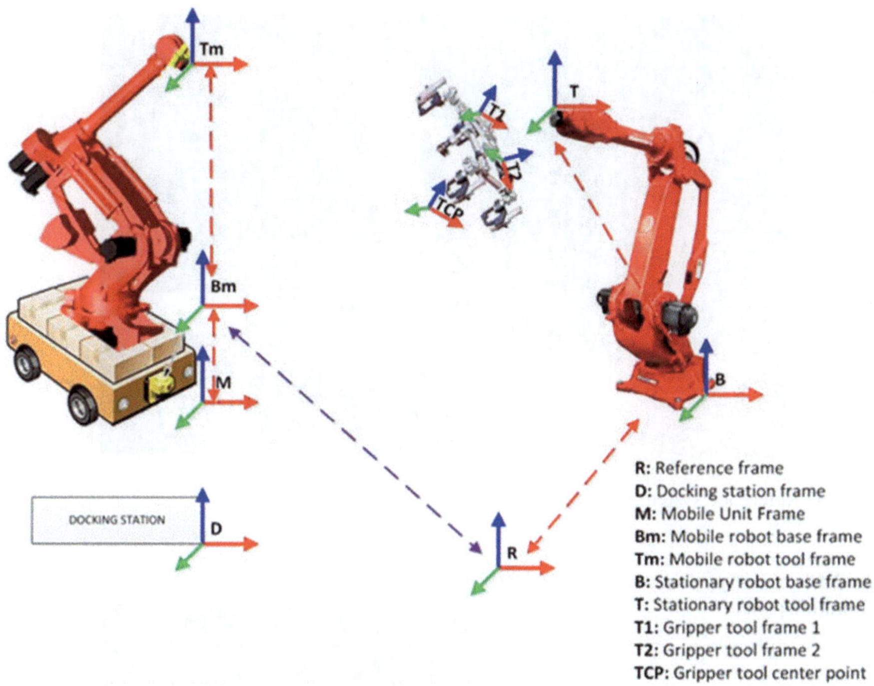

Fig. 2.8 End Effector Exchange Frames of reference

Furthermore, the mobile robot needs calibration for the definition of the relationships among the robot base frame, the reference frame and the tool center point. Finally, special attention will be needed at the calibration of the tool center point, as in this case the gripper has a double tool changer on it with two gripper tool frames, one for each tool changer head. The robots should be calibrated in such a way that they will understand the same tool center point on whichever tool changer head they are attached to (Figs. 2.9 and 2.10).

2.2.1.3 Robot Calibration and 3D Geometry Programming

When it comes to actual robot operating in such dynamic setup, it is necessary that it can identify the precise location of each frame discussed in the previous sections. This is feasible with the help of vision based systems and the following paragraphs explain the approach for achieving it. Briefly, there are two main calibration processes.

- Check and correct the robot position in respect to the working area every time the mobile platform is fastened to the docking station.
- Calculate the accurate position for the object located in racks and need to be grasped or handled by the robot.

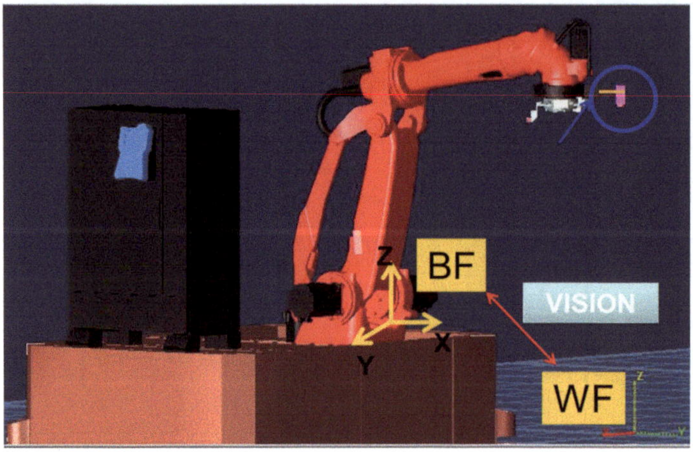

Fig. 2.9 Reference system transformation

Fig. 2.10 Vision system calibration panel

Robot-camera calibration: to produce measuring results in robot coordinates, normally it is necessary to perform a calibration procedure for the vision system using dedicated target and physically teaching points with the robot on it.

- **Camera calibration**: this setup involves mounting a camera on the robot arm in order able to provide results expressed in millimeters with respect to its reference system. At the same time the position of camera reference system with respect to the camera physical body must be defined.
- **Reference system transformation**: this step involves calculating the relation between the robot and the camera coordinate systems. Knowing the CAD data of the gripper and the camera position on it, it's possible to find the relation between

robot and camera coordinate systems, in order to obtain by the vision system the results in robot coordinates.

- **Measuring position teaching on robot**: after the docking station, the robot checks a fixed target, namely a well-known object on the mobile platform that allows to verify the general integrity of the robot-gripper system and then a target mounted on the working area to correct its base position. The use of fixed targets like perforated grid retro illuminated, could provide a simple and reliable method to perform the measuring task in industrial environments where the lighting are critical and variable, avoiding the use of external lighting on robot.
- There is a number of possible steps for the robot movement programming of each operation:
  - Performed by an operator, to minimize the complexity of the total amount of communication protocol and proceed with a first check regarding the general accuracy requested by each operation (target measurement for reposition on docking station, load area movements, etc.).
  - In this scenario the vision system is used to correct during the automatic cycle the robot base/frame and recreate the nominal conditions used in the movement teaching (1st time).
  - The robot movements are taught by the operator the 1st time, the programs are not store locally on the robot controller but on the ontology. They're passed in runtime in term of target position/task. The robot locally runs a client program that is continuously waiting for the next action to perform (movement, clamping, welding, etc.).
  - The working area is virtualized; models are prepared for the tools and mechanical parts. The robot working program is totally prepared offline; no human intervention is needed in any phase. The Robot receives the instruction for the next task in runtime by the ontology.

2.2.1.4 Physical Parts Recognition

Typically physical parts are stored in racks where robots can pick them up for further handling and assembly. However, in the generic case, parts may be located in non structured racks and can even be randomly placed. In this case, robots need to be equipped with parts pose identification system, which typically comprises one or more cameras as well as its accompanying recognition software.

After acquiring the images using the camera, an image processing algorithm attempts to match the shape of the image objects to their CAD counterparts in the images database. The edges from the picture are matched with the geometry of the parts and when they are coincident the position of the center of mass and orientation are determined. The method to guide the robot using vision in order to grasp parts on racks is greatly similar to the one discussed in Chap. 9 of this book.

2.2.2 *Line Level Control Logic*

This control approach is implemented in a number of steps. It mainly consists of a decision-making logic, which manages the actions of each resource of the manufacturing system and a communication infrastructure, which enables the flow of information between the various modules of the system.

The decision making logic has as its input the resources of the manufacturing system as well as the workload to be managed by these resources. In addition, the sequence constraints and the resources' suitability are given as an input to this model [5, 15]. The outcome of this process is a set of task assignments to manufacturing resources. Following, each task encapsulates a set of commands to be sent to the manufacturing resources. These commands are called Operations in this paper as shown in Fig. 2.12. Each Operation is executed by the resource where the task has been assigned to, until all the operations have been executed. In case of an unexpected event, the scheduling module is triggered to generate a new set of assignments for the manufacturing system's resources. Subsequently, the newly assigned tasks are executed as it has already been discussed. This approach is shown in Fig. 2.11.

2.2.2.1 Decision Making Method

Deciding which task will be performed by which resources, namely robot, mobile robot or grippers, is facilitated with the help of a four-level hierarchical model. This model helps to represent the resources as well as the workload, as shown in Fig. 2.12. The decision-making method is based on the approach that has been extensively discussed in [14]. The Factory represents the entire factory and includes a number of Assembly Lines. Each Assembly Line consists of a number of Work Centers which in turn, consist of a number of Resources. The Resources included in each Work Center are a sort of "parallel processors", namely they can "process" identical Tasks. Depending upon the assignment logic or the dispatching rules, a Task is assigned to one of the Work Center's Resources. In this particular application, the term Resource refers to a group of robots. With respect to the cell shown in Fig. 2.3, the Factory, the Assembly Line and the Workcenter levels are used to representing the entire cell. Then, each Robot is modelled as a single Resource, according to this model.

Corresponding to the facilities' hierarchy, there is also the workload's hierarchical breakdown. The Orders consist of Jobs, which in turn, consist of Tasks. The Orders correspond to the Factory and they are divided into Jobs, which are released to the Assembly Lines. A Job, based on its specification, can be processed only by one Assembly Line and it is thus released to the proper Assembly Line. The Tasks that are included in a Job can be again processed only by one Station and therefore, are released to the corresponding Stations. However, the Tasks can be processed by more than one of the Station's Resources and the assignment of a Task to a Resource is done with the help either of a complex decision making logic [15] or a simple dispatching rule. The assignment of the assembly tasks to the resources, results in a

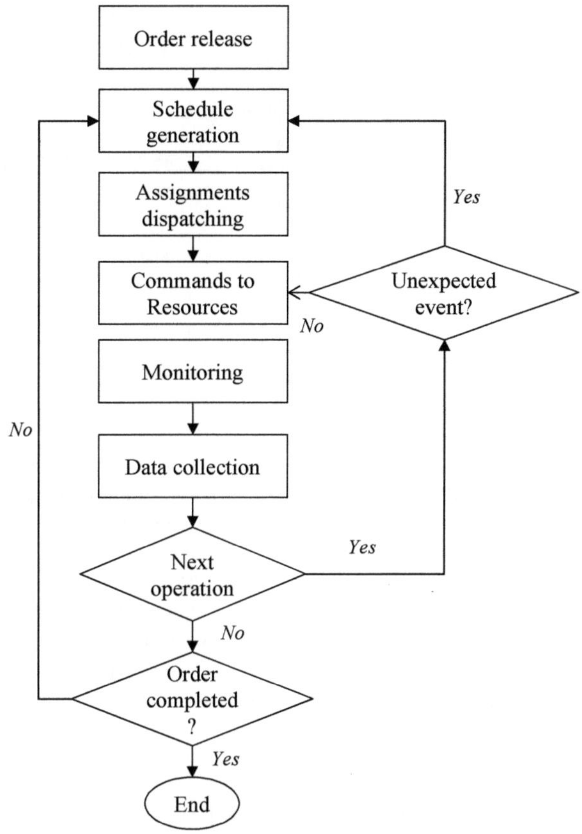

Fig. 2.11 Overall approach for assembly system reconfiguration

schedule for each resource of the cell and thus, a detailed plan and schedule for each resource is produced.

Furthermore, each Task consists of Operations which include the specific operations that each resource has to perform. Operations are divided into two categories, namely the Primitive operations and the Compound ones. The Primitive operations represent single operations such as a Robot moving from one location to another, a Gripper opening its clamps or a Gripper reconfiguring its structure. Beyond that, there are Operations that are more complex, such as a Navigation operation, the Invocation of a Vision system or a Gripper exchange operation. These operations are modelled as a collection of Primitives ones and are offered to the system's programmer in order to make for the system a more intuitive behavior program. Following this structure, commands are generated by the software system and are sent to the specific resources, namely to Robots, Mobile platform, Grippers and Cameras [12].

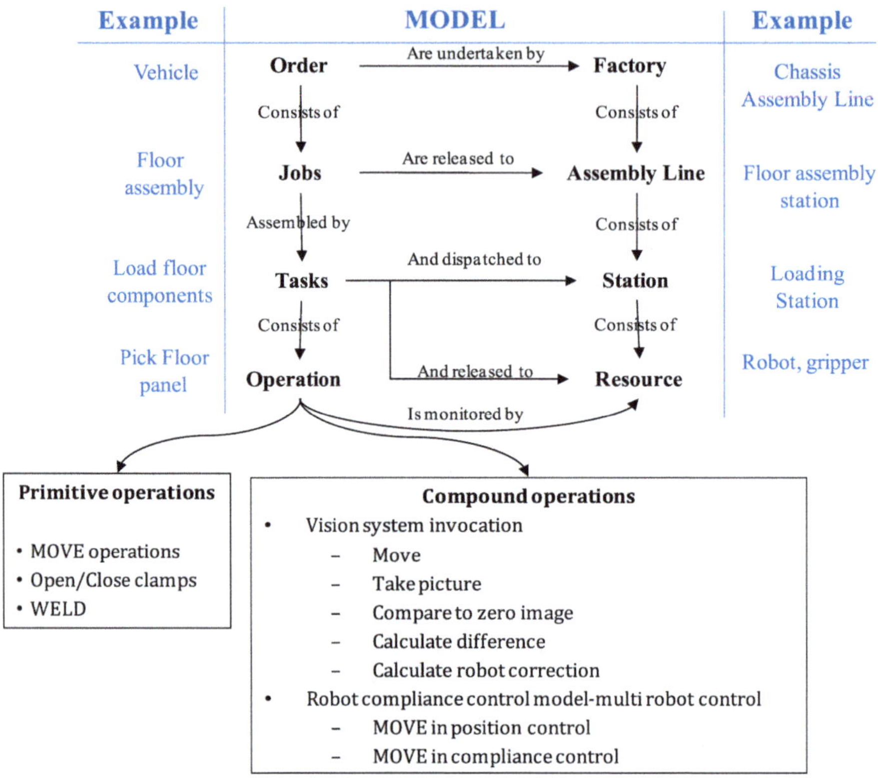

Fig. 2.12 Classification of resources and workload

2.2.2.2 Tasks Scheduler and Allocation to Robots

The following steps take place during the generation of the schedule, as shown in Fig. 2.13:

- The "Planning Creator" class constructs the scheduling problem to be solved, by loading data from the Ontology service.
- The "Dynamic Resource Fetching EndEffector Suitability Authority" class, is used to denying some preconfigured suitabilities. i.e.when a tool is not located at the station, then the static Resource cannot go and fetch it, only mobile robots can do it.
- Three criteria are used for the evaluation of each alterative as follows:
  - The "Dynamic Manufacturing Cost" is used for evaluating the cost of the system's reconfiguration alternatives. The cost is calculated based on the approach discussed in [16].
  - The "Dynamic Operational Time Calculator", calculates dynamically the operation time of tasks that cannot be pre-calculated, i.e. when a mobile resource

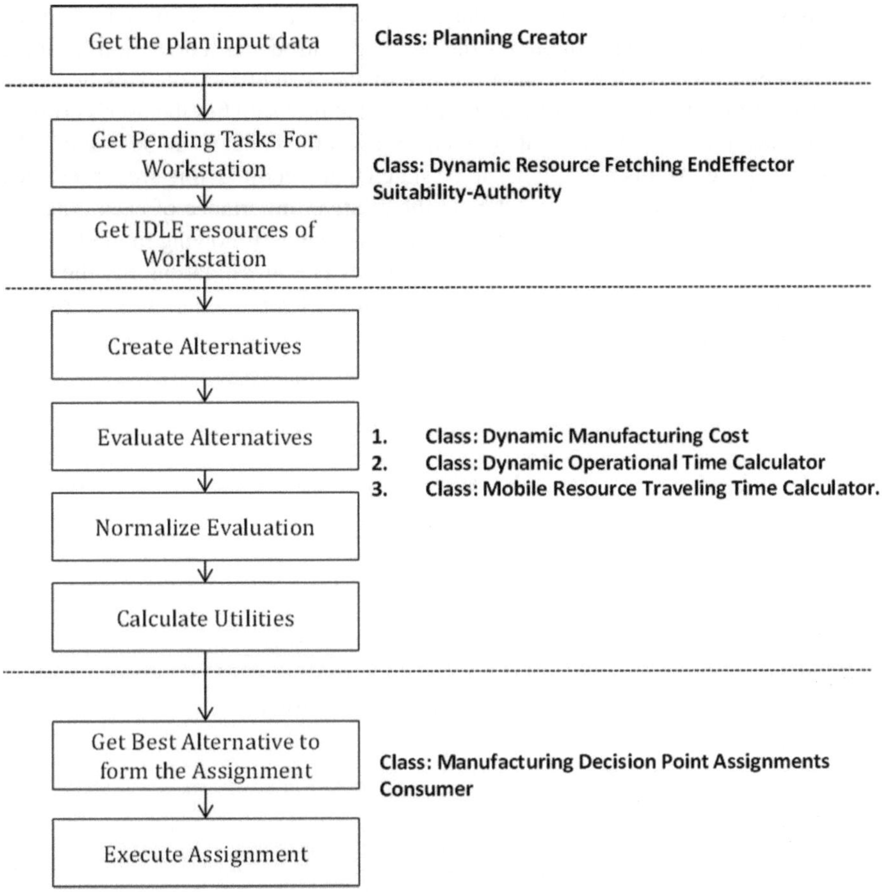

Fig. 2.13 Algorithm for generating assembly operation sequences

has to travel to fetch a tool, the tool position as well as the current robot position cannot be known a priori, so the time to retrieve the tool has to be calculated on the fly.

– The "Mobile Resource Travelling Time Calculator", calculates the travelling time required for a mobile robot to go from one place to another.

- Following their calculation, the values of these criteria are normalized, and weighted, so that a weighted utility value can be calculated. In order for the alternatives, based on the criteria values to be compared, a normalization of the values needs to take place. The normalization is carried out for criteria that should be maximized or (Eq. 2.1) minimized (Eq. 2.2) [1, 15]. The selection of the alternative that best combines the desired attributes is based on the total score (utility value) of each alternative, calculated as the sum of the products obtained by multiplying the normalized criteria values by a weight factor assigned to each criterion as shown in Eq. 2.3. The use of the utility value as a means of selecting the alternative that is most suitable for the end user, introduces a deviation from the predefined criteria values through the use of the weight factors. This however, allows for the identification of the alternative that best satisfies the user's criteria, even when a solution doesn't satisfy some or all of them, but it can be found in the solution space. At this point, it has to be stated that weight factors are represented on a scale between 0 and 1 and denote the relative importance of each criterion to the user. Currently, the user determines these factors following the judgement and experience of experts in the planning stage. The investigation into the ways that different weight factors may lead to a more efficient reconfiguration plan is proposed as future work.

$$\hat{C}_{ij} = \frac{C_{ij} - C_j^{min}}{C_j^{max} - C_j^{min}} \tag{2.1}$$

$$\hat{C}_{ij} = \frac{C_j^{max} - C_{ij}}{C_j^{max} - C_j^{min}} \tag{2.2}$$

where:

C_{ij} The criteria value of alternative i with respect to criterion j.
$\hat{C}_{ij}$ The normalized value of C_{ij}.

$$U_i = \sum_{j=1}^{n} w_c \hat{C}_{ij} \tag{2.3}$$

where: w_c is the criterion's weight factor and n is the total number of the criteria used. Alternatives with higher utility values are more preferable over the ones with low utility values.

- The "Manufacturing Decision Point Assignments Consumer" is used as follows. When an alternative has been selected for the formation of an assignment and its consumption, some extra steps should be performed in order for the factory context to be kept up to date. For example, when an end effector is transferred then its location has to be changed so as to be taken into consideration in the next decision point.

Regarding the step on the "creation of alternatives" as shown in Fig. 2.13, this work employed an intelligent search based algorithm. The algorithm is based on the depth of search concept, besides the number of layers for which the search method looks ahead. The main control parameters are the Decision Horizon (DH), the Sampling Rate (SR) and the Maximum Number of Alternatives (MNA). The generic tree of

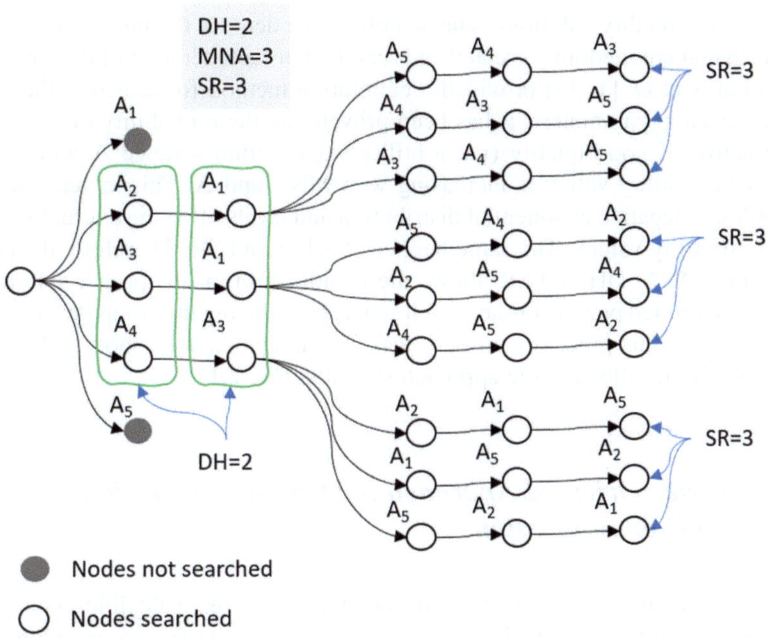

Fig. 2.14 Search tree example

Fig. 2.14, visualizes the use of the above parameters, while the nodes A1, A2... represent decision (assignment) points where a task is assigned to a resource. In essence, each node is a possible combination of a task with a resource (mobile unit, robot etc.) that is capable of executing the specific task. A branch which consists of such assignments (e.g. A1, A2 and A3) for all the tasks and resources, is considered being a complete reconfiguration alternative.

The algorithm uses the following steps to search for the solution:

Step 1: Starting at the root, create alternatives by randomly making assignments for all layers in DH, until MNA is reached.
Step 2: For each branch (alternative) in Step 1, create SR random alternatives (samples) until all nodes in the branch are searched.
Step 3: Calculate the criteria scores of all the samples belonging to the same alternative of Step 1.
Step 4: Calculate the branch's scores as the average ones achieved by its samples.
Step 5: Calculate the utility values of each alternative (branch).
Step 6: Select the alternative with the highest utility value.
Step 7: Store the assignments of the selected alternatives.
Step 8: Repeat steps 1 to 7 until an assignment has been completed for all the nodes of the selected branch.

MNA and DH control the breadth while DH the depth of search respectively. SR on the other hand, is used to directing the search towards branches that can provide higher quality solutions. The sampling rate denotes the number of samples per branch that are randomly created and searched in each iteration of the algorithm. Chryssolouris et al. [1, 15] provided a calibration method for selecting the values of these decision parameters. It has been proven that the probability of identifying an alternative of good quality (i.e. a utility value within a range Δ with respect to the highest utility value) is increasing with MNA and Δ. This increase follows more or less a negative exponential distribution and levels off at the alternatives with the highest utility values. The same happens for DH and SR. Therefore, the proper selection of MNA, DH and SR allows the identification of the proper solution by examining a limited portion of the search space, by reducing the computational time. The outcome of this process is a set of task assignments to the robots and grippers of the shop floor, following the approach shown in Fig. 2.11.

2.2.3 Service Oriented Approach for System Integration and Communication

In order for such a flexible system to operate, it requires a flexible way for the resources to communicate and interact. The aim is that a set of smart, interconnected resources be achieved, with embedded intelligence, capable of performing their operations and communicating with other resources. Each resource should be able to interact with other resources of the system, while from the information system's point of view, each resource software service should be able to communicate and exchange information with other services of the network. The outcome of this interaction should be an orchestrated cooperation of the machines towards implementing a flexible assembly system.

Each resource service includes three main layers, namely the Physical resources layer, the Logic layer and the Data repository layer as shown in Fig. 2.15. Moreover, each resource service consists of the Sensing module, the Control module and the Networking module. The Sensing module is responsible for sensing the process relevant parameters. Examples are the proper part grasping sensed by the gripper, the parts position, sensed by the camera system or the interaction with the environment or other robots, which is sensed by the robot. Based on these sensors' data, each resource performs local control functions. For example, when the robot arm mounted on the mobile robot, namely Robot 3, needs to pick up the gripper from Robot 2, Robot 3 will need to sense the interaction with Robot 2, to switch to a compliant behavior and perform the interaction. When messages have to be exchanged among the resources, it is done so via the Networking module. Moreover, data related to the control of the entire workflow as well as specific process data, are stored into the Data repository layer. For example, when Robot 3 needs to approach Robot 2 to pick up the gripper, this motion has to be implemented through a collision free path. In order for this behavior to be achieved, an intelligent motion planner, which is able to calculate the path, is called upon. This motion planner requires the station's CAD data as well

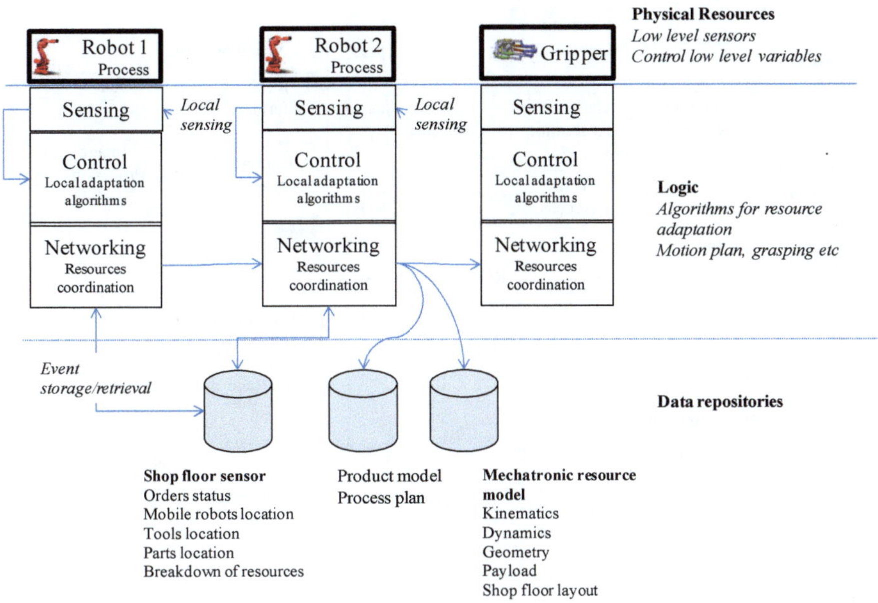

Fig. 2.15 A distributed architecture for resources coordination

as the robots' pose data at the station at a given point of time. These CAD data are available in the Data repository layer.

Furthermore, a data repository is used for the storage of the cell status as well as the currently executed operations, namely it stores the World model. When a robot or another resource has raised an unexpected event, this information is stored into the data repository, which knows at any given time, the cell's status. In this sense, the data repository acts as the shop floor "sensor", which is queried, on demand, by the other services.

Based on the service-oriented structure of the system, part of the entire system intelligence, is embedded in the various resources. The main benefit expected to be deriving from this approach, is the system's enhanced ability to reconfigure itself automatically. Using resource-level service-oriented architecture enables an open, flexible environment, where services can interact with each other from the lower levels of the resources' hierarchy up to the higher levels of the assembly line planning and control [9]. A single service bus, which is called "ROS Communication Framework", enables the manufacturing control system's traditional Hierarchical view to be transformed into a flat automation paradigm. In this framework, the communication is realized with the use of messages based on remote procedure calls (RPC) protocol which uses XML, namely the XML-RPC protocol. The structure of this integration middleware is shown in Fig. 2.16. In this approach, all the system's resources are able to communicate and exchange information either between themselves or with the information sources. The architecture integrates Robots, Grippers,

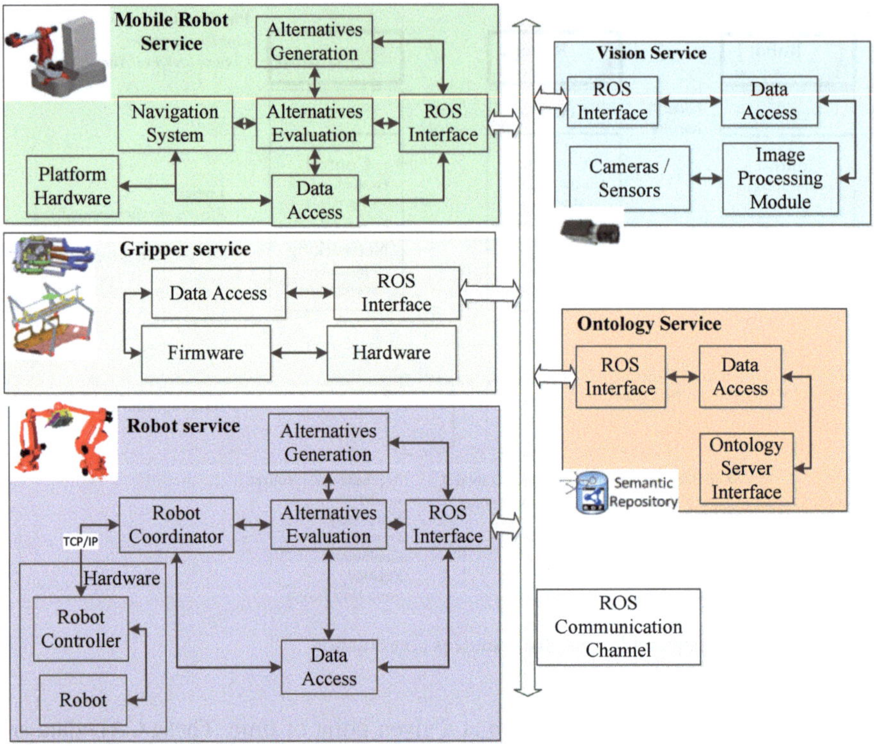

Fig. 2.16 System architecture services

Mobile platforms, Camera sensors and Data repositories which are implemented into the form of Ontologies and CAD data.

The block diagram of the entire architecture is shown in Fig. 2.16. All the resources exchange messages over the ROS communication channel. ROS stands for Robot Operating System and it has been used as the main communication middleware [17]. The ROS framework has been selected for the implementation of the communication platform, since ROS is a robotic framework oriented for implementing autonomous robot systems enabling a run time logic for shop floor control. Every resource in the architecture has a "ROS Interface" module and a "Data Access" module. The first module represents the implementation of the message exchanging mechanisms by utilizing the ROS framework, while the second is responsible for parsing the incoming messages to meaningful for the resource information or creating outgoing messages. The information of this level is consumed or produced for each resource by other software modules. The use of a common data model and a communication framework makes an open integration and communication architecture, able to integrate any existing or future manufacturing systems.

The second aspect of the integration and communication architecture is to enable the resources' autonomy, by avoiding the hierarchical control structures and so offer distributed control logic. In order for this aspect to be achieved, apart from the integration and communication framework, the architecture encapsulates the essential intelligence required by the resources in order to be autonomous. For this reason, there are two software modules, referring to the robot and mobile unit blocks and named 'Alternatives Generation' and 'Alternatives Evaluation'. The first module is responsible for creating alternative solutions when an unexpected event happens to the shop floor and the resources are required to make a decision. The alternatives created by this module are task assignments to resources: mobile unit moves to alternative docking station etc. depending on the unexpected event and the actions which are required to handle it. If there are more than one mobile robots, then all the mobile robots would be candidate for being assigned to the station that needs them. The second module evaluates the alternatives, selects the best one among them and distributes it to the relevant resources, through the ROS interface of the resource. In this way the resources are autonomous, able to deal with any unexpected situation that may occur in the shop floor and automatically reconfigure themselves.

These software services are discussed in more detail in the following sections. The interfaces of each manufacturing resource or services are described in detail. For each resource or service a table is provided describing the functions which the interface implements.

2.2.3.1 Ontology Service

The Ontology Service is composed of two software modules, namely the "Data Access" and the "Ontology Interface".

- The "Data Access" module, allows the communication of the Ontology Service with the other resources or services.
- The "Ontology Interface" module allows the communication of the Ontology Service with the "Ontology Repository". The communication mechanism between the "Ontology Interface" and the "Ontology Repository" web application is implemented using HTTP requests, while the communication between the "Data Access" and the "Ontology Interface" is implemented via function pointers. In this way, when a ROS message arrives at the "Data Access" module requesting some information from the "Ontology Repository", the "Data Access" module triggers the "Ontology Interface" module and retrieves the requested information from the "Ontology Repository".
  - The "Ontology Repository" software module is a web server application with an embedded semantic "reasoner". It serves the functionalities of storing semantic data and performing semantic queries upon them, using the embedded semantic reasoner to create the result sets. Moreover, it has the ability of using predefined semantic rules inside its reasoner.
  - The "Ontology Repository" contains the runtime data of the oprations running in the shop floor. These data are the online services and resources, participating in the platform, the shop floor pending tasks, the suitability between

the resources and the pending tasks. For this reason, the shop-floor ontology, describing all the appropriate semantic data has been developed.

- The "Ontology Repository" software module communicates with the rest of the platform, through the 'Data Access' module, as it is done in the robot architecture. In this case, the 'Data Access' module forms a set of services published to the platform as the 'Ontology Services'. The use of these services enables the participating resources to store and retrieve data from the Ontology Server.

The "Ontology Services" interface comprises the following methods:

- *registerResource* and *registerService:* The *registerResource* and *registerService* methods are called upon in order to inform the Ontology that a new resource or service has just been connected to the platform and should be registered. Every resource or service, participating in the framework, should call for the appropriate method to register itself to the platform. When registered, the Ontology knows that this resource or service is ready, online, to serve shop floor tasks. Only registered services are taken into consideration when a rescheduling is performed for the pending shop floor tasks. Furthermore, the Ontology Service, by analyzing the broadcasted messages, remains aware of the unexpected platform events and the resource or service break downs and updates the Ontology information respectively.
- *executeQuery (string SPARQLQuery)*: It allows the platform resources to query the Ontology and retrieve information from it. For instance, the resource that performs a rescheduling, queries the Ontology about the existing shop floor pending tasks, the online resources and their availabilities, the suitabilities among the resources and the pending tasks.
- *updateQuery (string SPARQLQuery)*: It allows the platform resources and services to update the existing information lying in the Ontology or feed it with new ones. For example, when the robot agent finishes the rescheduling task, it calls for this method so as to supply the Ontology with the new assignments and update the already existing ones.
- *retrieveSchedule*: This service is called for by every Robot's agent, after a rescheduling task has taken place. After calling these services, the Data Access module of the Ontology, will make a query at the Ontology Repository about the Robot's tasks and operations. A list of the assigned tasks will be returned as a response.

In Fig. 2.17, the Ontology Service integration and communication software architecture is shown.

The "Ontology Repository" is implemented as a web application and it is hosted on a web based server called Apache Tomcat [18]. An Ontology was implemented and stored into the "Ontology Repository" following a defined data model.

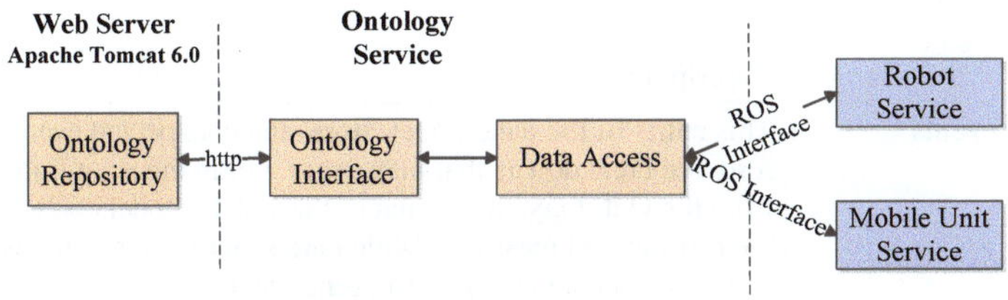

Fig. 2.17 Ontology service structure

2.2.3.2 Ontology Entities

The entities that have been implemented are described herein. This data model should encapsulate all the information required for enabling the resources and services to be autonomous and perform the decision making on their own. The shop floor is described both in terms of physical elements and in terms of operations. In the following sections, the elements are classified into categories and then the actual classes and their relationships are described. The shop floor is described both in terms of its physical characteristics and the resources lying in it, apart from the processing that takes place in it.

Entity	Description
Shop floor layout	The shop floor is composed of assembly lines and each assembly line has its stations, where the manufacturing resources are found
Docking stations	The docking stations are positions where the mobile unit can dock. For each docking station, information about the station where the docking station is placed exists in the data model. Moreover, information as to which types of robots can become docked into the stations will be included in the data model
Resources	A resource can be a gripper, a robot or mobile unit. Each resource contains information about the processes it is able to perform, its geometrical characteristics and its current position in the shop floor
Products—Subassemblies	The products and their subassemblies are also described in the data model. The geometrical characteristics of these components are included along with their welding spots. Moreover, for each subassembly the real time coordinates and orientation are included in the data model. These values are calculated, in real time, by the vision systems and are utilized by the network resources
Shop floor orders	The data model describes the manufacturing tasks using the order, job, task model
Scheduling	The data model contains information about the task scheduling within the shop floor. The assignments of the pending tasks to the resources are stored into the data model

(continued)

(continued)

Entity	Description
Shop floor events	This entity of the data model, stores the unexpected events, such as resource breakdowns that may occur within the shop floor. Apart from the Ontology, the format of these data model classes is used by the broadcasted messages, which are sent by the resources and services when an unexpected event occurs
Station geometry	In order for automated motion planning and collision avoidance capabilities to be carried out, it is required that a geometrical representation of the stations be made showing where this functionality is implemented. The geometrical representations are implemented using URDF formatted files, while the motion planning functionality is implemented using the MoveIt library [17]

In addition to the Ontology interface, the Ontology service is also able to receive messages broadcasted by other resources or services. In this way, it remains updated with the events occurring over the shop floor. If for example a resource breaks down, it will broadcast an appropriate message to inform the other resources and services about its failure. The Ontology service will also receive this message and update its data at the Ontology repository. Next, one of the resources will perform a rescheduling in order to assign the pending tasks of the broken down resource to another resource. To achieve this, this resource will exploit the reasoning functionalities of the Ontology service to retrieve the possible alternative resources that can perform these pending tasks. In this way, since the Ontology service is aware of the functioning and malfunctioning resources, it will be able to respond properly with the list of the possible alternative functioning resources.

2.2.3.3 Robot Service

The robot integration and communication software architecture involves the following two main computing components, as shown in Fig. 2.18:

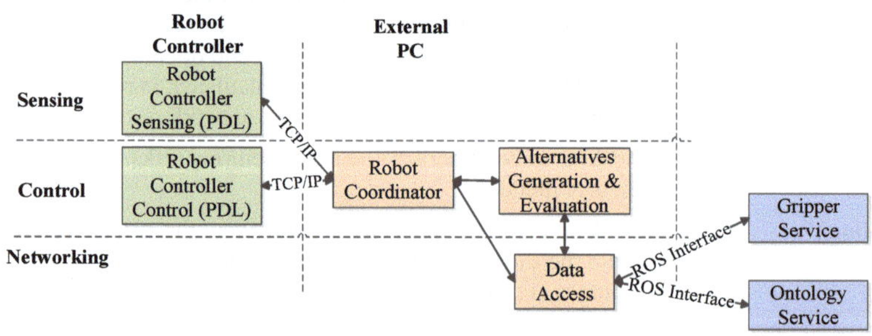

Fig. 2.18 Robot service modules

- The Robot controller. This controller runs a software written in the controller's language. This module is able to control the mechanical moves of the hardware robot as well as to provide sensing capabilities that are related to the robot's performance. This module communicates via its TCP/IP interface.
- The Robot service runs on a computer. It is responsible for communicating with the other services of the network, making decisions and sending the proper commands to the Robot controller.
  - The Robot service runs the "Robot Coordinator" module, which communicates with the Robot controller.
  - The "Robot Coordinator" is a software program that runs on the external PC, connected to the robot controller and being responsible for the robot's coordination.
  - The "Robot Coordinator" retrieves the manufacturing task descriptions from the Ontology Service through the "Data Access" module and the list of its assigned tasks either from the "Alternatives Evaluation" module or from the Ontology Service through the "Data Access" module. The data exchange between these software modules is implemented through function calls.
  - Moreover, the "Robot Coordinator" coordinates the robot's mechanical movements and monitors the execution of the manufacturing tasks by analyzing the task descriptions.
  - Furthermore, the "Robot Coordinator" monitors the robot and can communicate this information to the other resources or services, when necessary, i.e. in case of a resource breakdown or any other unexpected event.
  - The "Robot Coordinator", the "Alternatives Generation" and the "Alternatives Evaluation" modules, constitute the logic layer of the robot service.

The state diagram of the Robot service is shown in Fig. 2.19 and it works as follows:

- **Start Up**: When powered on the service enters at state STARTUP. At this state validates that all the robotic hardware as well as the networking works properly
- **Idle**: At this state it registers itself to the Ontology, it subscribes to the central ROS topic and waits for a start message to be published from the operator.
- **Negotiation**: When a "Start" message is published to the ROS topic, the robot service goes to the NEGOTIATION state where it exchanges negotiation messages with the other resources to select the resource which will perform the rescheduling.
- **Rescheduling**: After the negotiation, if it is decided that this resource is responsible for running the scheduler, it enters the RESCHEDULING state. At this state it performs the rescheduling of the manufacturing line by triggering an external java program for Alternative Generation and Evaluation.
- **Wait Schedule**: After the negotiation, all the other robots, apart from the one that will be responsible for performing the rescheduling, enter at the WAIT-SCHEDULE state, where it waits for being notified that the scheduling is over and it can query the Ontology for its service.

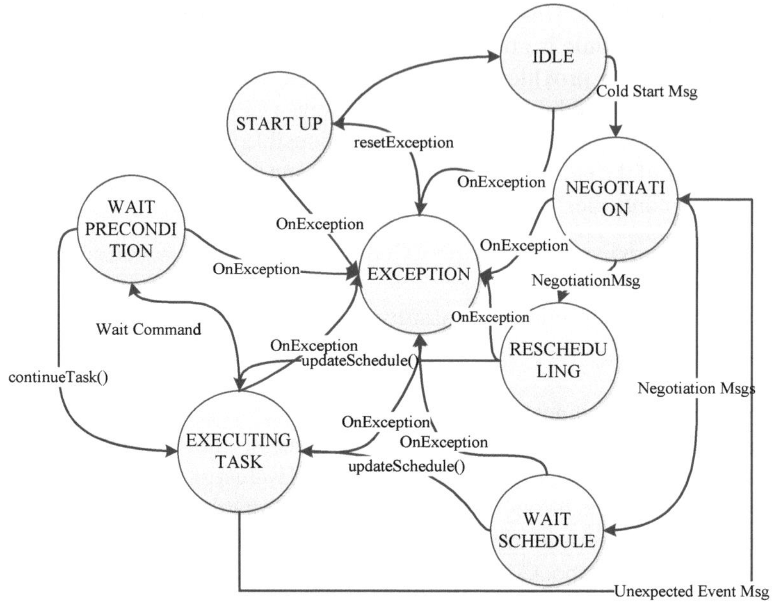

Fig. 2.19 Robot service state diagram

- **Executing Task**: When the rescheduling is performed and the new schedule is populate on to the Ontology Service the robot services are informed through the updateSchedule function and enter the EXECUTING TASK state. At this state they perform the pending tasks they are assigned to them.
- **Wait Precondition**: If one of the resources starts performing a task with precondition another pending task it enters the WAIT PRECONDITION state where it waits for the pre-condition task to finish and be notified by the corresponding resource.
- **Exception**: If in any of the states an exception happens, the robot service enters the EXCEPTION state while also publishing an unexpected event message at a ROS topic, where every other resource can receive it.

The interface of the robot resource is show in Table 2.2. This interface includes the messages exchanged among the manufacturing resources to inform each other that some manufacturing process has finished, the exchange of the robot user frames and the control of the tool changer.

The implementation of this interface utilizes the ROS framework and the C++ language. Further to this interface, the robot resource is able to broadcast ROS messages to the other resources and services to inform them about unexpected situations. One such example is when there is a failure at the robot arm hardware. In this case the Data Access software module of the Robot Service will broadcast an unexpected event message informing the other resource and services that there is an unexpected failure to this resources so that the other resources will decide to assigned

Table 2.2 Robot interface

Function name	Description
Update schedule	Triggered for informing the robot service that there are new task assignments for it. When this method is called, the robot service consumes the Ontology service interface to retrieve the new assignments using the 'executeQuery' method of the Ontology Service Interface
Continue task	Used for informing the robot resource that it should continue executing its next pending task. When called, the resources continues executing its pending manufacturing tasks
Get reference frame	Returns the reference frame which the robot is using
Update reference frame	Updates the reference frame of the robot
Get tool coordinates	Return the coordinates of the end effector attached to the robot. The coordinates will be based on the robot's reference frame
Get available tool changer head	Returns the free head of the attached to the robot dual tool changer
Release tool	Send the appropriate commands to the hardware robot to mechanically detach the tool changer

the broken resource's pending task to another resource capable of performing these manufacturing tasks.

2.2.3.4 Mobile Robot Service

The mobile unit's architecture is composed of two layers. In the first layer (Layer 1) the algorithms that deal with the platform's navigation are implemented and this layer has direct access to the hardware platform, its firmware and its sensor's I/O drivers. This layer provides both information on the navigation and on the state of the mobile unit to the second layer. The second layer (Layer 2) is composed of the decision-making modules, namely the Alternatives Generation & Alternatives Evaluation modules and the Data Access module. The Data Access module is responsible for the implementation of the Mobile Unit Interface, which is accessible from the other services of the network and it is further responsible for sending and receiving messages to and from the other network services. The implementation of the communication between the Layers 1 and 2 utilizes the ROS service mechanism (Fig. 2.20).

The mobile unit interface is described in Table 2.3. The three functions supported by this interface are implemented using the client server scheme of the ROS interface.

Further to the supported functions of its interface, the Mobile Unit is able to broadcast messages to inform the other resources about its state. The dock and undock actions will be followed by messages. In particular, every time the mobile platform undocks from a docking station or docks to one, a message is broadcasted to inform

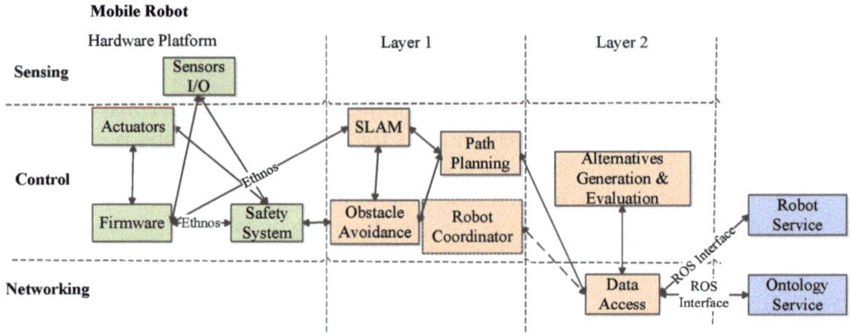

Fig. 2.20 Mobile Robot service modules

Table 2.3 Mobile unit interface

Function name	Description
Get current position	Returns the position of the mobile unit on the shop floor
Navigate to position	Commands the mobile robot to navigate to a new position
Get path info	Returns the distance that the mobile robot will travel as well as the estimated time for travelling this distance

the other resources and the Ontology service about this event. Moreover, when an unexpected event happens i.e. that the mobile platform has broken down due to a hardware failure, the mobile unit will broadcast an unexpected event message to inform the other resources and the Ontology service about its failure.

2.2.3.5 The Mobile Robot Relocation Task

Following the decision making logic that was previously described, a series of tasks have been assigned to each one of the resources. The following logic has been implemented in order to enable the relocation of a mobile robot on the shop floor, as shown in Fig. 2.21. Upon reception of an UNEXPECTED event on the ROS network, the scheduling logic that was previously discussed, assigns tasks to the various resources. In this particular instance, the mobile robot is assigned to move from its current location to a new one and take the gripper from the robot that communicated the UNEXPECTED event.

The mobile robot service logic, checks which end effector is required for the task that it has to perform. It checks if it already carries such an end effector and if not, it releases this tool. For instance, when the mobile robot is in the Loading area it carries the Dexterous gripper, which is not suitable for performing cooperative welding with Robot R3 in the welding area. So at this stage, Robot R1 releases the gripper it carries on its tool stand.

Next, the Mobile Robot R1 service, queries the Ontology in order to find out where from it can get the tool it needs for its task; the result is that the tool is mounted on

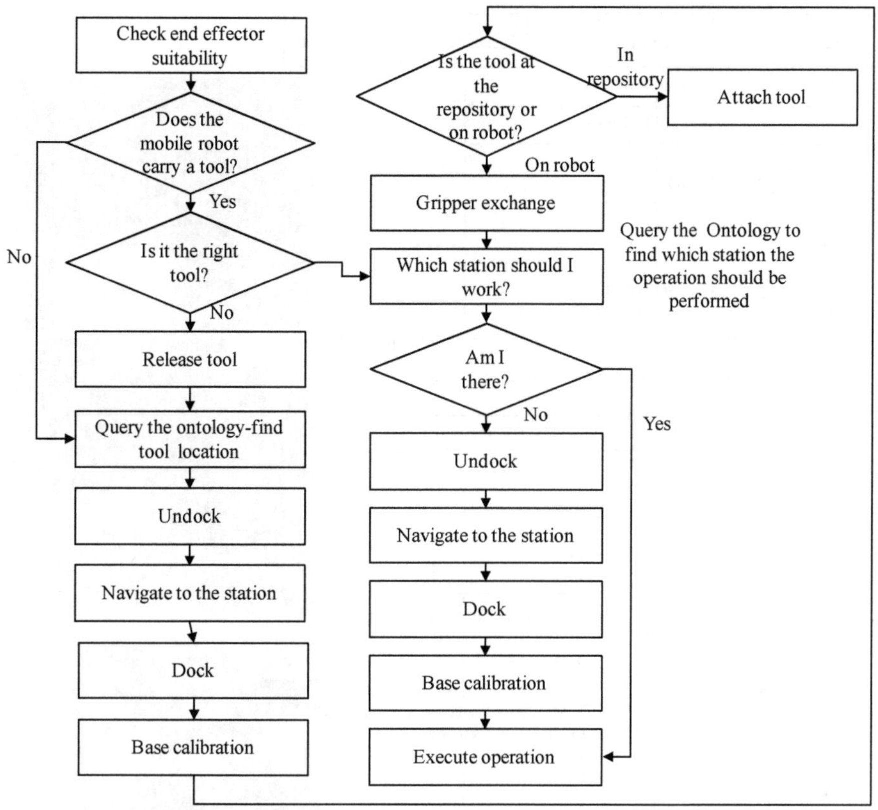

Fig. 2.21 Mobile robot task assignment and execution

the robot R2 that broadcasted the UNEXPECTED event. Based on that, the mobile robot R1 undocks, navigates to the station where R2 with the gripper are located and comes into dock.

2.2.3.6 The Gripper Exchange Task

At this stage, the image processing software is called upon for the calculation of the exact position and orientation of the mobile robot after docking, since such a calibration is required for minimizing alignment errors. Discussion on these errors is beyond the scope of this paper, but it is going to be done in the future.

In this new location, R1 and R2 exchange the gripper, following the logic that was shown in Fig. 2.22. After the gripper has been transferred from Robot R2 to Robot R1, Robot R1 can continue executing the operations that reside under the task assigned to it. In this case, these operations have to do with handling the gripper and the part in order for the Robot R3 to continue welding.

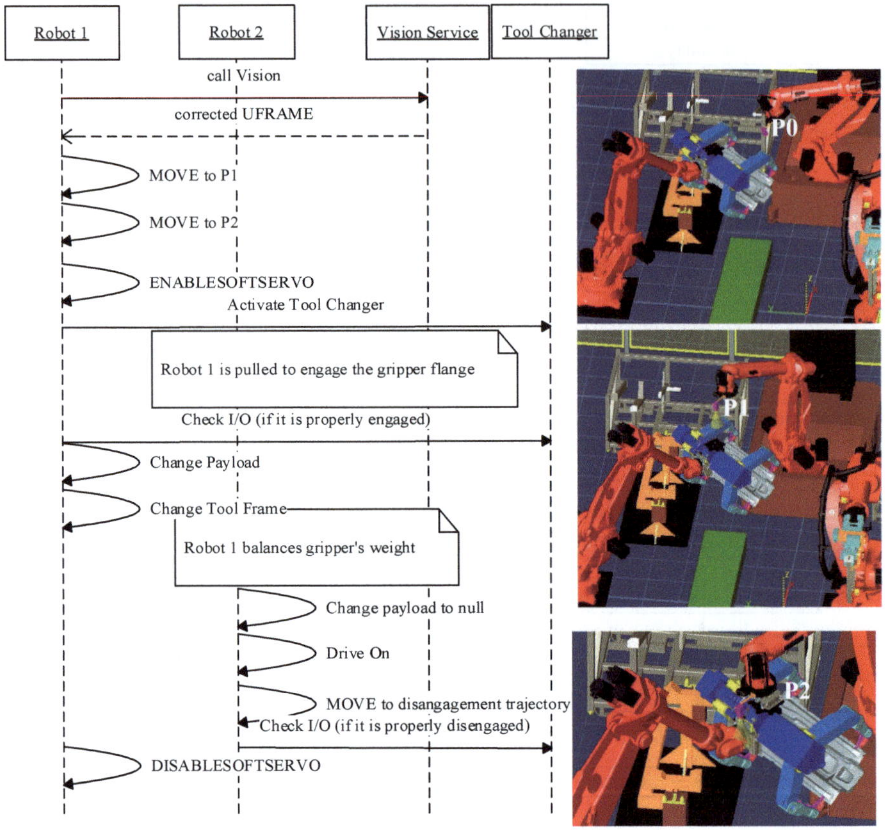

Fig. 2.22 Gripper exchange process flow

2.2.3.7 Gripper Service

Each gripper in such a flexible system, implements the control architecture shown in Fig. 2.23.

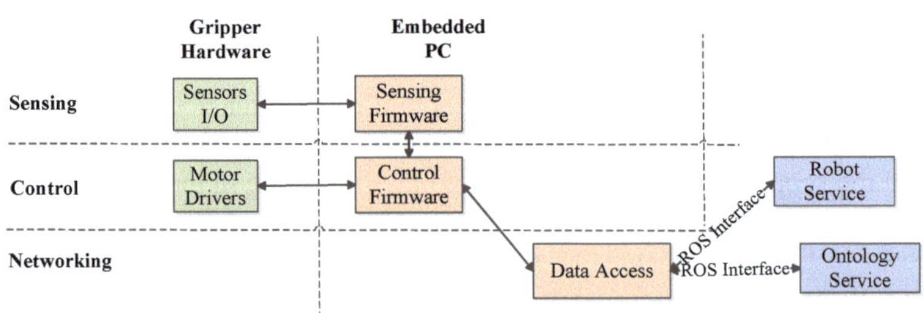

Fig. 2.23 Gripper service modules

The Gripper is composed of two software modules, the "ROS Interface" and the "Data Access" module being responsible for the. Apart from these two modules, there is also the "Firmware' module", which controls the hardware parts. Furthermore, it communicates with the "Data Access" module to receive commands contained in the messages that will be exchanged with the other resources or services.

The data from the sensors in the Gripper are read directly by the embedded PC unit, which will use the relevant sensorial information for the control and management of the Gripper. The kinematics of the Gripper and every relevant calculation runs on the embedded PC. The interfacing of the Gripper to the carrying robot is implemented through ROS.

The Data Access module serves merely as a data exchange unit or protocol/coding translation server, converting ROS data into register status. The physical link between the Data Access unit and the other services is implemented through Wi-Fi. It works independently from the mechanical plugging of the Flexible Gripper.

The interface of the Gripper can be seen at Table 2.4.

Further to its ROS interface, the Gripper broadcasts ROS messages to all of the resources in case of an exception. These messages show the hardware status of the gripper i.e. the power status, the clamps status or any other alarm. Similar to the Dexterous Gripper all these messages will be unexpected event messages and are handled by the logic and control layer of the resources.

Table 2.4 Gripper interface

Table	Description
Reconfigure	Passes from any current status to the status with arm configuration assigned
Close clamps	Implements the clamps closing functionality
Open clamps	Implements the clamps opening functionality. parking), performing a number of checks such as the tension of the backup battery is checked, if a work piece is hold, it is checked that the gripper is in closed status etc
Transfer	Prepares the gripper to a transfer from one resource to another one (either robot-robot or robot-tool stand)
Park configuration	Passes from any current status to parking configuration status (the parking configuration is with arms wrapped and clamps closed). All motors are powered; the joints are unlocked and unlock status is verified; the motors are operated to move the axes to the goal configuration; the joints are locked; a message of parked gripper is dispatched
Move arm	Allows an external ROS node to call this function and move each arm of the gripper separately on demand
Reconfigure	Informs the Gripper that it should reconfigure itself in order to be able to grab a specific subassembly part. The gripper must reconfigure itself according to the part which is to be picked
Aquire part	Closes the fingers for acquiring the part
Release part	Open the gripper fingers for releasing the part

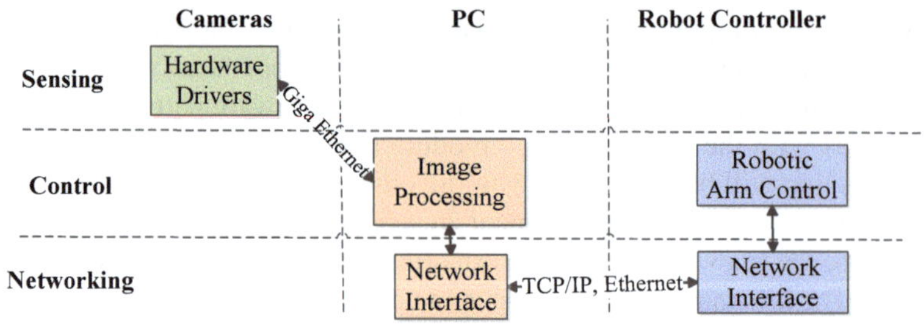

Fig. 2.24 Vision service modules

Table 2.5 Vision system interface

Function name	Owner module
Use visionservice	The identifier of the part to be grabbed is provided to this function. The Image processing system will then provide the correction of the user frame in order the robotic arm to be moved to the appropriate grasp position of the part

2.2.3.8 Vision Service

The vision system software service is composed of one camera on the robot's sixth axes that is connected to an external PC through a Giga Ethernet interface. The external PC hosts the Image processing algorithms, as a module capable of grabbing frames from the camera and processing them. The image processing module can perform image analysis and communicate with the other resources or services to exchange image processing result data, upon request, through the Ethernet protocol (Fig. 2.24).

Table 2.5 summarises the interface provided by the vision service.

2.2.3.9 Exception Handling Messages

A ROS topic is implemented and all the resources and services subscribe to this topic. Utilizing this topic, each resource or service is able to broadcast messages to all the other resources or services sending an unexpected event message. In order to perform exception handling, the unexpected event message has been defined. The structure of this message can be seen in Table 2.6.

2.3 Real World Implementation of the Robotic Cell

The real world implementation of the flexible system discussed above, has been implemented in the course of the European funded research project AUTORECON. The robots used were a COMAU NJ4 170 mounted on a mobile platform, which is

Table 2.6 Unexpected event message

Member	Description
Resource ID	The ID of the resource retrieved when registered to the Ontology Service
State	The state of the resource. Negative numbers are used for the state member of this message to declare that the resource or service has broken down. When such messages are received a dynamic rescheduling is performed to handle this situation
Description	The description of the exception

the Robot R1, a COMAU NJ 220 which is the Robot R2 and a COMAU NH4 200 is the welding robot R3. The mobile platform and grippers were built in this research by cooperating research partners. The mobile platform was also built in the context of this research. This flexible assembly cell was implemented in the premises of an automotive vehicle manufacturer (Fig. 2.25).

During the case study, the COMAU NJ4 Robot must firstly load 9 automotive parts of different size and geometry on a fixture. Having a dexterous gripper mounted on it, the robot moves, to the racks, grasps each part and the moves to the fixture in order to release it. Before approaching each part, the gripper must reconfigure itself for having the fingers at the right position in order to enter the holes and lock the parts. The reconfiguration is being achieved by calling the "reconfigure" ROS service, passing as argument the id of each part. After entering the holes, the gripper locks the parts with the "acquire_part" service call of the Dexterous Gripper Service (Fig. 2.23. Gripper service modules). Next, the Robot moves to the mobile fixture and places the part at the right position using the "release_part" ROS service call.

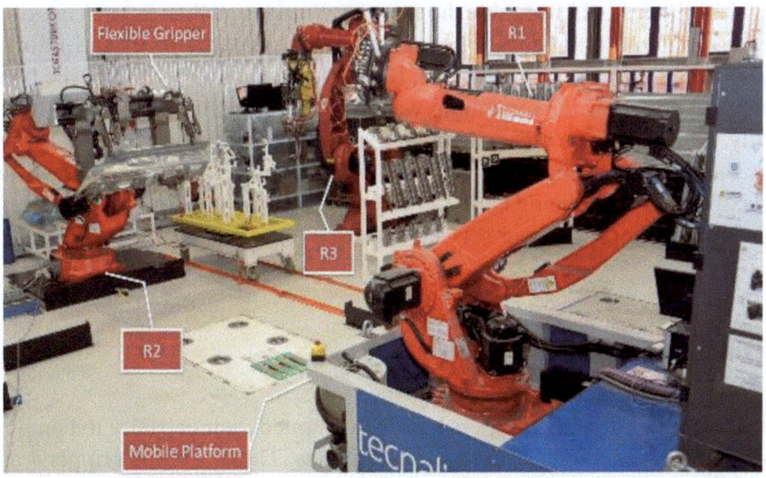

Fig. 2.25 Case study robotic cell

After all the parts were loaded on the fixture, the execution stops and a human operator enters the robotic cell in order to move the fixture to the welding area. After the fixture is moved to the welding area, a resume command is sent using the GUI to all the resources which enter into execution phase and continue their tasks. Initially, the welding R3 robot performs some tack welding (geometry spots) on the parts placed on the fixture. This welding task is consisted of several "move" and "weld" operations that are also stored at the Ontology repository and sent by the Robot Service to the Robot controller of the Welding Robot. The next task is a cooperative welding task, during which R2 handles the tunnel in different positions and R3 performs some welding operations.

After executing the above, a breakdown event of R2 is thrown by the robot service. This robot cannot continue the tasks assigned to it and the production line must reconfigure itself in order to take care of this unexpected event. A ROS message is transmitted and informs all the resources about the break down. Then, all the resources stop their work immediately after completing their current operation, they negotiate between each other and one of them is chosen to call the planning module in order to perform a rescheduling. The planning module again reads the remaining tasks and operations from the Ontology and, given that R2 is not available, it assigns its tasks to the only resource that can take its place: R1. So, after writing again the assignments in the Ontology repository, the resources are informed via the "retrieveSchedule" ROS service to retrieve their schedule and perform their work.

After querying the Ontology about the place where the job should be performed and about the tool that it needs, R1 undocks from the loading area docking plate, navigates to the welding area and docks to the docking plate at the welding area in order to take the flexible gripper from R2. Using its integrated vision system, R1 makes the fine tuning and enter into the free flange of the flexible gripper. Before entering, it notifies the R2 to enable the soft-servo mode and then enter drive-off state. After entering into the flange (tool changer), R1 locks the gripper and informs R2 to unlock using ROS service call. Then, R1 moves away and continues the handling task while R3 performs the remaining welding operations.

Finally, R1 releases the tunnel on the fixture and moves in order to pick up another part using a different configuration of the flexible gripper. The "reconfigure" service of the flexible gripper is being called in order to change the pose of the gripper arms and the "close_clamps" gripper service was used to grab and pick up the part.

2.4 Discussion

The introduction of flexible grippers has demonstrated their capacity to handle more than one part or even more than one subassemblies. The dexterous gripper was used for handling nine parts of different geometries, therefore, there is a high potential from this point of view. The placement of a robotic arm on a mobile platform has enabled the transfer of robot arms from one location in the shop floor to another. The reconfiguration potential was increased by this development resulting in a number of advantages.

The main advantages of assembly lines based on mobile robots are [19]: higher reconfigurability, reduced duration of breakdowns, lower commissioning time, higher reliability and flexibility, minimum need for human intervention due to their autonomous behavior and higher production variability. The findings of the case study from the automotive indicate that the addition of mobile robots increases the production volume of the line due to its higher response to breakdowns and the shorter period of time required for its reconfiguration. The mobility also results in the system's higher utilization and availability, thus rendering the line even more efficient.

Programming such reconfiguration, is a tedious process that has to be well prepared and verified, paying special attention to the interlocking and sequencing signals. This logic enables the performance of such reconfigurations on the fly, in an automatic manner. In this case, it would take around five working days to program the interaction and communication of the various resources, using a conventional programming approach. The specific method enables the programming of such communication in a matter of a few hours, ranging from four to eight. This corresponds to an 80% reduction for about four days, compared to that of the conventional approach.

References

1. Chryssolouris G (2006) Manufacturing systems: theory and practice, 2nd edn. Springer, New York
2. Scholz-Reiter B, Freitag M (2007) Autonomous processes in assembly systems. CIRP Ann Manuf Technol 56:712–729. https://doi.org/10.1016/j.cirp.2007.10.002
3. Koren Y, Heisel U, Jovane F, Moriwaki T, Pritschow G, Ulsoy G, Van Brussel H (1999) Reconfigurable manufacturing systems. CIRP Ann Manuf Technol 48:527–540. https://doi.org/10.1016/S0007-8506(07)63232-6
4. Michalos G, Makris S, Papakostas N, Mourtzis D, Chryssolouris G (2010) Automotive assembly technologies review: challenges and outlook for a flexible and adaptive approach. CIRP J Manuf Sci Technol 2:81–91. https://doi.org/10.1016/j.cirpj.2009.12.001
5. Michalos G, Makris S, Chryssolouris G (2015) The new assembly system paradigm. Int J Comput Integr Manuf 28:1252–1261. https://doi.org/10.1080/0951192X.2014.964323
6. Ranky PG (2003) Collaborative, synchronous robots serving machines and cells. Ind Robot: Int J 30:213–217. https://doi.org/10.1108/01439910310473915
7. Krüger J, Wang L, Verl A, Bauernhansl T, Carpanzano E, Makris S, Fleischer J, Reinhart G, Franke J, Pellegrinelli S (2017) Innovative control of assembly systems and lines. CIRP Ann 66:707–730. https://doi.org/10.1016/j.cirp.2017.05.010
8. Makris S, Michalos G, Eytan A, Chryssolouris G (2012) Cooperating robots for reconfigurable assembly operations: review and challenges. Procedia CIRP 3:346–351. https://doi.org/10.1016/j.procir.2012.07.060
9. Colombo AW, Bangemann T, Karnouskos S, Delsing J, Stluka P, Harrison R, Jammes F, Lastra JL (2014) Industrial cloud-based cyber-physical systems. Springer International Publishing, Cham

10. Leitão P (2009) Agent-based distributed manufacturing control: a state-of-the-art survey. Eng Appl Artif Intell 22:979–991. https://doi.org/10.1016/j.engappai.2008.09.005
11. ElMaraghy HA (2005) Flexible and reconfigurable manufacturing systems paradigms. Int J Flex Manuf Syst 17:261–276. https://doi.org/10.1007/s10696-006-9028-7
12. Malec J, Nilsson A, Nilsson K, Nowaczyk S (2007) Knowledge-based reconfiguration of automation systems. IEEE, pp 170–175
13. Tesla Gigafactory | Tesla. https://www.tesla.com/gigafactory. Accessed 30 Apr 2020
14. Michalos G, Sipsas P, Makris S, Chryssolouris G (2015) Decision making logic for flexible assembly lines reconfiguration. Rob Comput Integr Manuf. https://doi.org/10.1016/j.rcim.2015.04.006
15. Chryssolouris G, Dicke K, Lee M (1992) On the resources allocation problem. Int J Prod Res 30:2773–2795. https://doi.org/10.1080/00207549208948190
16. Michalos G, Makris S, Chryssolouris G (2008) An approach to automotive assembly cost modelling. In: Proceedings of the 2nd cirp conference on assembly technologies and systems. Toronto, CANADA, pp 478–487
17. Yoonseok Pyo DL, Cho H, Jung L (2017) ROS robot programming (English). ROBOTIS
18. Brown S (2001) Professional JSP 2nd Edition. Wrox Press Inc
19. Michalos G, Kousi N, Makris S, Chryssolouris G (2016) Performance assessment of production systems with mobile robots. In: Procedia CIRP. Naples, pp 195–200

Chapter 3
On the Coordination of Multiple Cooperating Robots in Flexible Assembly Systems Using Mobile Robots

3.1 Introduction

Typical manufacturing systems comprise rigid flow line structures by employing model-dedicated handling and transportation equipment of raw materials and components [1].

To address the need for handling market, product and plant variability, solutions that allow the production system to adapt to both planned and unplanned fluctuations and eventually achieve an autonomous operation of the complete system are required.

3.1.1 Manufacturing System Integration and Control

Starting from traditional production lines, it can be observed that they follow a rigid flow line structure with model dedicated transportation equipment, they conform to a fixed control logic and the signal based task sequencing creates need for high manual efforts in the case where changes are required. A single supervisor is responsible for planning and monitoring the system's operation, taking into account large pieces of information, such as process plans, due dates, processing times, set-up times, equipment status and the relations among these variables.

In Fig. 3.1, the representation of an assembly line with multiple stations and resources (R1, R2 etc.) is shown along with the tasks' breakdown into operations for each resource. The current practices involve the use of Programmable Logical Controller (PLC) signals to denote the start/stop of the operations, requiring a hard-coded approach that signifies high complexity and downtime in case of changes [1].

According to this logic, task sequences should be modelled in the form of PLC signals, as it is shown in Fig. 3.2.

S. Makris, *Cooperating Robots for Flexible Manufacturing*,
Springer Series in Advanced Manufacturing,
https://doi.org/10.1007/978-3-030-51591-1_3

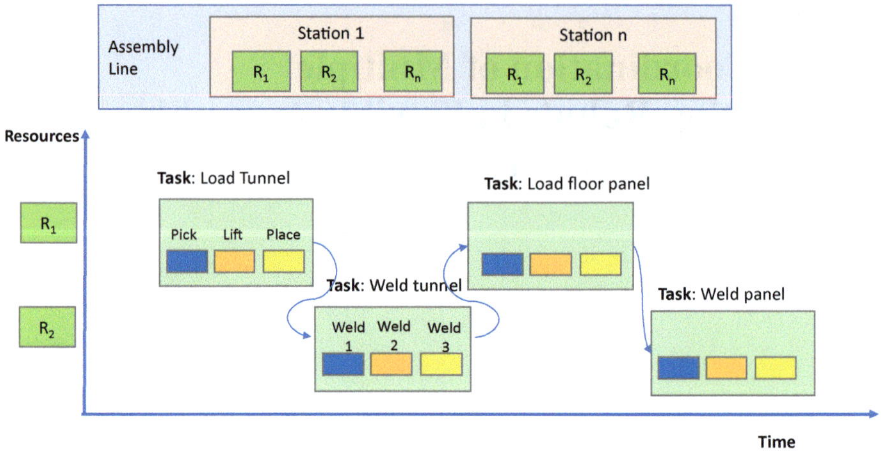

Fig. 3.1 Hierarchical model of production line and operations

The coordination of all the resources' tasks requires a substantial amount of signals and that this amount grows according to the amount of the tasks. This would mean that the PLC programmer should consider a very high number of signals in order to model all of them for the creation of a workflow, as shown in [1]. This is practically infeasible as the PLC programmer can neither identify all the possible alternative scenarios nor can he implement all of these signals in a PLC program that would control a flexible manufacturing system. Therefore, in the current practice, tasks are sequenced based on people's experience and there is no way to be automatically re-programmed. The only way that a sequence of tasks can be modified is by having these signals re-programmed. Alternatively, heterarchically controlled architectures lead to reduced complexity by localising information and control, for the reduction in software development costs by eliminating supervisory levels [2]. However, such a distributed system requires distributed communication architectures.

Based on the above, conventional production systems cannot follow the market needs, for fast introduction of new products or frequent improvement of the existing ones. New production systems need to exhibit attributes such as flexibility, reusability, scalability and reconfigurability [3–5]. The future vision is to have more autonomous and exchangeable production units, using highly interactive robotic structures, enabling random production flows and all that, implemented under a common and open communication architecture.

In order for these goals to be achieved, alterations of the production and logistics processes are required to enable the system's fast reconfiguration with minimal human intervention [6]. The current problems addressed by this approach may be summarized into the following:

- Reduction of hard wired control logic that allows limited or no reconfiguration capabilities and requires great effort in terms of human intervention. Activities such as those of scheduling, planning and programming of resources are now partially or individually automated. As a result, a significant reduction in the overall system reconfiguration time is expected.

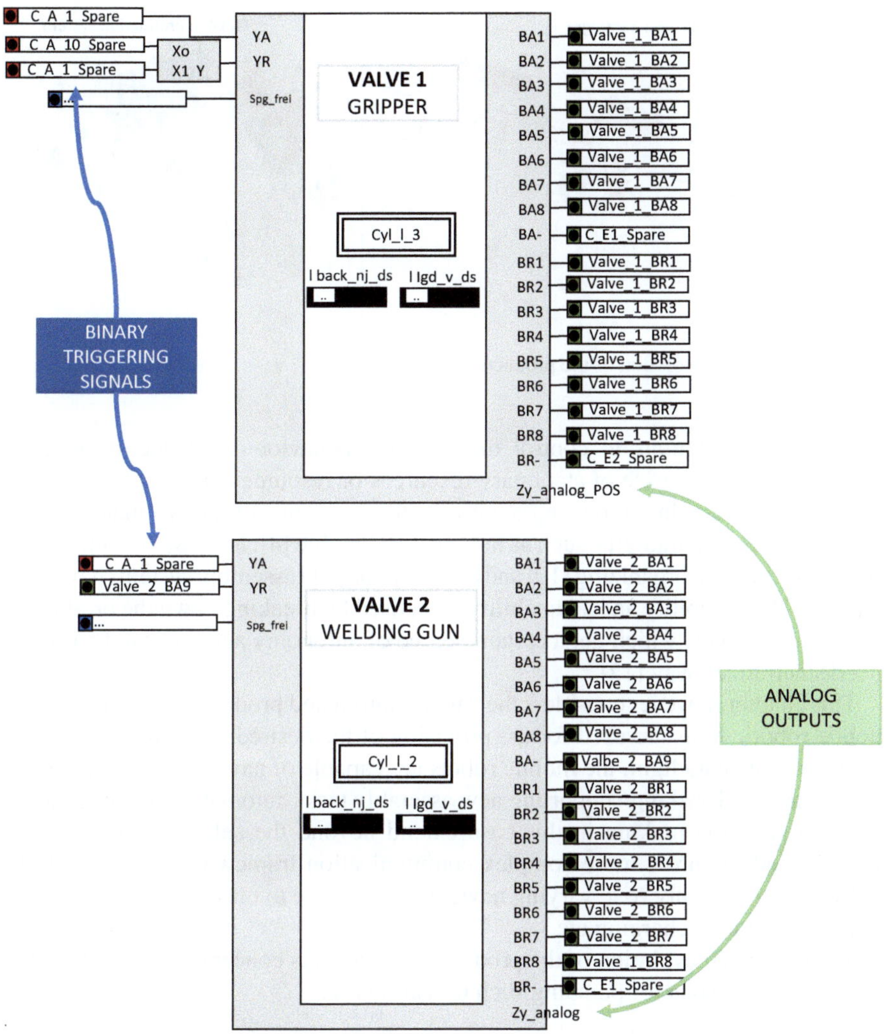

Fig. 3.2 Indicative example of signals required to control a gripper and a welding gun

- Reinforcement of random production flows through the use of mobile robots, eliminating the existing fixture based—static production paradigms that do not allow for changes in the production system structure.
- Autonomous behavior—planning of activities at multiple levels. Currently, autonomy is constrained by rules that are imposed by the strictly specified task execution routines for each resource. Robots however, can execute the same task, in a multitude of ways, but are now limited by the human dictated programming and planning. A significant reduction in programming efforts will be achieved.

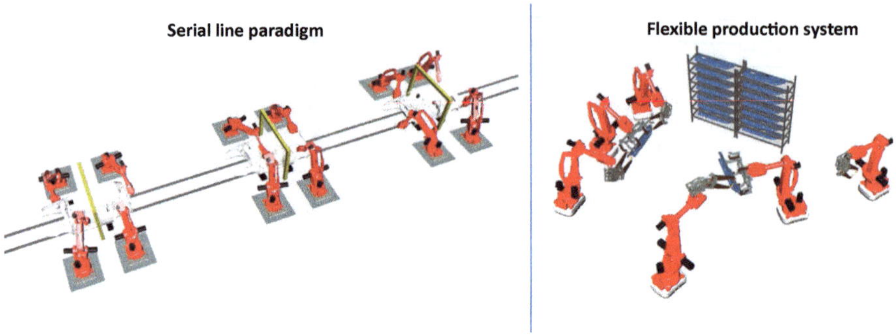

Fig. 3.3 Envisaged evolution of the production system

Up to now attempts to control the dynamic behavior of production systems have focused on the case of stationary resources or resources that can be relocated after rigorous planning, requiring a considerable amount of time. To manage these dynamics several paradigms such as holonic [7], flexible [4], lean [8], reconfigurable [9], evolvable, self-organizing [10] and autonomous [11] assembly systems have been realized. The flexibility and adaptability is realised by breaking down the production system in modules, which get a certain degree of autonomy and control themselves in a decentralized way [12].

This chapter however considers the case of automated production systems, where mobile robotic units are used for the provision of the desired reconfiguration capabilities. In this paradigm, the mobile robots are capable of navigating into assembly stations and undertaking/supporting new assembly tasks automatically. The requirement for decision making in this case extend beyond the selection of a suitable resource and require a more complex communication framework to deal with the real world uncertainty (e.g. varying navigation time due to obstacles, availability of mobile unit due to charging etc.)

The envisaged evolution of the production systems is conceptualized in Fig. 3.3. The main reasons for pursuing such transition are:

- To reduce the reconfiguration time required when introducing a new product or altering an existing ones
- To enable the implementation of random production flows by exploiting the flexibility of the robot and eliminating the fixed conveyors and fixtures
- To enable flexible and reusable tooling that can be used over multiple generations of products
- To achieve an autonomous behavior this reducing the need for human intervention in the case of breakdowns or unforeseen events
- And finally to reduce programming efforts by integrating or relevant information in repositories that are accessible through an open architecture.

3.1.2 Mobile Robots and Manipulators

Different attempts have been made so far to introduce mobile manipulators to industrial environments and exploit their flexibility potential [13]. The latest examples involve the introduction of a mobile manipulator for assembly applications [14] as well as the creation of an autonomous multi-purpose industrial robot [15]. Such robots mobile robots can overcome uncertainties and exceptions by using coordinated base and manipulator control, combined visual and force servoing, and error recovery.

The mobile robotic arm, developed by Henkel and Roth, can move at 1 m/s carrying a 10 kg payload robot [16].

High payload mobile robots for automotive Body in White (BiW) applications have been developed [17] while certain of them have become commercially available [18]. Nevertheless not many instances in high volume production have been identified. One of the most important problems of deploying mobile robots is that the environment around them is not static. Therefore, the mobile units should be capable of changing their path in case of any alterations in their surroundings [19–21].

The exploitation of the flexibility potential in systems that utilize mobile robots, signifies the definition and solution of a complex planning and scheduling problem. The first part of the problem deals with the production planning level (identification of tasks) and secondly, the scheduling part that deals with the assignment of these tasks to the resources [22]. Agent based systems have been the main research direction that has been followed towards addressing these problems [23, 24].

In most of the aforementioned cases, the mobile robots act as individual units that execute the pre-programmed tasks in the production schedule.

Most recently, the investigation of new assembly system paradigms that rely on mobile robotic units have come into focus [25]. It has been discussed that the combination of mobile robots with flexible and actively reconfigurable grippers results in production systems with higher performance in terms of reconfigurability, product variability and production capacity. The same authors have moved further to propose a decision making logic that can control mobile units for achieving a real near time reconfiguration. The application in a case of the automotive industry has indicated that higher resource utilization and line availability can be achieved [26].

The main difference with past attempts lies in the flexible nature of robots:

- Robots can carry out both processing and handling tasks and therefore, a large number of alternatives can be conceived and implemented involving decisions such as robot type selection, sequencing, motion planning etc. This is not the case with agents in Computer Numerical Controlled (CNC) machines that have several programs stored and the agents decide which one to execute on the basis of the pending operations. The structure of the system remains unaltered in this case.
- The coordination that is required between the resources is considered much more complex mainly due to the dynamic nature of the tasks (pick and place from unknown positions, navigation in the shop floor etc.). Higher level coordination mechanisms/services that allow the alignment between the planning level objectives and the unit operations need also be considered (vertical integration) as it has not been investigated for such types of resources.
- Agent based approaches, although are flexible in pursuing a smooth operation, they are not generic enough to support a dynamic operation by multiple, however dissimilar resources. New capabilities are sought through the use of open frameworks such as ROS [27, 28] including standardized interfaces, hardware and

software abstraction capabilities and decoupling of parameters, request/storage/ acquisition.

Summarizing the highlights in this chapter are the following:

- The introduction of a decision making logic which allows the different resource types to communicate and plan their actions in case of unexpected events
- The definition of an integration and communication architecture for implementing the transactions between mobile and stationary resources
- The implementation of the architecture using an open software platform (ROS) and the development of services for the negotiation and coordination between autonomous resources
- The application of the aforementioned architecture on a production scenario demonstrating the ability of the system to generate and execute alternative production plans.

3.2 Approach

The approach creates an architecture that would allow stationary and mobile robotic units to communicate with each other and negotiate the production plan that they need to implement. The importance of designing and simulating such system has been extensively demonstrated in [29]. In the same work a system capable of generating the assignment of operations to the proper robotic resources has also been discussed and is closely connected with the discussed approach. The main characteristics that justify the development of the architecture involve:

- Openness of the architecture: robustness and assures autonomous behavior in case of a failure.
- Flexibility: not unique to a particular robot or task.
- Dynamic operation: Unlike other resources, robots and especially mobile units require continuous data/status update during their execution.

In this context, a robot can be used in more than one production processes and in case of failure, another robot can undertake its task. The mobility enables the creation of random production flows since the production processes can be transferred to the shop floor. The building blocks and their implementation in such a system are presented in the following sections.

3.2.1 Data Model

A data model should include all the information needed to enable the resources and services to perform the decision making on their own i.e. autonomously [30]. This

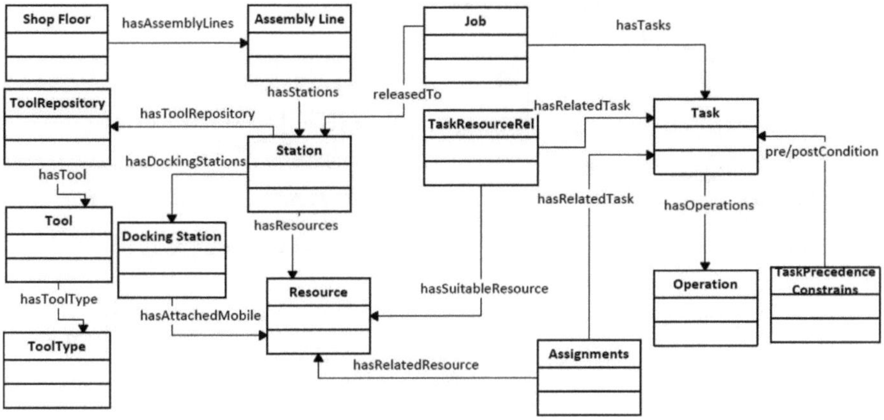

Fig. 3.4 Data model for the reconfiguration logic

information should depict the complete shop floor status, both in terms of physical elements and operations. The Unified Modeling Language (UML) schematic of the data model is shown in Fig. 3.4, where all classes are represented. Associations or composition between classes is also shown. For example, the association "hasDockingStations" denotes whether a station is equipped with an area where the mobile robot can dock and connect to external power [31]. The representation of these characteristics allows them to be automatically included in the decision making process.

3.2.2 Decision Making Triggering and Resource Negotiation Mechanisms

The decision making software module can be triggered after having received either an 'Unexpected Event' message. These messages are generated by each of the resources when there a need for reconfiguration is detected through the monitoring systems (Fig. 3.5). When each of these messages is broadcasted, the resources Networking module enters a negotiation phase, where they exchange with each other negotiation messages to decide which of all will perform the scheduling or rescheduling process. This allows the system to remain operational regardless of the breakdowns that it may have encountered, since each resource is capable of performing the scheduling process.

The approach discussed above has been implemented in the form of a software prototype, which comprises the following software packages: the Graphical User Interface—GUI, the reconfiguration/scheduling algorithms and the I/O module. The GUI and the reconfiguration/scheduling packages have been implemented in JAVA 6 language, while the I/O module utilizes the JAVA technology and the appropriate

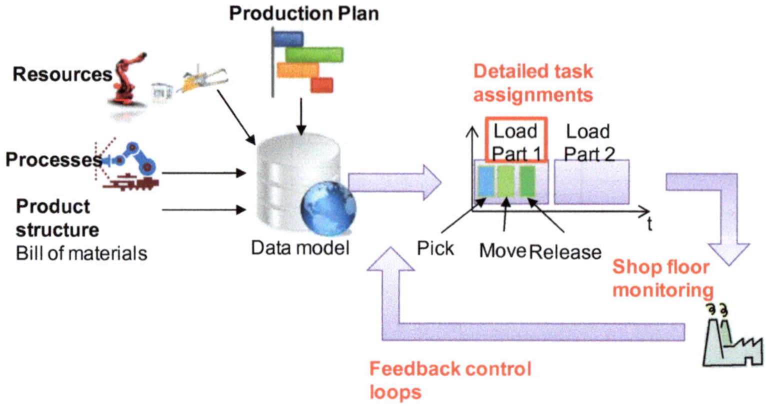

Fig. 3.5 Triggering of scheduling logic

libraries to support XML1.1 import and export capabilities. The system allows the user to construct a hierarchical model of the shop floor facilities and their workload. The facilities model includes the definition of the Plant, the assembly lines, the stations and the Resources. The user "fits" the workload model to the factory model, by specifying the *Orders*, the *Jobs* and the *Tasks*. Furthermore, the system allows the user to specify which Resources are suitable for the performance of each Task, the precedence relationships, the processing times and the setup times. The graphic user interface is driven by using the point and click operation for guiding the user through the modelling process (Fig. 3.6).

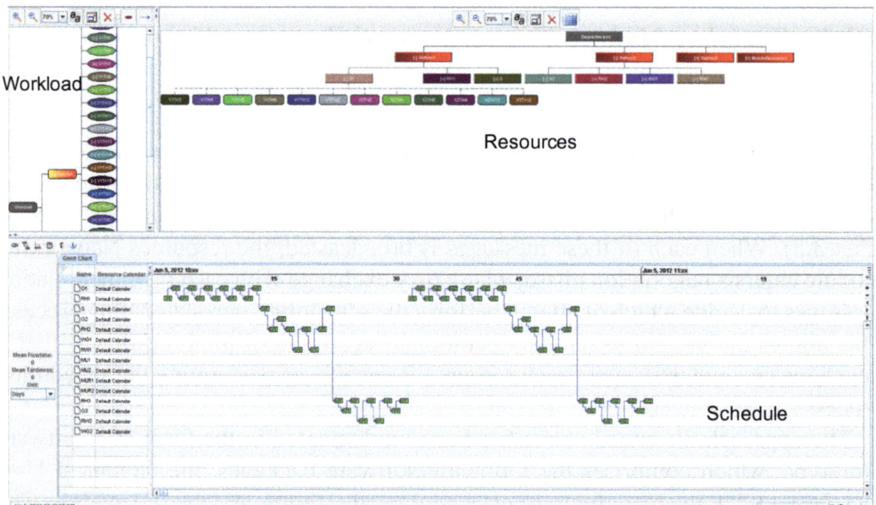

Fig. 3.6 Software tool for plant reconfiguration through task-resource assignment

The system uses event driven simulation for the operation of the assembly plant and the execution of the workload by the plant's resources. The simulation mechanism releases the workload to the Assembly lines and Stations, respecting the precedence relationships, which are defined by the user. In each station, an assignment mechanism decides which Task is to be assigned to which Resource. The assignment mechanism allocates the Resource available to a pending Task. The system simulates the operation of the production facilities either for a certain period of time (user specified) or until all the Tasks has been processed by the Resources. In either case, a detailed schedule for each Resource is produced in graphic format (Fig. 3.6).

The schedule produced is then communicated to the resources for further execution. In case of a change on the workload or on the status of the resources, the resources negotiate and initiate a new scheduling process resulting in an updated schedule.

3.2.3 Integration and Communication Architecture

The architecture is considered open when its specifications are public. This includes officially approved standards as well as privately designed architectures, whose specifications are made public. Such architectures enable the plugging of new modules and the entire system can be developed through the evolvement of the separate modules. The use of an open architecture has the advantages of: (a) Reduced cost, (b) Faster development, (c) Greater innovation potential and (d) Easier integration with existing systems.

Several attempts have been made recently for the development of standard robot software platforms with software development kits (SDKs) by robotics suppliers in order to simplify integration and robot development. The Robot Operating System (ROS) is an open framework for robot software development that aims to simplify the task of creating a complex and robust robot behavior across a wide variety of robotic platforms. Based on ROS, an architecture has been developed in this study for the integration and communication of mobile robotic units (Fig. 3.7).

The Line level control logic, implements the functionality of collecting information from the line stations and storing it to the database. The Ontology module is responsible for storing and retrieving data from the database. In addition, the Inference Engine is implementing certain rules that trigger the system's response. For example, when a resource breaks down, it reports its status to the Ontology and then the Inference Engine invokes the Scheduling module. The Scheduling module in turn, is responsible for generating a new schedule for the assignment of the remaining tasks to the system's resources. The outcome of this schedule is the system's reconfiguration.

Furthermore, the architecture implements the Unit level control, which is responsible for executing the production schedule and for monitoring and reporting the status of executing a task. Specifically, each resource, such as the Robot controller, is able to connect to the Ontology and retrieve its production schedule. Then, for each

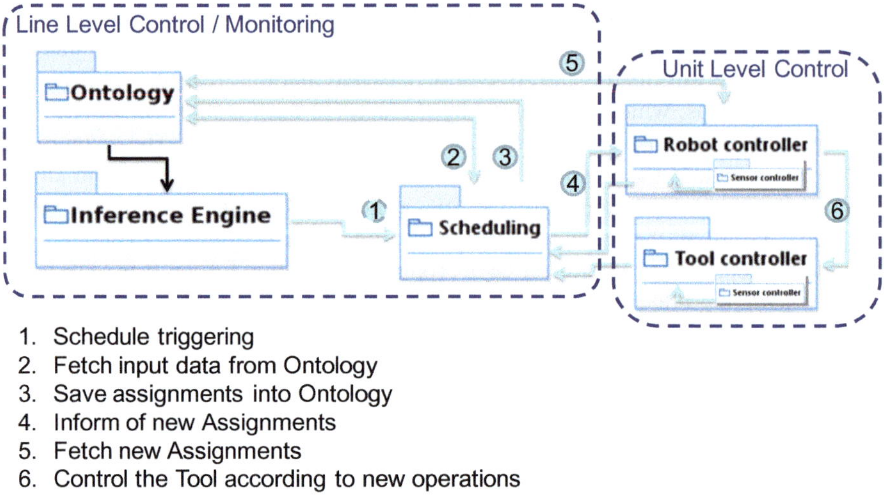

Fig. 3.7 Control architecture

task to be executed, they receive from the Ontology control commands, such as the Move ones for robots or for the gripper's, the opening and closing commands. As soon as a resource has detected an error, namely a resource breakdown, a message, which initiates the rescheduling process, is sent to the Ontology. The mechanisms that have been implemented for the development of this architecture are discussed in the following sections.

The communication model adopted in this work is the distributed control model shown in Fig. 3.8. Robot resources are modelled as services, following the service

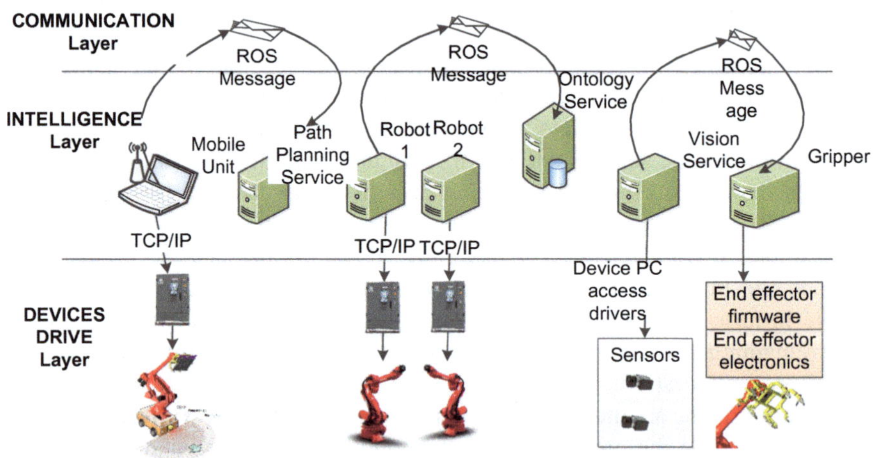

Fig. 3.8 Deployment of resources and control modules

oriented paradigm. Therefore, the communication between resources and services is achieved with the exchange of messages as shown in Fig. 3.8. In principle, each service publishes a set of functions in the form of an interface; these functions can be remotely called by another resource or service with the exchange of appropriate messages. These service interfaces, apart from the method's calls are able to broadcast messages or receive broadcasted messages. For the software implementation of the interfaces and the communication messages in this work, the ROS framework has been be utilized [32].

Every resource in the architecture has a "ROS Interface" module and a "Data Access" module. The first module depicts the implementation of the message exchanging mechanisms, through the ROS framework, while the second one will be responsible for parsing the incoming messages to meaningful for the resource information or for creating outgoing messages. The information of this level will be consumed or produced by other software modules, which enable each resource to communicate with other resources. The overall workflow is presented in Fig. 3.9.

The mobile unit itself has an interface and is able to communicate with the other resources by using the server/client or the publisher/subscriber model. The first allows the synchronous communication between the resources. In other words, a resource will have the role of the client, who can request something from the resource, playing the role of the server, and then waiting for the response. The second model (publisher/subscriber) is used for asynchronous topic-based communication. All the resources subscribe to a topic and everyone can publish messages. The messages are read by the subscribers and afterwards are parsed and translated into something useful by the resources interested in them. For example, when a robot breaks down, a message is published to the public topic and all the others can read it and do whatever is necessary in order for the broken down resource to be substituted by another mobile robot.

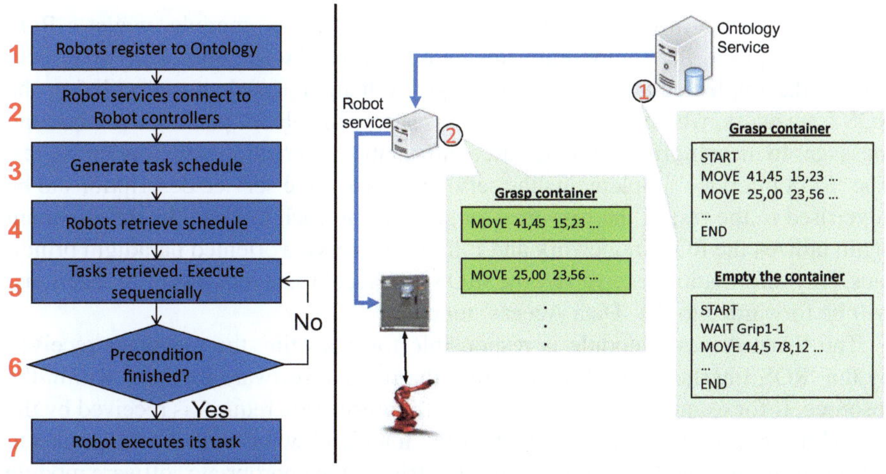

Fig. 3.9 Architecture and workflow concept

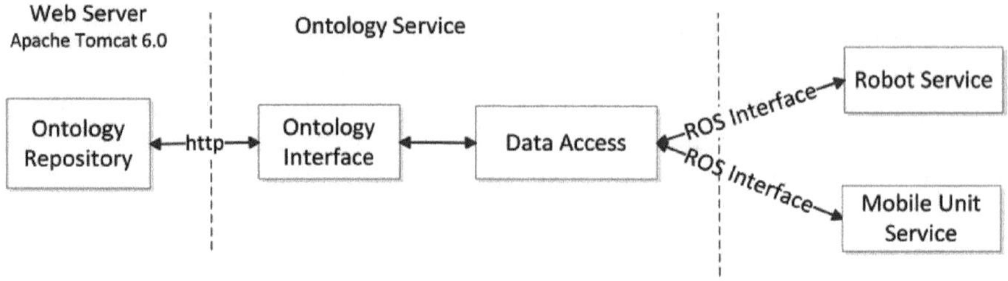

Fig. 3.10 Software architecture structure

Building on top of the ROS, the Ontology Service integration and communication software was developed. The Ontology Service is used for data management and storing services to the resources and services. A semantic repository, referred to as the "Ontology Repository" software module, is also utilized for these functionalities. In Fig. 3.10, the basic structure of the software architecture is shown.

The application and in this case, is hosted on an apache Tomcat server. To enable the proper semantic functionalities, the Jena framework is utilized for the implementation of the 'Ontology Repository'. The Ontology Service comprises two software modules, the 'Data Access' and the 'Ontology Interface'. The 'Data Access' module provides the communication among the Ontology Service and the other resources or services. The 'Ontology Interface' module enables the communication of the Ontology Service with the 'Ontology Repository'. In this way, when a ROS message, requesting information from the 'Ontology Repository', arrives at the 'Data Access' module, the latter triggers the 'Ontology Interface' module and retrieves the information from the 'Ontology Repository'. The communication between the "Ontology Interface' and the 'Ontology Repository' application is implemented using HTTP request, while communication between the 'Data Access' and the 'Ontology Interface' is implemented via function pointers.

Every resource in the architecture has a 'Data Access' module, whilst a ROS Interface exposed in this module runs as a separate thread. The 'ROS Interface' depicts the implementation of the message exchanging mechanisms utilizing the ROS framework, whereas the 'Data Access' is responsible for parsing the incoming messages to meaningful for the resource information or creating outgoing messages. The 'ROS Interface' implements the service calls and the service descriptions to be advertised to the rest of the resources or services by each module. Its development again utilizes the ROS framework and follows the service oriented paradigm principles. Any information received by the 'ROS Interface' (data requests or instructions) will be forwarded to the 'Data Access' module.

The 'Data Access' module is responsible for analyzing the messages received by the 'ROS Interface' and for triggering the relative software modules within the resource. If for example, a message for a path information request is received by the 'ROS Interface' of the Mobile Unit, the 'ROS Interface' will pass the message data to the 'Data Access' module and the latter will trigger the appropriate software module within the Mobile Unit internal control systems. After the pieces of path information have been calculated, the 'Data Access' module sends them to the relevant resource.

3.2.4 Mobile Robot Control Services

The Mobile Unit Software modules are developed in such a way so as to perform the entire control and navigation of the Mobile Unit. All modules included are shown in Fig. 3.11. In order to be integrated into the platform, the Mobile Interface was developed following the client–server ROS model.

The most important functionalities to be handled through the communication architecture involve:

- Retrieve Mobile Unit Position: The Mobile Unit interface provides information about its current position so that it can be used for task allocation
- Navigate to position: the Mobile Unit interface is used to passing a command to the Mobile Unit Navigation system in order for it to move to a new position
- Retrieve Information about a Path: the Mobile Unit should provide information about the selected paths between two positions. Path distance and estimated arrival are used by the planning algorithm
- Leave/Arrive at a docking station: the Mobile Unit is able to push/update information about its docking or undocking to the Ontology service.

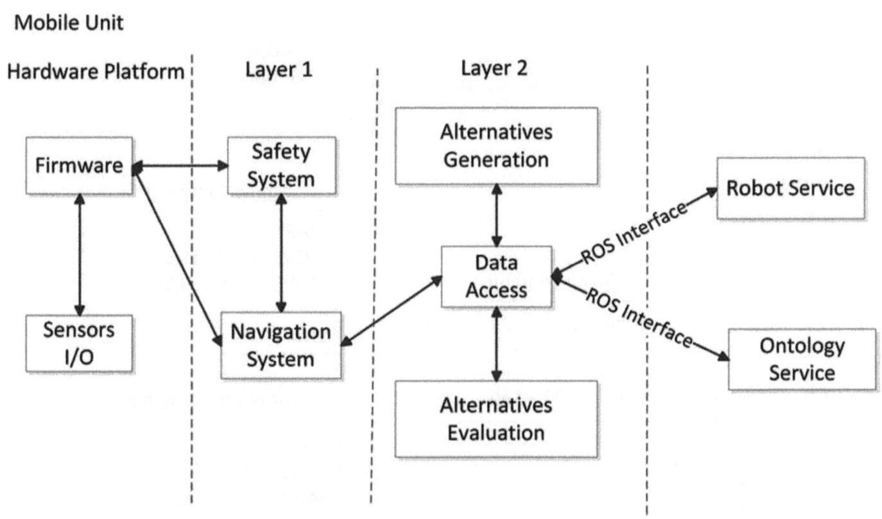

Fig. 3.11 Mobile unit local architecture

3.2.5 Execution Software System Implementation

A main bottleneck in implementing a flexible system, is the complexity of the programming of the resources, since they need to perform various tasks and in a variety of sequences that complicates the programming. In this sense, this work has developed a strategy, implemented with a software module for the reduction the of the robots' required programming. The design of the logic that will be used for the creation of new or the alteration of the existing robot paths and programs will be added to this section. The main focus is to enable resources such as mobile robots that have not been programmed for a specific task, to communicate with the other resources and obtain routines, created specifically for them and the task at hand.

The strategy for resource autonomy is to avoid manual reprogramming or teaching. Instead, use task modeling/description within the Ontology and apply frames of reference on the logic level of the robot resources.

On the one hand, the task is instantiated in the ontology and described as a set of operations, shown in Fig. 3.12. Each operation describes robotic arm mechanical moves in a parametric way and provides information that will allow the logic level of the robot resources to calculate the parameters and coordinate the mechanical moves.

On the other hand, the robot resources have a logic level which retrieves the robot resource schedule (assigned tasks and task's operation) when necessary (i.e. when a mobile robot docks to a new station) from the Ontology Service calling the 'executeQuery' function. The logic level then processes the operations of the resource

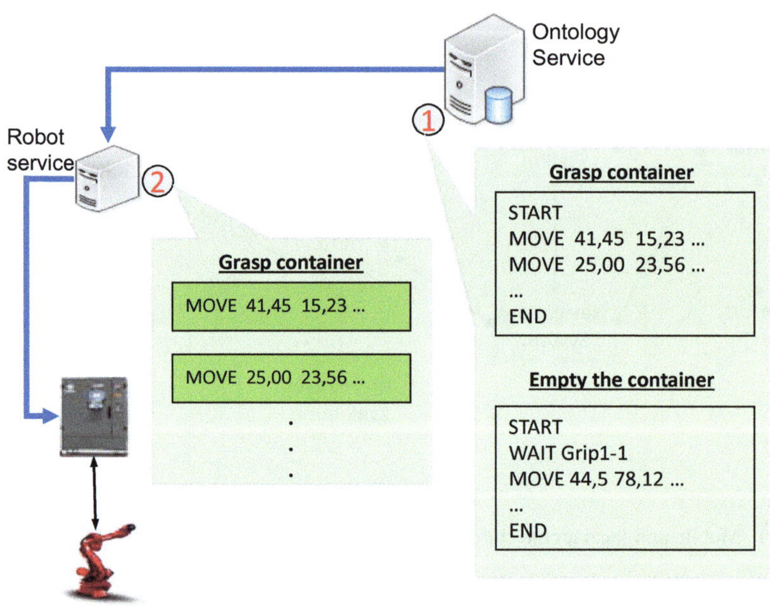

Fig. 3.12 Task model for enabling execution by robot

assigned tasks, calculates the operation parameters consuming other services whenever required such as a vision service and finally, monitors and coordinates the robot's mechanical moves. The communication between the resources and the services utilizes the communication architecture and the defined resource or service interfaces as described in this section. In order for the parameter calculation to be enhanced, the initial construction of each task's operations will utilize frames of reference.

In particular, the initial definition of the operations will be based on the utilization of the reference frame (R), which will be a fixed point within the station. Each robot, either stationary or mobile will be calibrated in order to translate its tool central point (TCP) moves according to this reference frame (R). In the case of the stationary robots, the Robot Controller can perform the necessary calculations.

3.3 Case Study

A case study was setup with the use of a mobile robot in order for the aforementioned functionalities in a manufacturing system to be demonstrated. The scenario involves two Comau® Smart5 Six fixed robots and one Robotino® mobile robot, as shown in Fig. 3.13. The stationary robots are used to sorting small parts, in this case, plastic shaver handles, while the mobile robot transfers new parts into a small container that the robot on the right (R2) picks and places on the conveyor in front of it.

The testing workload included only one job that comprised several tasks to be undertaken by the mobile and the stationary robots respectively. The hierarchical breakdown is shown in Fig. 3.14.

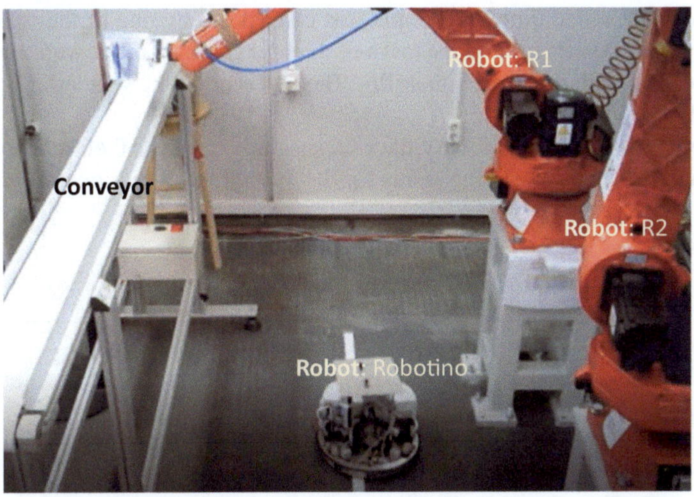

Fig. 3.13 Case study physical layout

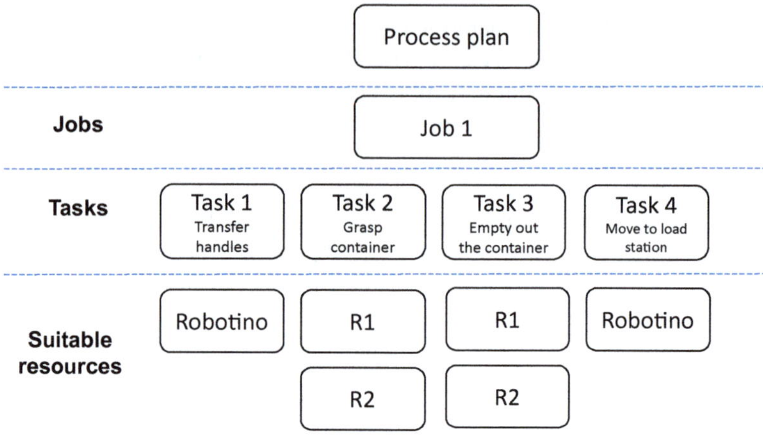

Fig. 3.14 Workload breakdown

Tasks 2 and 3 can be assigned only to R1 or to R2 or be performed by the cooperation of the two fixed robots. The scenario foresees a breakdown of robot R2, which is communicated to the ontology. As a result, a negotiation between the services of all the resources takes place and the task of unloading the container from the robot is assigned to the other robot (R1). The mobile robot is also automatically instructed to move in front of R1 so as for the unloading to take place.

As described in Sect. 3.2, ROS was used in order to implement the communication among the resources. At the system's startup, all the PCs services that were directly connected to the robot controllers, registered to the Ontology. The breakdown signal was transmitted from the service that was connected to the stationary robot. The execution of the scheduling algorithms was carried out by one of the robots (after negotiation between them) and the tasks of the broken down robot were re-assigned to the other one with the same suitability. The outcome of the scheduling process is graphically shown in Fig. 3.15.

All robots, mobile and stationary, have successfully retrieved their schedule, whilst the commands entered a waiting status for the execution of their operations.

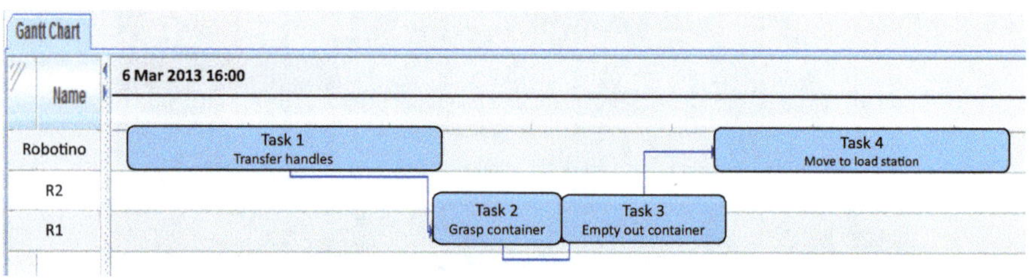

Fig. 3.15 Scheduling algorithm result for the case study

The RobotinoView® software suite was used for the navigation and interfacing of the mobile robot through the WiFi network. This software enables the intuitive creation of programs through the use of function blocks that allow logic, mathematics and arithmetic operations. A camera and a line detector function block were used for the navigation program (guiding line in Fig. 3.13). Its final position is dependent on the fixed robot that will execute the gripping task.

At the beginning of the execution process, a signal was sent from ROS to Robotino to define the new (updated) final position and pose of the mobile robot. Firstly, from the starting position following its path, Robotino transferred the container, which included the plastic handles. Upon resuming this position, Robotino via the RobotinoView, signaled to ROS the task's completion. The next task being the gripping of the container by the robot was then initiated. After the robot had performed the third task and emptied the container above the conveyor, it also signaled the end of the task. Finally, ROS sent a signal to Robotino and the last task (return of mobile robot to starting position) was executed by the mobile unit. The small scale testing proved that all technologies could be integrated and work efficiently. The verification in large scale assembly environments (such as the automotive) and diverse applications (such as consumer goods industry) is an ongoing study.

3.4 Discussion

This paper discussed the integration and communication architecture for the efficient utilization of an autonomous mobile robot, in cooperation with other fixed robots inside the production system. The architecture is open and enables the integration of multiple resources through the use of ontology and service technologies. The hardware and software architectures that the robotic resources need to comply with for the exploitation of this method, have also been presented. The implemented systems are able to:

- generate task assignments for the robots (mobile and stationary) within the production environment
- sequence these tasks automatically without the use of hard coded PLC approaches
- coordinate the operation of the resources for the automatic execution of these tasks.

With this method's application to a case study, a feasible reconfiguration plan, allowing the replacement of a malfunctioning robot by an adjacent one, without any human intervention, was generated. The benefits of adopting such technologies over the traditional control techniques mainly lie in the shortened reconfiguration time as well as in the reduction of the efforts, required for the commissioning of new resources.

Employing mobile robots imposes further challenges to be addressed. Future research should focus on the standardization of hardware (electrical and mechanical) and software interfaces in order for a seamless 'Plug & Produce' behavior of the resources to be achieved. Particularly, the following items are considered requiring further investigation:

- A mobile robot should be able to navigate autonomously; therefore, such autonomous motion will need to be modeled in the central ontology that models the skills of the robots.
- Communication capabilities of a mobile robot need to be investigated, since such a robot should communicate over a wireless connection.
- Further investigation is required on the issues of the robots' accuracy and the local autonomy. Robots require that information be exchanged about their working frames, a process that is not always accurate and therefore, further investigation should be made.
- Finally, the method's expansion to account for processes that humans and robots can cooperate in the same workspace, is also an open challenge [33].

References

1. Makris S, Michalos G, Chryssolouris G (2012) Virtual commissioning of an assembly cell with cooperating robots. Adv Decis Sci 1–11
2. Duffie NA, Prabhu VV (1994) Real-time distributed scheduling of heterarchical manufacturing systems. J Manuf Syst 13(2):94–107
3. Michalos G, Makris S, Papakostas N, Mourtzis D, Chryssolouris G (2010) Automotive assembly technologies review: challenges and outlook for a flexible and adaptive approach. CIRP J Manuf Sci Technol 2:81–91
4. Chryssolouris G (2006) Manufacturing systems: theory and practice, 2nd edn. Springer-Verlag, New York
5. Paralikas J, Fysikopoulos A, Pandremenos J, Chryssolouris G (2011) Product modularity and assembly systems: an automotive case study. CIRP Ann Manuf Technol 60:165–168
6. Reinhart G, Tekouo W (2009) Automatic programming of robot-mounted 3D optical scanning devices to easily measure parts in high-variant assembly. CIRP Ann Manuf Technol 58:25–28
7. Zhao F, Hong Y, Yu D, Yang Y, Zhang Q (2010) A hybrid particle swarm optimisation algorithm and fuzzy logic for process planning and production scheduling integration in holonic manufacturing systems. Int J Comput Integr Manuf 23:20–39
8. Houshmand M, Jamshidnezhad B (2006) An extended model of design process of lean production systems by means of process variables. Robot Comput Integr Manuf 22:1–16
9. Koren Y, Shpitalni M (2010) Design of reconfigurable manufacturing systems. J Manuf Syst 29:130–141
10. Ueda K, Hatono I, Fujii N, Vaario J (2001) Line-less production system using self organization: a case study for BMS. CIRP Ann Manuf Technol 50:319–322
11. Scholz-Reiter B, Freitag M (2007) Autonomous processes in assembly systems. CIRP Ann Manuf Technol 56:712–729
12. Valente A, Carpanzano E (2011) Development of multi-level adaptive control and scheduling solutions for shop-floor automation in reconfigurable manufacturing systems. Manuf Technol 60:449–452
13. Angerer S, Strassmair C, Staehr M, Roettenbacher M, Robertson NM (2012) Give me a hand—the potential of mobile assistive robots in automotive logistics and assembly applications. In: 2012 IEEE international conference on Technologies for Practical Robot Applications (TePRA), pp 111–116

14. Hamner B, Koterba S, Shi J, Simmons R, Singh S (2010) An autonomous mobile manipulator for assembly tasks. Auton Robots 28:131–149
15. Hvilshoj M, Bogh S (2011) "Little Helper"—an autonomous industrial mobile manipulator concept. Int J Adv Rob Syst 8:80–90
16. Henkel and Roth Mobile Robot (2013) https://www.henkel-roth.com/en/performances/mobile-robot.html.
17. URL AUTORECON (2013) [Online] Available: https://www.autorecon.eu
18. KUKA omniMove (2015) https://www.kuka-omnimove.com/de/
19. Yu-Qing CHEN, Yan ZHUANG, Wei WANG (2007) A dynamic regulation and scheduling scheme for formation control. Acta Automatica Sinica 33:628–634
20. Volos ChK, Kyprianidis IM, Stouboulos IN (2013) Experimental investigation on coverage performance of a chaotic autonomous mobile robot. Robot Auton Syst 61:1314–1322
21. Posadas JL, Pérez P, Sim JE, Benet G, Blanes F (2002) Communications structure for sensory data in mobile robots. Eng Appl Artif Intell 15:341–350
22. Giordani S, Lujak M, Martinelli F (2013) A distributed multi-agent production planning and scheduling framework for mobile robots. Comput Ind Eng 64:19–30
23. Posadas JL, Poza JL, Simó JE, Benet G, Blanes F (2008) Agent-based distributed architecture for mobile robot control. Eng Appl Artif Intell 21:805–823
24. Shena W, Hao Qi, Wanga S, Lia Y, Ghenniwac H (2007) An agent-based service-oriented integration architecture for collaborative intelligent manufacturing. Robot Comput-Integr Manuf 23:315–325
25. Michalos G, Makris S, Chryssolouris G (2014) The new assembly system paradigm. Int J Comput Integr Manuf. https://doi.org/10.1080/0951192X.2014.964323
26. Michalos G et al (2015) Decision making logic for flexible assembly lines reconfiguration. Robot Comput Integr Manuf. https://doi.org/10.1016/j.rcim.2015.04.006i
27. URL ROS (2013) [Online] Available: https://www.ros.org/
28. Quigley M, Gerkeyy B, Conleyy K, Fausty J, Footey T, Leibsz J, Bergery E, Wheelery R, Ng A (2012) ROS: an open-source robot operating system. In: ICRA workshop on open source software
29. Michalos G, Kaltsoukalas K, Aivaliotics P, Sipsas P, Sardelis A, Chryssolouris G (2014) Design and simulation of assembly systems with mobile robots. CIRP Ann Manuf Technol 63(1):181–184
30. Yang QZ, Miao CY, Zhang Y, Gay R (2006) Ontology modelling and engineering for product development process description and integration. In: 2006 IEEE international conference on industrial informatics, pp 85–90
31. Secchi C, Bonfe M, Fantuzzi C (2007) On the use of UML for modeling mechatronic systems. Autom Sci Eng IEEE Trans 4:105–113
32. Makris S et al (2012) Cooperating robots for reconfigurable assembly operations: review and challenges. Procedia CIRP 3:346–351
33. Morioka M, Sakakibara S (2010) A new cell production assembly system with human–robot cooperation. CIRP Ann Manuf Technol 59:9–12

Chapter 4
Cooperating Dexterous Robotic Resources

4.1 Introduction

Higher demand for new products, tailor made to customer needs and likes, require flexible robotic systems to adapt to the production changes. Perception technology play an important role to this as well, while it has already been used for logistic operations in warehouses and in quality control activities [1]. World's economic crisis which demands for more lightweight and flexible grippers, that would be integrated with versatile automation solutions, able to perform more functions than simple grasping and holding during handling [2–4]. More specifically, delicate handling operations cannot be easily performed by the automation solutions due to the difficulty to address many problems. On the one hand, parts with deterministic feeding method require less flexible gripping solution while on the other hand bin picking case requires a gripper that could accurately grip items from variable positions, angles etc. Additionally, restrictions on the design and selection of a gripper are higher when bigger speed, acceleration, reorientation etc. is needed during the part manipulation [5]. Another obstacle is the futuristic design that more and more is introduced in the modern products. In this case, industrial design favors aesthetics and ergonomics over design for handling. This obstacle is usually overcame by using the most dexterous resource, human being and his hands in the production [6]. However, in order for industrial robotics [7] to be employed in such kind of operations an effective way of interaction with the product is also required.

In the past years, have been developed cooperating dexterous robotic systems to increase the flexibility of industrial assembly operations [8]. At the resource level, this flexibility has been achieved in multiple forms: control strategies for flexible lines [9], reconfigurable robotic systems [2, 10], autonomous mobile manipulators [11, 12], cooperating robots [3, 13] and advanced perception/sensing technologies [14]. The term "end effector" refers to the device or tool attached to the end of a robot manipulator, allowing it to act on the product. Tools serve, mainly, part modification purposes [15] while gripping devices are integrated to perform handling operations on

S. Makris, *Cooperating Robots for Flexible Manufacturing*,
Springer Series in Advanced Manufacturing,
https://doi.org/10.1007/978-3-030-51591-1_4

the product. The efforts aim towards the development of low-cost grippers that present dexterity combined with high speeds and gripping forces, yet robust and simple. To satisfy these requirements, two distinct paths have been followed for the design of robotic hands. The more industrially oriented, focuses on more robust, simple and low-cost solutions, while sacrificing dexterity. The second approach aimed to achieve higher speed, dexterous grippers on the expense of high complexity and costs.

The first approach has been pursued by most industrial gripper suppliers. Schunk has developed two finger [16] and three finger grippers [17]. On a similar concept, RobotIQ has developed two-finger grippers, with one controlled DoF and a three finger with four actuators [18]. The Barrett hand can also be placed in this category [19]. It consists of 8 joints, which are controlled by 4 servomotors, is lightweight, powerful and dexterous in precise motions. On the other hand, it is limited by its relatively low number of independent DoFs. In a similar direction, the LARM Hand IV [20] is a three-finger gripper, actuated by four motors embedded in the palm of the hand. Furthermore, the Tuat/Karlsruhe Hand [21] is a five-finger hand that consists of only one active DoF. The University of Nazarbayev also presented in 2013 a five-finger gripper whose controlled DoFs had been kept to four by creating a dependency in the movement of its last three fingers [22]. Prensilia Srl. developed a five-finger hand, called Azzurra IH2, with five DOFs, while its two last fingers open and close simultaneously [23].

Highly dexterous grippers consist of a high number of DoFs and mostly use the multigrasp hand paradigm, while being relatively complex and expensive. The Stanford Hand/JPL [24], Utah/MIT dexterous hand [25], and the fourteen degrees-of-freedom Robonaut hand for space operations [26] were among the first high-DoF hands to be developed. The BUAA/Beijing University [27] consists of 4 fingers with a total of 16 DoFs and the NTU hand [28] consists of a five-finger design, which accounts for an overall of seventeen DoFs. In the Gifu Hand II [29], and III [30] the sixteen actuators required for the movement of its 20 joints have been incorporated inside the fingers as opposed to many of the other grippers. One of the most widely known hands for its dexterity, is the Shadow C5 Dexterous Hand [31], developed by the Shadow Robot Company that can reproduce the 26 degrees-of-freedom of the human hand by using 23 DoFs, controlled by remotely placed motors or pneumatic tendons. In the dexterous gripper category belong "in-hand" manipulators. Although this subject still undergoes research in terms of human capabilities [32], it already applies to a variety of end effectors. Bio-inspired gripper designs are common nowadays, trying to mimic both human dexterity [33] and tactile sensing [34].

Another concept adopts Bernoulli's principle and has been applied in a suction-operating gripper [35, 36]. High airflow speed between the item and the gripper surface results to local drop of pressure, something that pushes the item towards the gripper [37]. A new trend of adhesive grippers is also presented and has found some practical applications [38]. A layer of pins on the outer surface of the gripper jaw, that have wearable adhesive toppings, can firmly hold the product for all the time needed. Replacement of the adhesion layer is achieved through UV light, which destroys the light-sensitive sheathing [39]. Apart from these, if the handling material is soft or porous, needle grippers have also been manufactured. This has immediate effect on the food industry, especially on raw meat processing [40]. Electrical principles are also widespread in gripper designs. If the voltage drop between the grasping surface and the product is high enough, attractive electrical forces keep these two stuck or even re-orientate the product [41]. Magnetic forces are likewise able to pull and hold the product. Such grippers require extensive calculations [42] to identify the optimum design that permits the magnetic flux density to be intense between the gripper jaws. Last but not least, more unconventional techniques have been implemented. Ashkin [43] has proposed a set of optical tweezers that exploit particle-scale forces, generated by a single optical beam, that are able to control particles of micrometer down to nanometer dimensions inside a liquid solution. The above method may serve Biology purposes, since it can be a tool to handle organic cells. This idea has been evolved and single optical fiber is now used as the medium to transmit the beam [44].

The structure of this chapter is the following. Following the literature review elaborated in this section, presenting similar developments as well as their drawbacks, in the field of dexterous cooperating robots, a control logic is discussed that coordinates the cooperation of two robots that share the same working space. Next, perception algorithms play an important role in the cooperation activities among robot resources and for this reason a section is dedicated on such algorithms. Furthermore, another aspect that plays important robot on dexterous robotics is the use of dexterous tools. This section will be dedicated on the design and implementation activities performed to create two type of dexterous grippers. The first one focuses on dexterous manipulation activities, where a number of different objects, in terms of geometry and size, can be manipulated, while the other one focuses on high-speed handling activities of similar objects with complex geometries. Last but not least, the final chapters focus on two use cases where the aforementioned technologies have been applied.

4.2 Robot to Robot Cooperation

In modern industrial environment, it is quite often that more than one robotic solution should be used in the same workspace and perform a specific process. For this reason, a number of sensors and smart control systems are usually used to coordinate them and ensure their smooth collaboration. In this section a sensorless smart algorithm will be presented to achieve this and ensure the task synchronization among the different resources.

In order to achieve more than one robot cooperation on the same cell, a software platform, namely Synchronization and Control (S&C) Framework, has been implemented. The S&C framework is used as the basis for message exchange between the resource drivers and the task executors, libraries responsible to track the tasks for each resource, using ROS nodes. As a result, multiple robot nodes have been created to facilitate the communication between the robots as well as the other resources. Each task executor sends the respective robot instructions to the specific robot, ensuring the correct order of task execution. Nevertheless, the execution of a certain task list is not enough. There should be established a mechanism that certain tasks should

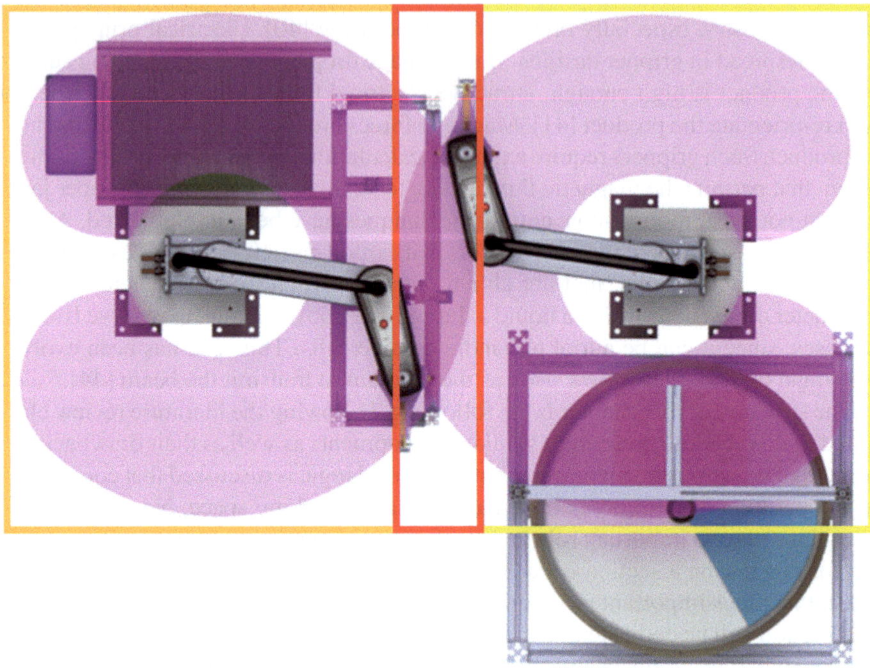

Fig. 4.1 Robots' common workspace

not be executed at the same time by the two robots or on the other side, that two tasks start in parallel. For this reason, two synchronization mechanisms have been introduced, namely LOCK and BARRIER mechanism.

Regarding the first mechanism, namely LOCK, the principle is quite simple. Each robot has a specific workspace: In Fig. 4.1, the workspace of the left robot is inside the orange/red box while for the right robot, the workspace is the yellow/red box. It is obvious that the red box is a common workspace among the two robots and if both enter it there will be a collision. For this reason, if for a task that needs to be executed, the robot has to enter or pass from the red box, it is claimed as lockable. When a lockable task begins from the first robot, it grabs the "lock" preventing the second robot to start his own lockable task, since the lock is missing. Upon the completion of the task from the first robot, the "lock", which is an algorithmic flag in the code, is released and the second robot is able to capture it and begin its task.

Inside the S&C framework has been included another synchronization mechanism called barrier. The idea of this feature is to allow the programmer to ensure that a group of operations that need to be ran by the different robots, as well as by multiple resources, can start simultaneously. In other words, the programmer is able to define group of tasks that all should be executed and completed until the group finishes and the next group of operations will start. This prevents the starting of the following group of operations unless all the operations from the previous group have finished.

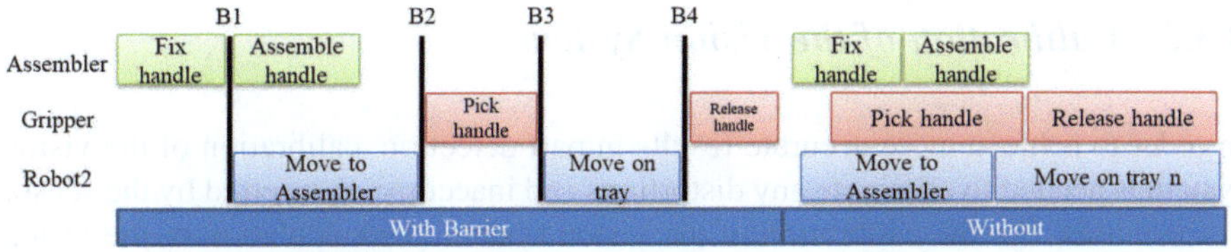

Fig. 4.2 Overview of barrier mechanism

This operation is better visualized in Fig. 4.2. On the left side is shown how the operations by different resources are grouped together, start simultaneously in each group and the next group waits for all operations from the previous group to be concluded. On the right side is shown how the operations are ran randomly, without specific order and not synchronized.

4.3 Robotic Perception

The manipulation of all industrial products should be based on its 3D coordinates, thus, the identification of the position and orientation of the objects along all axes is required (x, y, z, e, a, r).

This section will describe a hybrid algorithm, which is based on two processing stages. The first step is dedicated to calculating rough position and orientation data from a 2D vision technique. The second step is the generation of the object's 3D coordinates, using information from the object's CAD files. As a result, even with the part's partial identification (e.g. rotation over the vertical and horizontal axes/pose) the estimation of all the coordinates of the points of interest can be provided.

4.3.1 Training of Vision Algorithm for Object Detection

The initial step in the set-up phase is the training of the algorithm for the new objects. Object's geometry, such as its dimension, area, maximum length, is the key focus of this procedure. Firstly, the operator places a single object on a surface under the camera's field of view, activating the vision algorithm to capture an image. This image is processed and 2D object's characteristics are calculated and stored under a name by the operator. This process is repeated multiple times to gather a substantial amount of data. Once the data is gathered, their characteristics are filtered via the RANSAC algorithm and after that, the mean value is saved. Lastly, for each object, the operator provides information about the height (on z-axis) and the nominal position of the object.

4.3.2 Calibration of the Vision System

In order to achieve more accurate results in part detection, calibration of the vision system is needed to eliminate any distortions and inaccuracies inserted by the sensor or the lenses. For the calibration of the vision system, the calculation of the vision system's intrinsic parameters, such as distortion coefficients, the camera focal length (f_x, f_y) and position of the optical center in the image plane (c_x, c_y), takes place. The calibration model, based on the OpenCV library [45], takes into account both the radial and tangential factors. Radial is the distortion caused by the lens "barrel effect". This effect is lower in the lens center and becomes very strong on the edge of the lens. The second distortion is the tangential, which is caused due to the lens not being parallel to the camera sensors. Then, the correlation of pixel coordinates (X_{pixel}, Y_{pixel}) with the camera frame, on the camera sensor, which is located in the (X, Y), is based on Eq. 4.1.

$$\begin{aligned} X_{pixel} &= f_x X + c_x \\ Y_{pixel} &= f_y Y + c_y \end{aligned} \tag{4.1}$$

Another calibration procedure that needs to be performed is that of calculating the coordinate system of the camera to robot frame. For this again a standard procedure of OpenCV calibration algorithm has been used. The transformation from one frame into another is possible provided that the relative position from one frame to the other is known. Generally, this matrix has 12 parameters that need to be calculated, of which 9 are dedicated to rotation ([R]) and 3 to the translation vector ([T]), therefore, the general matrix is in the form of $T = [(R_{3\times3}\ T_{3\times1})]$. For the simplification of the set-up phase instead of the robot base frame, a user frame has been used on a conveyor surface. The user frame is defined in such way, in parallel to the vision frame, to accurate presume that there is only a sifting on the x axis as the conveyor moves the object in space. Therefore, the transformation is described by Eq. 4.2 where the X_{offset} is the one to be calculated.

$$T = \begin{bmatrix} X_{offset} & 0\,0\,0 \\ 0 & 1\,0\,0 \\ 0 & 0\,1\,0 \end{bmatrix} \tag{4.2}$$

For the calculation of the transformation matrix between the vision system and the robot frame, the idea is that data will be retrieved for the same object in relation to both frames.

- The object was placed on a position near the robot's user frame (Object position 1).
- The position of the object's point of interest in relation to the user frame R1 was measured manually (Pr1).

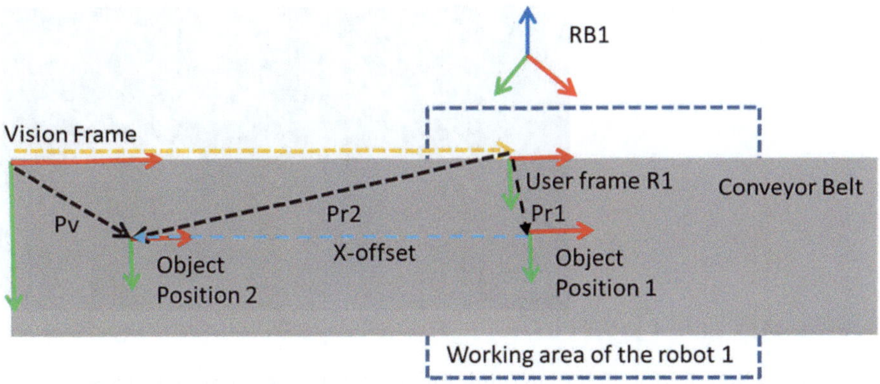

Fig. 4.3 Frame position calculation

- The motion tracking measured the x-offset (blue arrow) of the object that was moving on the conveyor inside camera's field of view.
- The position of the point of interest, in relation to the vision frame, was calculated by using the vision system algorithms (Pv).

  The frames used in the above calculation are depicted in Fig. 4.3
  The Eqs. 4.3 and 4.4 describe the connection between the measurements:

$$P_v = X_{offset} + P_{r2} \tag{4.3}$$

$$P_{r2} = P_{r1} + X_{conveyor} \tag{4.4}$$

This process was iterated several times. The results of the process were collected in two matrixes. The first matrix contained the measurement of the vision system (P_v), while the other contained the measurement of the same object coordinates in relation to the robot frames (P_r). Due to error in the measurements, the final equation is 5, where X_{offset} refers to the unknown variables.

$$E = P_v - \left(P_r + X_{conveyor} + X_{offset}\right) \tag{4.5}$$

4.3.3 Region of Interest (RoI) Identification

The procedure aims to identify the area within the image that potentially includes the desired objects. A robust way of achieving this is by using the high contrast filter between object's color and the conveyor surface. This step is essential because

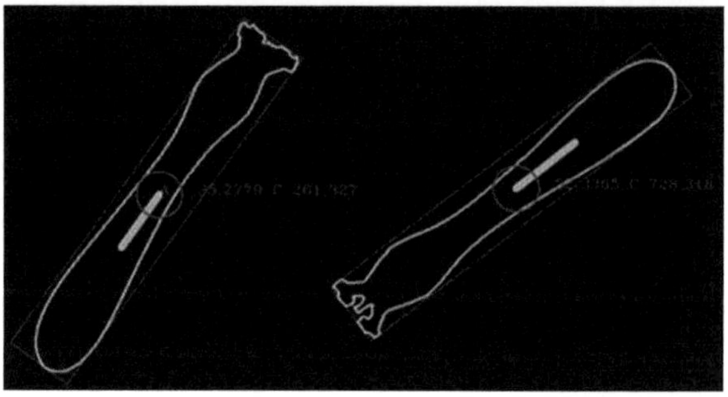

Fig. 4.4 Object's position defined coordinates

the image is acquired without any high-quality lighting system and therefore, any shadow or light variation may affect the identification process.

The next step is to calculate all the characteristics of all objects in the image. Any parts other than the object are removed from the image. For all the objects found in the previous step, a rough estimation of their orientation takes place. To find an estimation of the rotation, there is an attempt to fit an eclipse in the object's contour. A rough estimation of the object's rotation is the orientation of the eclipse. For the estimation of the eclipse orientation, the OpenCV function, which is based on this algorithm has been used. After the orientation has been found, the algorithm calculates the minimum area rectangle.

The next step is the object's classification. There are several poses of the object that are stored during the calculation phase and used for comparison reasons against the acquired image. The comparison between the two objects is based on the Hu invariants. The result is the identification of the template which is similar to the object. Next, the algorithm tries to locate the offsets on both X and Y axes of object's point of interest which will optimize its accuracy. At the beginning, the algorithm calculates an approximate position of the object. Next, the algorithm tries to optimize the vertical axis. After that the object's position in the vision frame system is generated as displayed in Fig. 4.4.

4.3.4 Hybrid 3D Vision Algorithm

As explained in the previous paragraphs, due to the fact that a 2D vision system is used, the coordinates of the object points, along the z-axis, are not calculated. In order for this problem to be bypassed, during the calibration phase, this information was extracted from the CAD file and was stored into the software tools for the different points of interest. Therefore, the algorithm supplements the 2D coordinates with the CAD info and can provide a complete set of coordinates to guide the robot.

Another important part of the identification is that the object is placed on the surface of a moving conveyor. This makes its relative position to be defined in relation to the conveyor frame. Since the conveyor is continuously moving, all the calculated data need to be shifted through time. To simplify the problem, the vision system's coordinates are chosen in such way, having the X axis parallel to the direction of the moving conveyor belt. As a result, only the x position needs to be considered as a variable. A conveyor tracking device is attached onto the conveyor belt. The conveyor tracking calculates the motor's rotation that is dedicated to rotating the conveyor belt. Using the encoder provided values, the position of the handle on the conveyor can be predicted. In this way, it is possible to calculate the position of each object that has passed through the camera frame, even if it is not visible through the camera. For this reason, the vision system should acquire the encoder's data and store them along with the object's data.

4.3.5 Machine Learning Method

Object recognition can be also performed through the use of Convolutional Neural Networks, namely CNNs. CNNs are similar to neural networks, consisting of a multi-layer complex, that is trained using labelled images by the human. After this process, CNN provides as output the respective information from an unknown image. Each layer processes a part of the input image, putting a label on this part and a score which defines the confidence of this labelling. Following multiple iterations among the different layers, the labels with the higher scores are combined, reaching to a final conclusion. This process is faster and more precise that normal neural networks.

Regarding the CNN-based visual recognition, in the training phase, multiple groups of predefined images are used to create multiple classes, one per group, based on a specific characteristic. This would allow the classify an unknown image, that contains this unique characteristic, to its respective group/class. Once a big data set of images is used, to create a robust classifier, the CNN is ready to accept an unknown image to categorize it. The output of this process is a set of scores that reflect the matching rate of the specific image to each class. Assuming that the classifier is smart enough to perform the classification activity accurately enough, the title of the class with the biggest score, is the label that matches to the input image [46, 47].

In order to increase the accuracy of the aforementioned system, the camera configuration should be the same in both training and classification phases. Additionally, the camera should be static and the object background should be as clear as possible. This would increase the image matching scores and algorithms accuracy.

4.4 Dexterous Manipulation

A common requirement that exists in the manufacturing industries is the part handling with different characteristics. The key advantage of the manual assembly lines is the handling capabilities that human operators have. They are able to grasp, pick-up and orient objects with different geometries, material textures and size. This capability cannot be easily automated or transferred to a robotic system and therefore complex gripping systems are created, trying to reach human level.

An electromechanical gripper will be described in the next sections, starting from its design through its realization. This gripper is capable to pick-up objects from a flat surface, that have random position and orientation in 3D space, as well as non-symmetric geometry. Its key characteristics is the high number of DoF, its simple design, high capabilities of manipulation and grasping.

4.4.1 *Design of the Gripper*

The presented gripper has been designed with specific requirements in mind, that arise from the manufacturing sector, such as the body in white from the automotive, where parts vary in geometry and weight. These requirements refer to high-speeds and versatility characteristics that are needed in industry. At the same time, the gripper should have a lean design for robust operation and simple maintenance.

Having the above in mind, a gripper with three fingers and eight degrees of freedom has been designed, achieving high re-configurability for grasping multiple positions/geometries. Additionally, the design of the two fingers enables to the gripper to enlarge its grasping capabilities as discussed in [48], while a hand with such capabilities would need at least nine DoF to be able grasp a randomly positioned object. Nevertheless, this gripper has been restricted to eight DoF, reducing the rotational ability of the middle finger, simplifying the design (Fig. 4.5). This limitation can be easily replaced by the robotic manipulator, upon which the gripper would be attached.

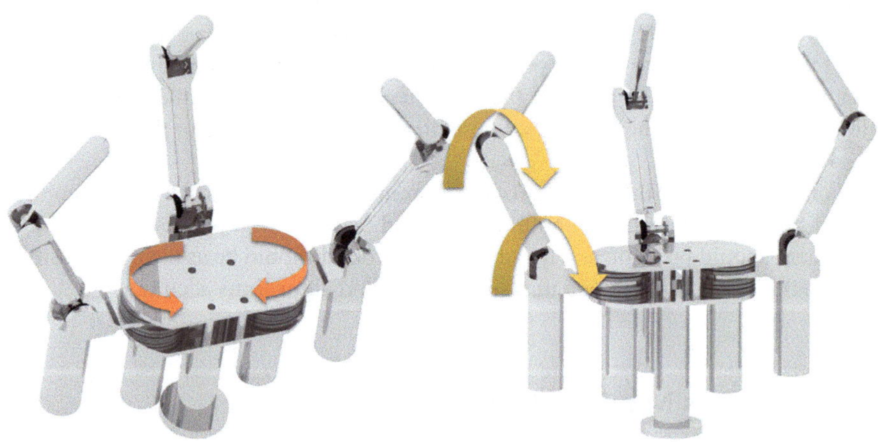

Fig. 4.5 Gripper with 3 fingers and 8 DoF freedom

Regarding its dynamics, the full movement of each joint, namely a rotation of 180°, can be executed in 0.1 s while another important aspect for high-speed applications has also been taken into consideration. This aspect is the gripper weight, which has been limited to 5 kg, being able at the same time to grasp objects weighing a few kilograms.

Lastly, control aspects have been taken into consideration, creating a central master node to track and control all motors at the same time. Additionally, it is able to track all joint activity and real time message exchange between the motors and the main controller.

4.4.1.1 Design Notion

The notion that has been followed during the design phase of the gripper is the capability to achieve a wide range of grasping motions. In general, there are two ways of categorizing the grasping of an object. Either by the relative positioning of the fingers, namely centripetal or parallel, or by the grasping mode that is achieved, namely encompassing or pinching.

The final result on the implemented gripper is demonstrated in Fig. 4.5. Regarding the aforementioned grasping categories, in this gripper for the centripetal grasp, all fingers are positioned around a central point.

For the parallel grasping, the fingers are positioned in parallel relative to each other. As far as the grasping modes are concerned, the difference lies on the precision of the grasping task performed as explained in [49].

4.4.1.2 Mechanical Notion

Following the simplistic design and high-speed movement, lean motors and transmission mechanisms have been selected. The component that supports the two side fingers is fixed directly on the motor output shaft, located on the gripper base. This ensures the rotational movement of these fingers. In order to achieve a frictionless movement, the finger assembly is supported by single direction thrust bearings, placed above and below of the finger.

The rest of motors used for the other joints, have been inserted in the finger body, forcing the conversion of the rotational axis, from vertical to horizontal (relative to the motor's output shaft).

In order to achieve this, two straight bevel gears have been used, as shown Fig. 4.6, where the one has been positioned on the motor output and the other perpendicularly to the first one, mounted inside a ball bearing. At the center of this bevel gear, the joint axle motion is fixed by another ball bearing in the other end.

Although the movement is responsive, the direct connection described above can be a weak point, since the motor can be damaged in case of high load on the fingers.

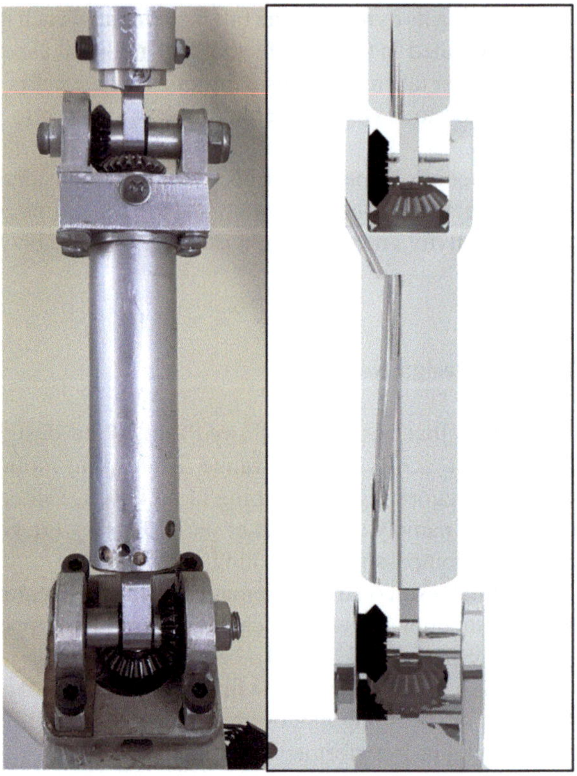

Fig. 4.6 CAD and real joint views—axle to bevel assembly

4.4.1.3 Kinematics

The typical approach of solving the kinematics of a mechanism is using either forward or inverse kinematics calculations. The math on the above methods can be correct, in reality in the physical conditions they could be unrealistic. For this reason, each solution should be tested first for its viability.

In the case of the gripper, each finger has been considered as a serial manipulator with the first joint axis as base. Following the execution of forward kinematics of the manipulator, the transformation of each joint based on the previous one is calculated. In other words, the matrices that contain the coordinates of all the joints starting from the base have been defined and following a merging into one, it is possible to extract the final transformation matrix that shows the relative position of the fingertip based on the base joint. Similar mechanism structure and inverse kinematics has been discussed in [50]. Figure 4.7 shows on a simple visualization the kinematic model with the link and joint frames of a 3-DoF finger, provided by the Mobile Robot Programming Toolkit [51].

Fig. 4.7 Kinematic model of the left three-DoF finger

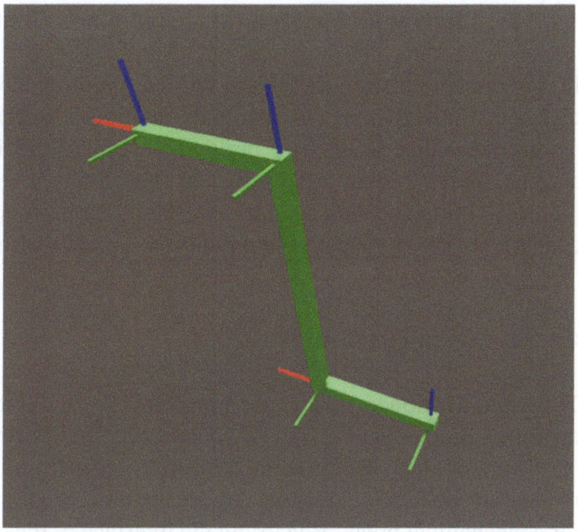

4.4.2 Implementation

Figure 4.8 displays the implemented gripper holding different parts using its fingers. As it is shown, in order to grasps different parts, commonly used in the automotive industry and its suppliers, asymmetric grasping positions have been used.

Since each finger has its own independent kinematic chain, due to the individually actuated joints, non-symmetric configurations can be achieved, both in terms of finger joints and palm dimension, defining abduction and adduction movement. This level of versatility can realize multiple configurations and perform gripping operations that were not feasible.

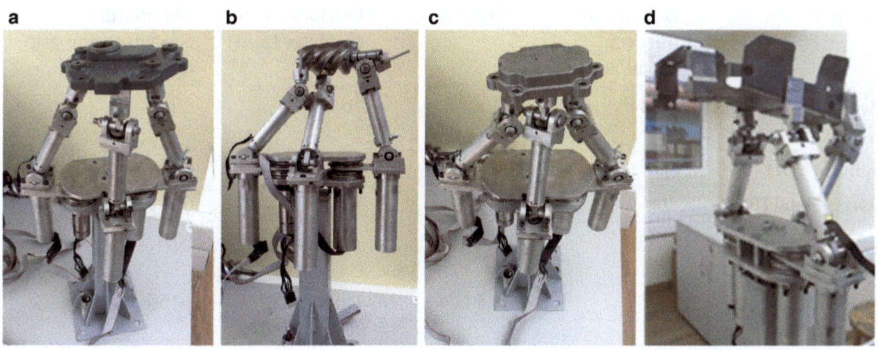

Fig. 4.8 Grasping positions holding **a** heat shield, **b** manifold, **c** car computer, **d** car floor part

4.4.2.1 Electronics for Motor Activation and Control

In this implementation, brushless DC servomotors have been selected, due to their high speed and efficiency in a compact solution. This enables the gripper to satisfy the requirements discussed in the previous section of precise positioning, speed control and maximum torque, to be achieved. Those motors have planetary gearheads and high-resolution absolute encoders. As far as the motors that control the fingertips are concerned, they are smaller and therefore less powerful compared to the others.

4.4.2.2 Gripper Control

The aforementioned motors are controlled by actuation controllers using different operation modes, namely speed, position, speed or acceleration curve and voltage or current control. The motion controller, apart from the aforementioned operation modes, can be used as splitting device, enabling the motor stoppage without the need for a mechanical break. Regarding the interface of the controller, the communication protocol that has been used is the Controller Area Network bus (CAN), which is similar to the RS232 but supports higher transfer rates, of up to one Mbps. On the other side, the physical connection of the CAN bus requires in both ends one resistor of 120 Ohms, while all the devices are connected in parallel and are consider as independent nodes. In other words, a common bus is used for all the motion controllers, while individual buses are used to connect the motors with their own controllers.

Having presented in this section all the information regarding the design and development of the reconfigurable gripper, a short comparison is presented in Table 4.1 five with other similar solutions that have been found in the literature.

Table 4.1 Specification comparison among different gripping solutions

Gripper	Degrees of freedom (Total/actuated)	Weight (kg)	Grasped object diameter (mm)	Maximum encompassing grip (N)
Robotiq—3 Finger Gripper [52, 53]	10/4	2.3	20–155	100
Schunk SDH [54]	8/7	1.95	17–215	N/A
Barret Hand [55, 56]	8/4	1.2	Up to 240	60
LARM Hand IV [57]	9/3	1.5	10–100	N/A
Proposed gripper	8/8	3.5	Up to 215	>150

4.5 High Speed Handling

As already described above, industrial requirements dictate the need for dexterous and flexible grippers able to handle a variety of parts. In the previous section such gripper has been described, able to manipulate different parts with relatively big geometries and stiff density, that are used in the automotive industry. Nevertheless, there are cases, where smaller parts with a more rubber behavior need to be handled. In these cases, handling speed plays a more important role while the friction provoked by the rubber-like plastic material should be taken into consideration. Such gripper is described in the sections below, starting from the design and motion analysis stage up to the implementation.

4.5.1 *Motion Analysis*

A design principle that usually has to be taken into account is "minimalism"; meaning a device should not consist of convoluted motions in overall [58]. Under this scope the addressed problem can be formulated as follows. A robotic arm approaches the item, comes in contact, grasps and holds the part, rotates it with respect to one or more specific degrees of freedom and finally releases it. Some states may be served by the same sub-system of the mechanism, while others includes both alignment and rotation, meaning that at least two degrees of freedom must be controlled over the object. Supposing that the alignment is just a shift on any of the three axes, a linear translation must be obtained. Considering that the degrees of freedom are controlled sequentially, this translation should be able to fetch the item directly to a pivot point, ready for rotation. Keeping in mind that items of varying geometries should be addressed, the whole mechanism should be able to adjust itself to such changes.

With the correct transmission (reduction, rotational to linear movement transition and vice versa), a single source of power may be able to manipulate the necessary degrees of freedom, but such a method is not preferred because it increases the problem complexity. Keeping things simple, a dedicated actuator should not only serve each required degree of freedom but also be of the same type, so that no power conversion between linear and rotational forms takes place.

4.5.2 *Grasping Mechanism*

Having specified the correct collection of actuators, and a set of jaws or fingers is required in order to grasp the object. This system must not only be able to hold the object securely but must also be able to align it with pivot points that will perform the object's rotations along the different axes.

Fingers should not be mechanically coupled with each other but behave independently. This allows potential asymmetry in finger design, meaning that fingers may not be the same, depending on the item. There are various concepts concerning the grasping motion of the jaws; either both of them move (active closure) or one is fixed while the other pushes the item against it (passive force closure) or even a mixed combination of those (hybrid closure) [59]. In this case, active closure is chosen, due to the parametric nature of the proposed solution. Another important fact about the fingers is that the fingertip is required to move over the "xy" plane. This set-up yields perfectly to the designed motion principles.

4.5.3 Rotating Mechanism

The subsystem mentioned in the previous section referred to the gripper fingers gripping and aligning the portion. A rotational motor may be used to rotate the component along any axis. The fingers must however ensure that the element is properly aligned with the pivot point. Nevertheless, the entity must be un-grasped and a portion that will retain is needed to execute the rotation. This part will be the only fixture between the object and the mechanism at the time of rotation and it must also be the linking part between the rotating point and the object. The object can be pushed inside the rotating part along the "Z" axis while it is still held by the fingers and can be released afterwards. This emphasizes on the design of this "third" finger, because an improper fixture would easily lead to item dropping. The coupling surface of the finger should pair with the object contour, resulting to the maximization of the contact area and therefore, maximization of the grasping force [4]. An additional factor to be considered is that the average distance between the finger and object surface should be minimized.

4.5.4 Constraints

Although the scale of the proposed mechanism is case-dependent, there are measurements which can be put within a set, guaranteeing both the device's effective operation and preventing conflict between moving parts and the target. The following constraints should be taken into account:

- Firm holding of all possible sides of the object on the "xy" plane.
- Grasping of all sides of the object on the "xy" plane.
- No collision will take place between the fingertip and the object during rotation.
- All the required x, y, z-distance and rotation can be performed.
- Ensure successful disengagement.

Another important dimension is the finger width. This has immediate relationship with the bending strength of the finger, according to the material strength. An essential requirement is that the displacement of the finger in the tip area should be minimized during grasping, allowing more precise control and centering of the object with respect to the pivot point.

4.5.5 *Integration and Networking*

Another factor is the nature of the communication that will be formed, in order to control the whole process. Broadcast communication networks have a lot of documented advantages over point-to-point ones [60], with the most basic of them the reduced wiring. In addition, broadcast networks can provide helpful troubleshooting of the relation developed between the nodes as well as painless introduction of new nodes or reconfiguration of old nodes into the current circuit:

- Reduced wiring is always desirable due to the reduction of the overall cost but also of the wiring complexity that will take place in a device that will be held by a robotic arm.
- Troubleshooting and error detection is vital.
- Reconfigurability is of equivalent importance, since the proposed device's actuators are highly depended on the product specifications.

4.5.6 *Hardware and Software Implementation*

For the modelling, design and construction of the gripper, a complex shape product from the consumer good industry (shaver handle) was used. The initial implementation of the gripper was based on this part, while a variety of similar shape and size products can be manipulated after making some easy adaptations.

4.5.6.1 Mechanism Design

At the first phase of the implementation, the gripper was modelled using a 3D CAD software. Following the motion analysis above and the specific gripper implementation, the details of each handling state are described shortly below:

1. Grasp
2. Align while grasped
3. Engage with rotary part while aligned and grasped
4. Ungrasp while engaged with rotary part
5. Rotate while engaged with rotary part
6. Re-grasp and keep aligned while engaged with rotary part

7. Disengage from rotary part while grasped and aligned
8. Release from mechanism by ungrasping.

As it can be seen from the list above, not every state involves object motion with respect to the gripper. The object is affected once by a linear shift with respect to y-axis (Step 2) and once by a rotation around z-axis (Step 5). This table verifies that there is no redundant motion concerning the product and leading to time being wasted. Based on the previous analysis, the internal parts of the mechanism should be able to accomplish 1 rotational and 7 linear movements.

The couple of the front grasping fingers is one of the most vital parts of the design. To maximize the grasping effect, a fingertip should achieve high coefficient of friction between the object and itself. An indispensable condition for efficient grasping, is to include the center of mass of the object inside the grasped volume, to prevent any turning during lifting. The object's width diminishes while moving from back to front side in an area around its center of mass. This indicates that in order to grasp the item, the fingertip should follow the same geometric pattern. Although a linear interpolation has been approximated as far as the grasping surface is concerned, there still remains a problem of surface coupling between the fingertip and the item. Due to the different curve that the fingertip will be expected to couple with per orientation, a soft material is needed such as polyurethane.

The finger of the engaging mechanism, unlike the grasping one, has only one part that holds the shaver handle, meaning that there has to be a kind of smart fixture between the two of them. An underactuated finger would embrace the handle during the relative pushing between them, but such a scenario increases the complexity of the mechanism even more. By studying the rear side of the handle, where the finger is supposed to approach four available sides for contact can be identified. The linear motor fetches the finger to the grasped object, until engagement is accomplished. Then the rotational motor revolves the finger along with the object up to a predefined angular position. After the two fundamental elements were designed, they have to be combined into the greater mechanism, using the drilled assembly holes on the fixed frame.

4.5.6.2 Construction

All parts described above were constructed and put together. The metallic parts were CNC aluminum cut while the 3D-printed ones were made out of ABS material. The final gripper has been integrated on a robot and its grasping and manipulation capabilities have been tested as demonstrated on Fig. 4.9.

4.5.6.3 Programming

In order to control effectively the device, a custom interface has been created, which will be able to synchronize the motion of all components and effectively use the

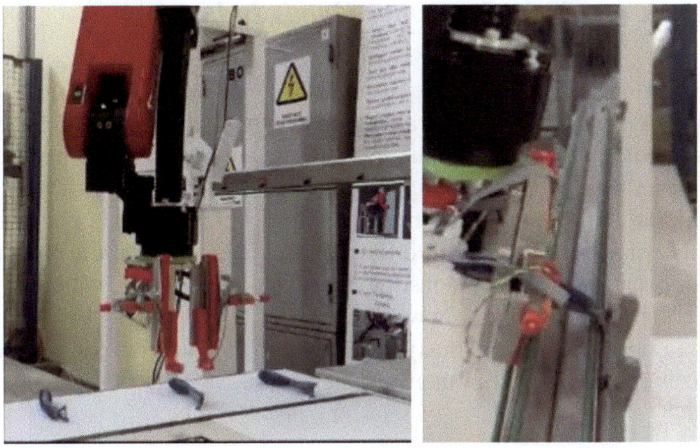

Fig. 4.9 Final gripper attached on robotic arm manipulating handles

sensors' feedback. The motors' controllers comply with CANOpen, a high-level communication protocol in terms of the OSI Model. In order to control the whole CAN network, hexadecimal message frames will be sent across the devices. The controllers of the motors provide messages for all the available states that they can be in, as well as which is the sequence to follow for a state change that includes several steps in-between. The custom graphics user interface that has been implemented using C# is depicted in Fig. 4.10.

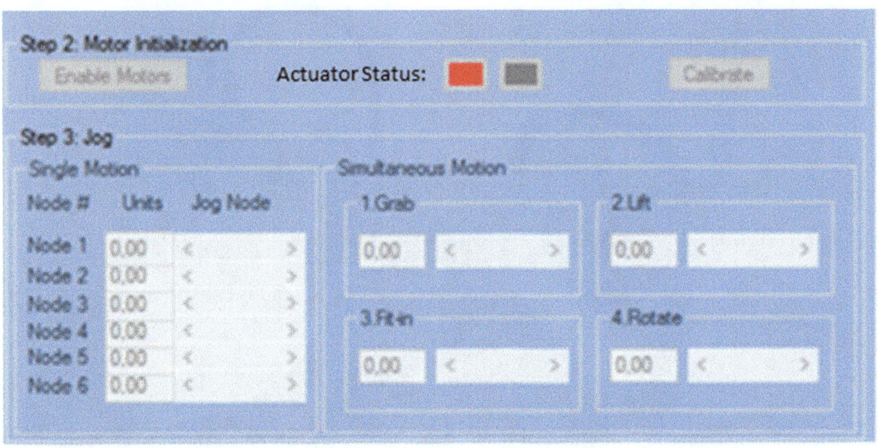

Fig. 4.10 Custom interface for gripper control

4.5.7 Use Case from Consumer Goods Industry

The aforementioned vision, flexible gripping systems as well as the robot cooperation technologies have been integrated together to create a production station that assembles handles with heads of shavers. Shavers are in general composed of parts with complex shape geometry, weighing around 20 grammars and come in many models and variation. Its main characteristic however is that it is not symmetrical along all its axis and it can lay in different orientations when left randomly on a surface. An overview of the demonstrator is given on Fig. 4.11, while the different areas that comprise the cell are shown on Figs. 4.12, 4.13 and 4.14.

Each area operates autonomously and they are all synchronized with the use of Synchronization and Control Framework algorithm.

- Area 1:
  - Detection of the position and orientation of the handle on vibratory surface with the use of vision system
  - Pick, manipulation (orientation correction) and placement of the handle in the Assembly area.
- Area 2:
  - Assembly operation using dual platform assembly device.

Fig. 4.11 Demonstrator of the consumer goods pilot case

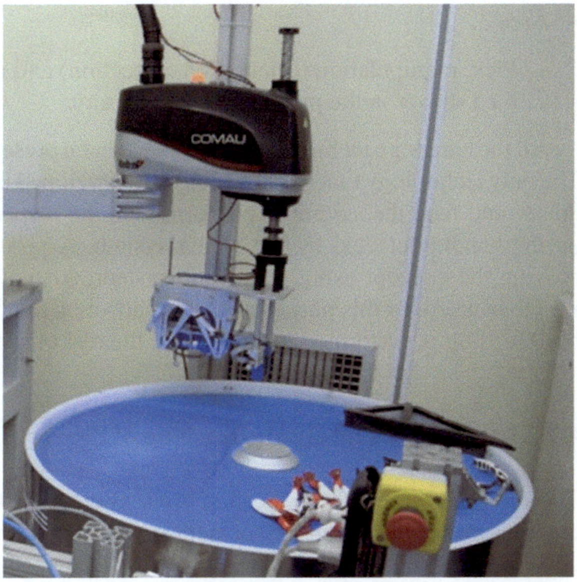

Fig. 4.12 Area 1. Detection—feeding area

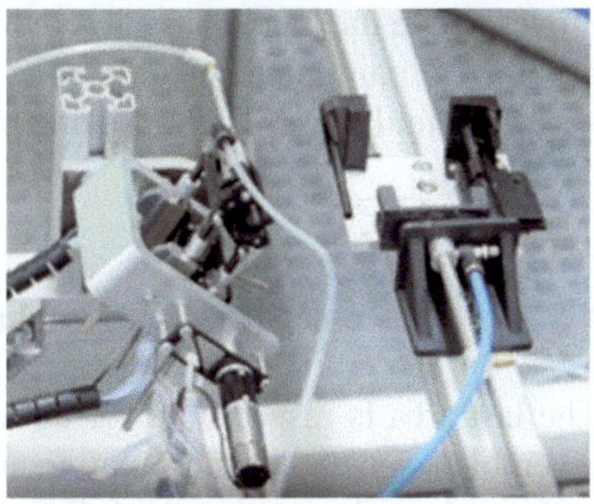

Fig. 4.13 Area 2. Assembly area

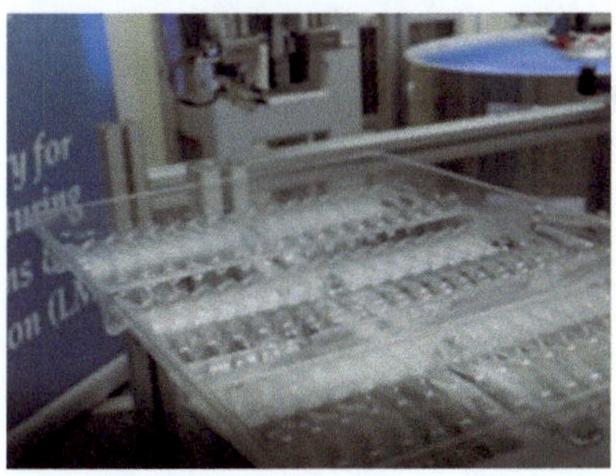

Fig. 4.14 Area 3. Packaging area

- Area 3:
  - Pick, manipulation (orientation correction) and placement of the fully assembled shaver in the primary packaging tray.

All the hardware technologies used, have been presented in detail in TRL5 (D5.3). The only technology that has not been fully developed is the head feeding mechanism that would feed the assembly platform with the respective shaver head, depending on the handle detected by the vision system. A prototype has been implemented proving the concept using pneumatic system. Also, a robot has been examined as idea to be used for this purpose. As it is shown on the final demonstration video, the head feeding is performed manual while the rest of the system is fully autonomous.

4.5.7.1 Processes Involved: Handling of Shaver Handles

The main functionality of the first area, after correct detection is ensured, is the picking, manipulation and placement of the handle on the assembly mechatronic device. The hardware used during this phase is the first robot and the flexible gripper. Based on the initial pose of the handle, as detected by the vision system, the robot moves above the handle and together with the gripper approaches the handle, grasps it, retracts, manipulates it and moves to the area 2 to place it on the assembly device with its default pose. The following pictures (Figs. 4.15 and 4.16) will give an overall perspective.

4.5.7.2 Processes Involved: Packaging of Shavers

The last area is responsible for three main sub processes. The picking of the fully assembled shaver, its manipulation and its placement in the packaging tray. All those processes are depicted at the following figures (Figs. 4.17 and 4.18).

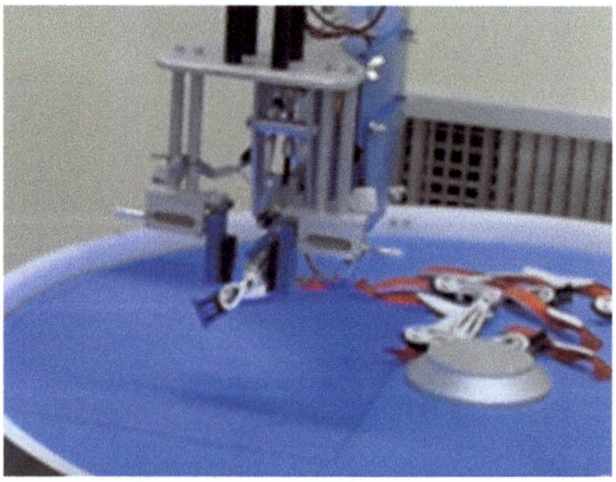

Fig. 4.15 Handle picking and manipulation

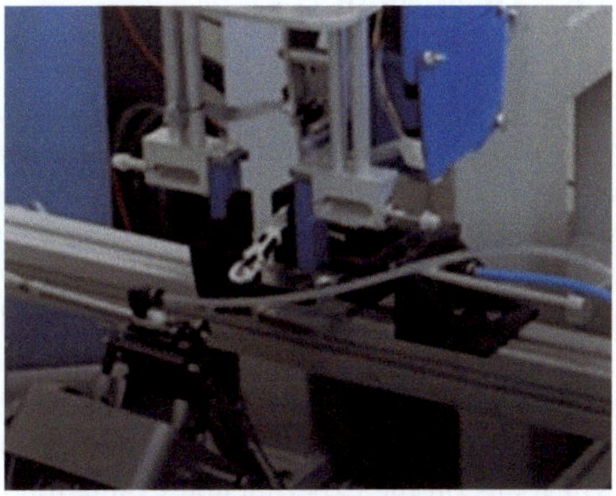

Fig. 4.16 Handle placement on assembler

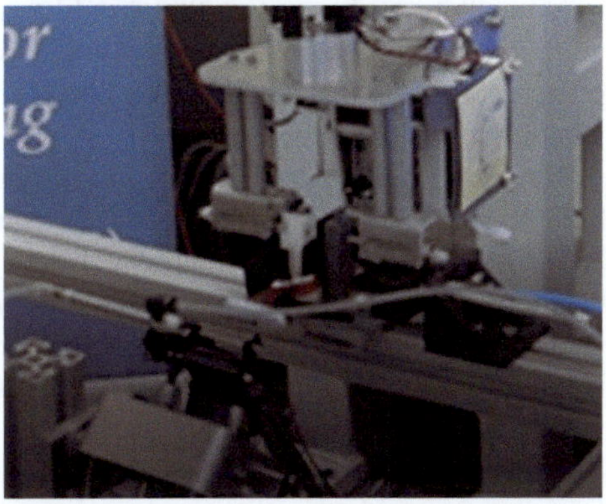

Fig. 4.17 Picking and manipulation of assembled shaver

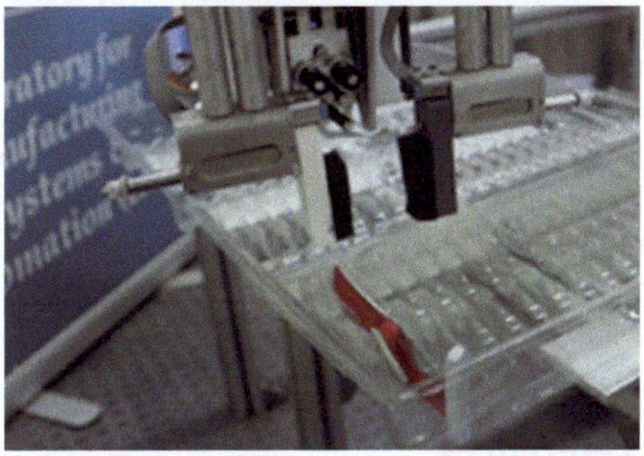

Fig. 4.18 Shaver placement in tray

The hardware used in this area is the second SCARA robot and a second (identical to the first) with a 6 DoF gripper attached to it.

4.6 Discussion

In this chapter, two similar reconfigurable grippers have been presented, throughout the different stages of implementation. The first one is a modular reconfigurable anthropomorphic gripper with multiple fingers that can execute high speed movements and grasp components weighing a few kilos, while keeping its design simple and its dimensions relatively compact. The gripper has been tested for its capability of implementing on-demand reconfiguration of its fingers through the developed control modules. Different grasp types (pinching, enclosing etc.) have been successfully tested and more are considered possible with the use of customized fingertips.

Following the design and the construction of the first prototype, future work will focus on the experimental evaluation to ensure that it can sufficiently achieve the force/torque and speed values considered in the design stage. Furthermore, the implementation of a system that would provide feedback to the gripper's executed motion, and correct any mistakes, should be the next goal. Finally, automatic grasp planning algorithms should be developed to facilitate the automated and optimized grasping of different objects.

Regarding the second gripper, the proposed handling concept has been proven both applicable and efficient. There is, though, a lot of room for improvement to be reconsidered in future implementations. By pushing the item inside the engaging mechanism, there is a chance that it slips from the grasped fingers, something that leads to its unpredictable positioning inside the mechanism. A way to prevent such behaviors is to break down the direct engaging motion into smaller steps, between which, additional individual re-grasping motions are inserted.

The technique mentioned above, aims at using the terminal switch inside the back finger, to sense whether the item has slipped or not. Different type of sensors such as analogue optical proximity sensors may also be considered to control more efficient the relative position of the part and the finger. Additionally, further redesign to allow industrial applications that can meet production requirements is also necessary. The use of material other than plastic that are less prone to wear, is also an open issue.

In order to further characterize the capabilities of the proposed concept, it should be tested for a variety of different objects. Nevertheless, even though the manipulated object was small, a holistic mechatronics approach was required to decently adopt the grasping concept to it.

References

1. Brumson B (2011) Robots in consumer goods, robotic industries association. Available at: https://www.robotics.org/content-detail.cfm/Industrial-Robotics-Industry-Insights/Robots-in-Consumer-Goods/content_id/2701.
2. Michalos G, Makris S, Papakostas N, Mourtzis D, Chryssolouris G (2010) Automotive assembly technologies review: challenges and outlook for a flexible and adaptive approach. CIRP J Manuf Sci Technol 2:81–91. https://doi.org/10.1016/j.cirpj.2009.12.001

3. Papakostas N, Michalos G, Makris S, Zouzias D, Chryssolouris G (2011) Industrial applications with cooperating robots for the flexible assembly. Int J Comput Integr Manuf 24:650–660. https://doi.org/10.1080/0951192X.2011.570790
4. Fantoni G, Santochi M, Dini G, Tracht K, Scholz-Reiter B, Fleischer J, Kristoffer Lien T, Seliger G, Reinhart G, Franke J, Nørgaard Hansen H, Verl A (2014) Grasping devices and methods in automated production processes. CIRP Ann 63:679–701. https://doi.org/10.1016/j.cirp.2014.05.006
5. Fantoni G, Capiferri S, Tilli J (2014) Method for Supporting the selection of robot grippers. Proc CIRP. 21:330–335. https://doi.org/10.1016/j.procir.2014.03.152
6. Kuo L-C, Chiu H-Y, Chang C-W, Hsu H-Y, Sun Y-N (2009) Functional workspace for precision manipulation between thumb and fingers in normal hands. J Electromyogr Kinesiol 19:829–839. https://doi.org/10.1016/j.jelekin.2008.07.008
7. Panel GB, Sanderson A, Wilcox B. WTEC Panel report on international assessment of research and development in robotics
8. Chryssolouris G (2006) Manufacturing systems: theory and practice. Springer, New York
9. Krüger J, Wang L, Verl A, Bauernhansl T, Carpanzano E, Makris S, Fleischer J, Reinhart G, Franke J, Pellegrinelli S (2017) Innovative control of assembly systems and lines. CIRP Ann 66:707–730. https://doi.org/10.1016/j.cirp.2017.05.010
10. Pellegrinelli S, Pedrocchi N, Tosatti LM, Fischer A, Tolio T (2017) Multi-robot spot-welding cells for car-body assembly: design and motion planning. Robot Comput Integr Manuf 44:97–116. https://doi.org/10.1016/j.rcim.2016.08.006
11. Michalos G, Makris S, Chryssolouris G (2015) The new assembly system paradigm. Int J Comput Integr Manuf 28:1252–1261. https://doi.org/10.1080/0951192X.2014.964323
12. Makris S, Michalos G, Eytan A, Chryssolouris G (2012) Cooperating robots for reconfigurable assembly operations: review and challenges. Proc CIRP 3:346–351. https://doi.org/10.1016/j.procir.2012.07.060
13. Makris S, Michalos G, Chryssolouris G (2012) Virtual commissioning of an assembly cell with cooperating robots. Adv Decis Sci 1–11. https://doi.org/10.1155/2012/428060
14. Tsarouchi P, Matthaiakis S-A, Michalos G, Makris S, Chryssolouris G (2016) A method for detection of randomly placed objects for robotic handling. CIRP J Manuf Sci Technol 14:20–27. https://doi.org/10.1016/j.cirpj.2016.04.005
15. Guillo M, Dubourg L (2016) Impact and improvement of tool deviation in friction stir welding: Weld quality and real-time compensation on an industrial robot. Robot Comput Integr Manuf 39:22–31. https://doi.org/10.1016/j.rcim.2015.11.001
16. Schunk parallel gripper. Available online: https://schunk.com/de_en/gripping-systems/category/gripping-systems/schunk-grippers/parallel-gripper/
17. Schunk centric gripper. Available online: https://schunk.com/br_en/gripping-systems/category/gripping-systems/schunk-grippers/centric-gripper/
18. Adaptive Robot Gripper 2-Finger. Available online: https://robotiq.com/products/adaptive-robot-gripper/.
19. Townsend W (2000) The BarrettHand grasper—programmably flexible part handling and assembly. Ind Robot 27:181–188. https://doi.org/10.1108/01439910010371597
20. Carbone G, González A (2011) A numerical simulation of the grasp operation by LARM Hand IV: a three finger robotic hand. Robot Comput Integr Manuf 27:450–459. https://doi.org/10.1016/j.rcim.2010.09.005
21. Fukaya N, Toyama S, Asfour T, Dillmann R (2000) Design of the TUAT/Karlsruhe humanoid hand. In: 2000 IEEE/RSJ international conference on Intelligent Robots and Systems (IROS 2000) (Cat. No.00CH37113). IEEE, Takamatsu, Japan, pp 1754–1759. https://doi.org/10.1109/IROS.2000.895225.
22. Kappassov Z, Khassanov Y, Saudabayev A, Shintemirov A, Varol HA (2013) Semi-anthropomorphic 3D printed multigrasp hand for industrial and service robots. In: 2013 IEEE international conference on mechatronics and automation. IEEE, Takamatsu, Kagawa, Japan, pp 1697–1702. https://doi.org/10.1109/ICMA.2013.6618171.

23. Robotic Hands (Self-contained) | Prensilia s.r.l. Available online: https://www.prensilia.com/index.php?q=en/node/40. Accessed on 21 July 2017.
24. Salisbury JK, Craig JJ (1982) Articulated hands: force control and kinematic issues. Int J Robot Res 1:4–17. https://doi.org/10.1177/027836498200100102
25. Jacobsen S, Iversen E, Knutti D, Johnson R, Biggers K (1986) Design of the Utah/M.I.T. dextrous hand. In: 1986 IEEE international conference on robotics and automation. Institute of Electrical and Electronics Engineers, San Francisco, CA, USA, pp 1520–1532. https://doi.org/10.1109/ROBOT.1986.1087395
26. Lovchik CS, Diftler MA (1999) The Robonaut hand: a dexterous robot hand for space. In: 1999 IEEE international conference on robotics and automation (Cat. No.99CH36288C). IEEE, Detroit, MI, USA, pp 907–912. https://doi.org/10.1109/ROBOT.1999.772420.
27. Zhang Y, Han Z, Zhang H, Shang X, Wang T, Guo W, Gruver WA (2001) Design and control of the BUAA four-fingered hand. In: Proceedings 2001 ICRA. IEEE international conference on robotics and automation (Cat. No.01CH37164). IEEE, Seoul, South Korea, pp 2517–2522. https://doi.org/10.1109/ROBOT.2001.933001
28. Lin L-R, Huang H-P (1998) NTU hand: a new design of dexterous hands. J Mech Des 120:282–292. https://doi.org/10.1115/1.2826970
29. Kawasaki H, Komatsu T, Uchiyama K, Kurimoto T (1999) Dexterous anthropomorphic robot hand with distributed tactile sensor: Gifu hand II. In: IEEE SMC'99 conference proceedings. 1999 IEEE international conference on systems, man, and cybernetics (Cat. No.99CH37028). IEEE, Tokyo, Japan, pp 782–787. https://doi.org/10.1109/ICSMC.1999.825361
30. Mouri T, Kawasaki H, Ito S (2007) Unknown object grasping strategy imitating human grasping reflex for anthropomorphic robot hand. JAMDSM 1:1–11. https://doi.org/10.1299/jamdsm.1.1
31. Dexterous Hand—Shadow Robot Company. Available online: https://www.shadowrobot.com/products/dexterous-hand/
32. Palli G, Ficuciello F, Scarcia U, Melchiorri C, Siciliano B (2014) Experimental evaluation of synergy-based in-hand manipulation. IFAC Proc Vol 47:299–304. https://doi.org/10.3182/20140824-6-ZA-1003.00784
33. Ma RR, Dollar AM (2011) On dexterity and dexterous manipulation. In: 2011 15th international conference on advanced robotics (ICAR). IEEE, Tallinn, Estonia, pp 1–7 https://doi.org/10.1109/ICAR.2011.6088576
34. Mattar E (2013) A survey of bio-inspired robotics hands implementation: new directions in dexterous manipulation. Robot Auton Syst 61:517–544. https://doi.org/10.1016/j.robot.2012.12.005
35. Michalos G, Dimoulas K, Mparis K, Karagiannis P, Makris S (2018) A novel pneumatic gripper for in-hand manipulation and feeding of lightweight complex parts—a consumer goods case study. Int J Adv Manuf Technol 97:3735–3750. https://doi.org/10.1007/s00170-018-2224-2
36. Hawkes EW, Christensen DL, Han AK, Jiang H, Cutkosky MR (2015) Grasping without squeezing: shear adhesion gripper with fibrillar thin film. In: 2015 IEEE international conference on robotics and automation (ICRA). IEEE, Seattle, WA, USA, pp 2305–2312. https://doi.org/10.1109/ICRA.2015.7139505
37. De Meter EC (2004) Light activated adhesive gripper (LAAG) workholding technology and process. J Manuf Process 6:201–214. https://doi.org/10.1016/S1526-6125(04)70075-4
38. Lien TK (2013) Gripper technologies for food industry robots. In: Robotics and automation in the food industry. Elsevier, pp 143–170. https://doi.org/10.1533/9780857095763.1.143
39. Biganzoli F, Fantoni G (2008) A self-centering electrostatic microgripper. J Manuf Syst 27:136–144. https://doi.org/10.1016/j.jmsy.2008.11.002
40. Roy D (2015) Development of novel magnetic grippers for use in unstructured robotic workspace. Robot Comput Integr Manuf 35:16–41. https://doi.org/10.1016/j.rcim.2015.02.003
41. Pettersson A, Davis S, Gray JO, Dodd TJ, Ohlsson T (2010) Design of a magnetorheological robot gripper for handling of delicate food products with varying shapes. J Food Eng 98:332–338. https://doi.org/10.1016/j.jfoodeng.2009.11.020
42. Read S, van der Merwe A, Matope S, Smit A, Mueller M (2012) An intuitive teachable micro material handling robot with Van der Waals gripper design and development. In: 2012 5th

robotics and mechatronics conference of South Africa. IEEE, Johannesberg, South Africa, pp 1–6. https://doi.org/10.1109/ROBOMECH.2012.6558469
43. Ashkin A, Dziedzic JM, Bjorkholm JE, Chu S (1986) Observation of a single-beam gradient force optical trap for dielectric particles. Opt Lett 11:288. https://doi.org/10.1364/OL.11.000288
44. Zhang Y, Zhao L, Chen Y, Liu Z, Zhang Y, Zhao E, Yang J, Yuan L (2016) Single optical tweezers based on elliptical core fiber. Opt Commun 365:103–107. https://doi.org/10.1016/j.optcom.2015.11.076
45. https://opencv.org/. Last accessed on 15 Nov 19
46. Krizhevsky A, Sutskever I, Hinton GE (2012) ImageNet classification with deep convolutional neural networks. In: Pereira F, Burges CJC, Bottou L, Weinberger KQ (eds) Advances in neural information processing systems, vol 25. Curran Associates, Inc. , pp 1097–1105
47. Zeiler MD, Fergus R (2013) Visualizing and understanding convolutional networks. CoRR. abs/1311.2901
48. Luo M, Carbone G, Ceccarelli M, Zhao X (2010) Analysis and design for changing finger posture in a robotic hand. Mech Mach Theor 45:828–843. https://doi.org/10.1016/j.mechmachtheory.2009.10.014
49. Cutkosky MR (1989) On grasp choice, grasp models, and the design of hands for manufacturing tasks. IEEE Trans Robot Automat 5:269–279. https://doi.org/10.1109/70.34763
50. Lu Z, Xu C, Pan Q, Zhao X, Li X (2015) Inverse kinematic analysis and evaluation of a robot for nondestructive testing application. J Robot 1–7. https://doi.org/10.1155/2015/596327
51. MRPT—Empowering C++ development in robotics. Available online: https://www.mrpt.org/
52. Robotiq 3 Finger Adaptive Robot Gripper. Available online: https://robotiq.com/products/3-finger-adaptive-robot-gripper
53. Franchi G, Hauser K. Technical report: use of hybrid systems to model the RobotiQ adaptive gripper. Available online: https://www.cs.indiana.edu/ftp/techreports/TR711.pdf
54. SCHUNK Dextrous Hand 2.0 (SDH 2.0) Available online: https://www.nist.gov/sites/default/files/documents/2017/05/09/9020173R1.pdf
55. Lauzier N (2012) Barrett Hand vs Robotiq Adaptive Gripper 2012. Available online: https://blog.robotiq.com/bid/52340/Barrett-Hand-vs-Robotiq-Adaptive-Gripper
56. Barret hand. Available online: https://www.barrett.com/products-hand-specifications.htm
57. Larm Hand/ARMAND. Available online: https://www.mechanimata.it/flyers/ARMAND.pdf
58. Bicchi A (2000) Hands for dexterous manipulation and robust grasping: a difficult road toward simplicity. IEEE Trans Robot Automat 16:652–662. https://doi.org/10.1109/70.897777
59. Yoshikawa T (2010) Multifingered robot hands: Control for grasping and manipulation. Ann Rev Control 34:199–208. https://doi.org/10.1016/j.arcontrol.2010.09.001
60. Nof SY (ed) (2009) Springer handbook of automation: with 149 tables. Springer, Berlin

Chapter 5
Cooperative Manipulation—The Case of Dual Arm Robots

5.1 Introduction

The evolution of the market today and the steady rise of competition increase the pressure on industry to gain a larger market share [1]. It is difficult to exclude the automotive industry from this race. Particularly because the variety of products is increasing, the paradigm mass customization [2, 3] dictates the need for manufacturing processes to be designed to meet individual needs [4]. Several performance factors need to be examined and optimized in terms of metrics such as cost, efficiency, quality and flexibility for higher levels of automation.

Integrating a dual arm robot in a production cell may have significant impact in both small and larger industries. The approach of integrating simple and lean gripping devices in the two arms is of high importance, while the concept of dual gripper in a single arm has been successfully tested [1]. The dual arm robot capabilities for synchronized and coordinated motions are expected to increase dexterity and flexibility in manufacturing [2, 3]. Inspired by human task execution, the introduction of dual arm robots in human based assembly lines is a promising concept [4]. So far, their presence in industrial applications is restricted, despite their great advantages, such as dexterity, flexibility, space saving, decreased complexity of tools and gripping devices [5].

The human-based assembly lines in particular contain operations involving flexibility which can only be performed by two-handed means. These are also classified as bi-manual activities [5–7]. The implementation of dual arm robots in assembly lines, inspired by human task results, poses a double novelty [4]. On the one side, a typical manual assembly cell's attempted automation is a task on its own. On the other side, it is important to choose a human-like robot to handle tasks that require limited space and that both arms work in cooperation. Several research attempts have already addressed the idea of using multiple cooperating robots for assembly operations [8]. The use of such robots in assembly lines seems exciting, as it enables multi-tasking, as well as space and cost efficiency by eliminating fixtures and clamping devices [9].

S. Makris, *Cooperating Robots for Flexible Manufacturing*,
Springer Series in Advanced Manufacturing,
https://doi.org/10.1007/978-3-030-51591-1_5

Last but not least, the fact that dual arm robots emulate the structure of the human body has made designing and programming assembly operations simpler and more intuitive [3, 17].

In operations typically performed by humans, this chapter focuses on the mechanical and programming dimensions of using a dual arm robot. The case chosen includes a dual arm robot architecture and pre-assembly programming for a vehicle dashboard. The first step is to raise the traverse from the loading area and position it on the assembly table. The body computer of the car is then grasped, installed and set on the traverse.

5.2 State of the Art Solutions

Research and industry have shown significant interest in dual arm robots, due to their advantages described above [10, 11]. Examples of dual arm robot platforms that have been used mainly for research purposes such as Smart Dual arm and Amico (COMAU) [12], Yumi (ABB) [13], NASA dual arm robot, Dual arm robot series (YASKAWA—Motoman) [14], Adroit Manipulator arm [15], Baxter and later Sawyer (Rethink Robotics) [16].

The main research areas related to dual arm robots include the control methods for bi-manual operations, the motion planning tools and the simulation of both arms motions, the intuitive robot programming methods, the human robot interaction techniques and so on. Some of these areas have been investigated also for single arm robots, where research challenges arisen. An example is the intuitive robot programming techniques, such as programming by demonstration, learning and instructive systems gain space [17, 18]. Such a programming system has been developed in X-act project as well [4, 19]. Additionally, the human-like structure of dual arm robots enables also their close collaboration with humans. In this direction, several research works have focused on enabling human robot collaboration, investigating issues such as interaction and safety [20–23]. Related standards that X-act has been based on and contributed as well are the ISO 10218-1 and ISO 10218-2 for robots safety requirements and the ISO/TS 15066 regarding the collaborative workspaces requirements.

Because of their advantages such as agility, versatility, space saving, reduced tool complexity and gripping tools, both academia and industry have shown significant interest in dual arm robots [4, 5, 10, 11]. The main advantages of the dual arm robots compared with single arm robots are interesting, since a dual arm robot:

- requires less floor space than two single arm robots;
- has larger workspace compared to two fixed single arm robots;
- allows significant simplification on the gripping devices, tooling and fixtures compared with single arm robots;
- enables the easier programming compared with two separate single arm robots with individual controllers.

The fact that dual arm robots resemble the human body structure, makes the design and programming of assembly operations easier and more intuitive. Due to their higher agility compared to traditional single-arm robots, the human-like dual-arm robot capabilities for synchronized and controlled movements are expected to increase production flexibility [2, 3]. Based on their configuration, they can be used to perform tasks in the same way as humans do with both hands, so integrating dual-arm robots into human-based assembly lines is a promising prospect [4].

5.3 Approach

The proposed versatile cell allowing human-robot collaboration involves a dual-arm robot of industrial size, operating with the human in shared or separate workspaces. This cell offers increased flexibility for the following reasons. At first, the dual arm robot mainly involves the use of two arms, which makes it possible to manipulate more dexterously, save space, increase work space, reduce the complexity of the tools and fixtures compared to that of single arm robots [1, 4]. Second, requiring a person to assemble more complex parts, such as wire harness assembly, requires intelligence and dexterity to further increase the versatility of the cell.

The requirements' analysis given the manual assembly processes was the basis for the design of a robotized assembly. The most important requirements are the product quality, the minimum modifications in the current production line, cost efficiency, the process cycle time, working conditions and ergonomics, safety, overall production cost reduction, maintainability and space saving. The human motions can be divided into single arm actions, as well as bi-manual actions, according to the classification described in [14]. The bi-manual operations can be considered as both the coordinated movement of the two arms, referred to in this work as "COOP," as well as synchronized movement when both arms move independently and synchronized, referred to as "SYNC" movements.

The dual arm robot movement capabilities can cover most human movements, both single and bi-manual. In the case study on the automotive sector, discussed in this chapter, both forms of motions were found and presented in [14]. There are 22 bimanual operations between them, 16 of which are SYNC and 6 are operations on the COOP arms. In comparison, only eight single-arm operations are used. This is also a rationale to guide the dashboard assembly cell robotizing towards the use of a dual arm robot. Bi-manual operations can be called both the coordinated movement of the two arms, referred to in this work as "COOP," and synchronized movement when both arms move separately and synchronized, referred to as "SYNC" movements (Table 5.1).

In the design and selection of tools and grasping devices as well as in the workstation layout, the study of tasks in "SINGLE", "SYNC" and "COOP" movements has an important role to play. A dual-arm robot allows the entire workspace around it to be used, as opposed to other approaches for single-arm robots.

Table 5.1 Single and bi-manual operations of dashboard automotive use case [19]

Dual arm robot tasks	Number of SINGLE arm operations	Number of SYNC arm operations	Number of COOP arm operations
Pick up traverse	0	4	1
Place traverse	0	4	1
Pick up body computer	0	3	2
Place body computer	1	2	2
Pick up screw driver	2	1	0
Screwing process	3	0	0
Place screw driver	2	1	0

The fenceless flexible cell approach is based on the use of a safe 3D camera, called SafetyEYE [1]. Using this device, it is possible to define multiple configurations of virtual fences that can be used within the same cell and can be changed with the push of a button. Three different safety configurations were specified in the proposed flexible cell. The first refers to the configuration of the robot task, where the human and the robot operate in separate workspaces; the second refers to the con-figuration of the cooperative task, where the human and the robot work in a shared workspace, but on different tasks; and the third refers to programming configuration, where the human coexists only for programming tasks.

The yellow zones refer to the warning signals so that if they are crossed by a human or an obstacle, a light warning may eventually decrease the speed of the robot if necessary. When enabling the red zones, also known as hazardous zones, the robot drivers are switched off automatically for safety reasons. There is only an alert zone in the programming mode, which helps the person to work closer to the robot by using the teach pendant to program it.

5.4 Industrial Relevance and Examples

The vehicle dashboard demonstrator is the case study, which the proposed solutions have been applied to [18], having originated from the final assembly line of an automotive manufacturer, performed manually. The selection of a COMAU dual arm robot was based on the handling of heavy and complex geometry parts, as well as flexible parts from different materials such as metal and plastic. The first stage consists of raising the traverse out of the loading area and placing it on the assembly table. The body computer of the car is then grasped, mounted and placed onto the

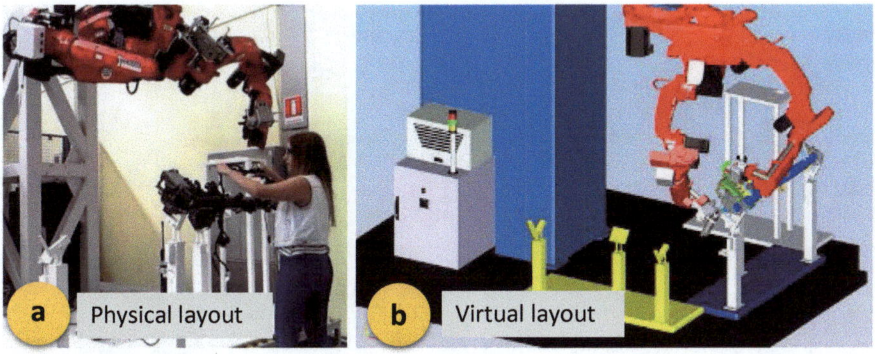

Fig. 5.1 Flexible layout [1]

traverse. Installation of cables, air-conditioning unit etc. takes place near this station. The flexible layout is illustrated in Fig. 5.1.

5.4.1 Heavy Part Grasping with a Dual Arm Robot

The product specific grippers that are integrated in the Smart dual arm robot allow the manipulation of traverse, without high cost. In the case that a commercial gripper should be selected, a high payload would be required (~ 12 kg). This would be a restricted factor for integrating it in robot, as the gripper itself would have a significant load (each arm can load 10 kg). The first step of the assembly process refers to the pick and place task of the dashboard traverse (Fig. 5.2). This task was carried out using the mechanical fingers of the dual-gripper concept.

The selected solution allows the manipulation of traverse with mechanical low-cost fixtures. The integration of low-cost solution for robot tooling is even easier in a dual arm. The concept of using a commercial screwdriver with low cost, lifted by the PG70 fingers is adopted in this case. A construction for lifting the screwdriver is used, without requiring the need for tool changer that would increase the cost. The grippers designed in such a way, so as to allow the handling of more than one objects.

5.4.2 Parts Grasping and Screwing Processes

The second task of the dual arm robot is to fasten the fuse box with the use of screws, where the first robot arm carries the fuse box and the second handles the screwdriver and the screws. This is an example of tasks requiring the use of both arms during the process. The rotation of the Torso axis to the base of the Body Computer is the first step and is followed by the lifting of the Body Computer from its base, as well as

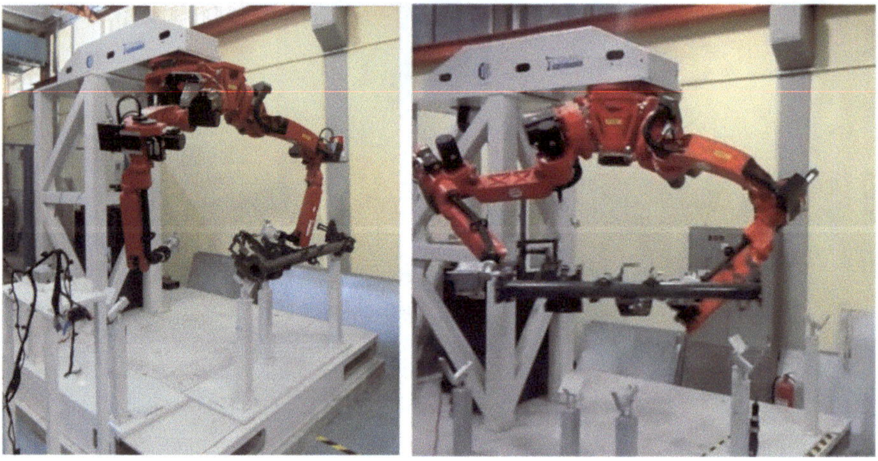

Fig. 5.2 Dashboard traverse pick and place task

the placing of the Body Computer on the traverse, and the extraction of both ARMs (Fig. 5.3).

During the Screwing Process, the Body Computer is correctly aligned with the holes in the traverse in order to be fixed. ARM1 is lifting the Screwdriver from its base and screws the bolts. The screwdriver is returned to its base and the two ARMs are re-positioned in a neutral position. These steps are visualized in Fig. 5.4.

Fig. 5.3 Dual arm robot pick and place task

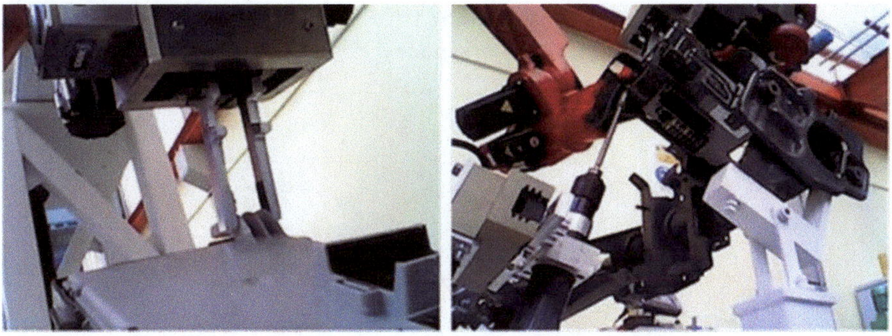

Fig. 5.4 Screwing task

5.5 Discussion

The use of dual-arm robots was assessed on the basis of the distinction between single and bi-manual behavior. The results show that this type of robot is suitable for these tasks as it has the ability to control the performance of single and bi-manual tasks. The final product quality—in this case the car improvement compared to the manual assembly line – is also an important issue that can be solved through this workstation, as a human being is more likely to make assembly mistakes. In addition, the working conditions and improvement in ergonomics can be accomplished through the new design, as the weight of the traverse for a human operator in the line is rather high.

If a dual arm robot is used, greater use of the workspace is permitted, higher levels of robot functionality as well as simpler arms programming and coordination. This is compatible with the development of algorithms in the robot's control system, which can easily perform bi-manual actions. The robot workspace can also be expanded as the robot has an external rotary axis that allows the robot to rotate 180° around its base. In this case, the cooperative, coordinated and single-arm motions workspace occupies the robot cell's larger area.

The use of two single arm robots was an alternative approach to designing this unit (Fig. 5.5). In the same figure, the region where the two separate robots will operate is visualized within cooperation. The cooperative working space is the purple spherical shape and, in comparison with the dual arm robot, is obviously restricted. In the same figure, a red dotted line can display the workspace of Robot 1 as well as the non-reachable red-colored areas. Areas 1–3 are the areas that need to be reached to complete the dashboard assembly scenario. Both robots can access Area 1, while Area 2 cannot be entered by robot 1. It is also clear that in this situation, when attempting to move the traverse from area 1 to area 2 the two robots' collision.

Another idea was the use of a single arm robot, with a product-specific gripper to lift the traverse. The gripper would be pretty heavy and complicated to manage the traverse in this method, and it may not be used to grasp another component. In this case, the cost would be significantly increased and an advanced screwing system would be needed. The dual arm robot has supported the possibility of somewhat

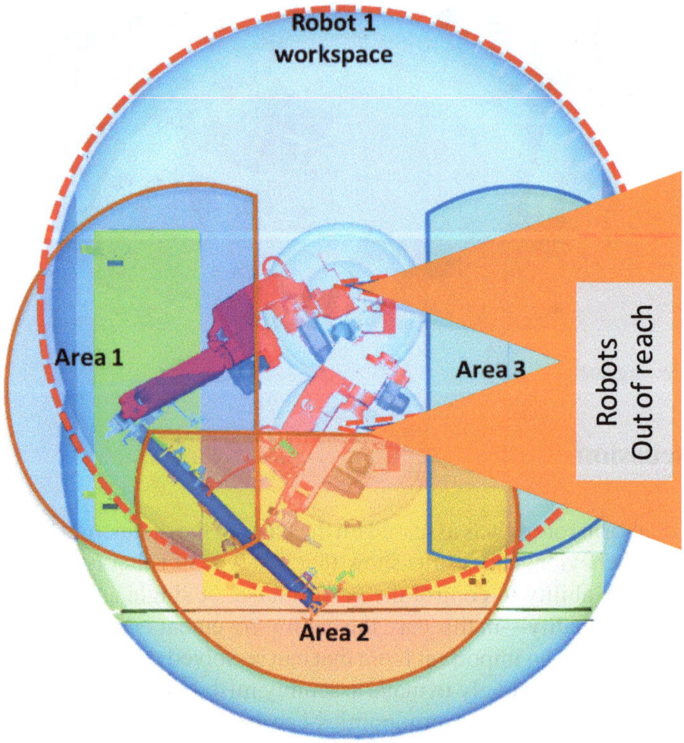

Fig. 5.5 Limited workspace of two separate robot arms [19]

complex parts being handled in a dexterous manner and using lower complexity tooling compared to this approach.

While great progress is being made and there is broad interest in dual arm robots, multiple problems still need to be addressed. The dual arm robots have made it possible to substantially reduce the number and complexity of the tools and grasping devices, while also helping to save space. To this purpose, a significant challenge influenced by human performance is the design and selection of the correct solutions.

The problems for single and dual arm robots are similar in terms of the intuitive interfaces and interaction mechanisms for robot programming (see Chap. 15). In addition, the incorporation of advanced vision skills and situational awareness in semi-structured or fully unstructured industrial environments is in the future research plans. The need for a standardized universal integration architecture remains an obstacle, allowing the implementation of different robot cell behaviors to achieve extensibility and flexibility.

Aspects of safety namely how safe the person is and behaves, or what sensors can be used to bring people and robots closer together, should be investigated as well. The 3D sensor integration ensured safety and workspace sharing, having limitations in bringing a human and a robot working close at the same time. Integrating new safety sensors such as visual-based sensors, safety mats and skins is one way to overcome the limitations of 3D sensors and enable people and robots to work closer in the same workspace. In conclusion, with the implementation of a dual arm robot, the proposed solution allowed versatility in assembly cells working close to a human.

The results of this study were important in promoting cell and system structure, enhancing ergonomics aspects, saving space and reducing the complexity of the devices. The approach for automatic robot motion generation, based on the proposed hierarchical model, contributes to simplifying robot programming. It also becomes simpler and more time-efficient to reconfigure a cell system. Interaction mechanisms allowed human and robotic activities to be synchronized during their execution, while security is assured in a fenceless environment that allows for coordination with the communication system. Last but not least, this study has shown simpler solutions focused on advanced ICT technology, even in the case of complex robot systems.

References

1. Makris S, Tsarouchi P, Matthaiakis A-S, Athanasatos A, Chatzigeorgiou X, Stefos M, Giavridis K, Aivaliotis S (2017) Dual arm robot in cooperation with humans for flexible assembly. CIRP Ann 66:13–16. https://doi.org/10.1016/j.cirp.2017.04.097
2. Chryssolouris G (2006) Manufacturing systems: theory and practice, 2nd edn. Springer, New York
3. Smith C, Karayiannidis Y, Nalpantidis L, Gratal X, Qi P, Dimarogonas DV, Kragic D (2012) Dual arm manipulation—a survey. Robot Auton Syst 60:1340–1353. https://doi.org/10.1016/j.robot.2012.07.005
4. Makris S, Tsarouchi P, Surdilovic D, Krüger J (2014) Intuitive dual arm robot programming for assembly operations. CIRP Ann 63:13–16. https://doi.org/10.1016/j.cirp.2014.03.017
5. Surdilovic D, Yakut Y, Nguyen T-M, Pham XB, Vick A, Martin-Martin R (2010) Compliance control with dual-arm humanoid robots: design, planning and programming. 2010 10th IEEE-RAS international conference on humanoid robots. IEEE, Nashville, TN, pp 275–281
6. Wang L, Mohammed A, Onori M (2014) Remote robotic assembly guided by 3D models linking to a real robot. CIRP Ann 63:1–4. https://doi.org/10.1016/j.cirp.2014.03.013
7. Reinhart G, Tekouo W (2009) Automatic programming of robot-mounted 3D optical scanning devices to easily measure parts in high-variant assembly. CIRP Ann 58:25–28. https://doi.org/10.1016/j.cirp.2009.03.125
8. Makris S, Michalos G, Eytan A, Chryssolouris G (2012) Cooperating robots for reconfigurable assembly operations: review and challenges. Procedia CIRP 3:346–351. https://doi.org/10.1016/j.procir.2012.07.060
9. Krüger J, Schreck G, Surdilovic D (2011) Dual arm robot for flexible and cooperative assembly. CIRP Ann 60:5–8. https://doi.org/10.1016/j.cirp.2011.03.017
10. Rehnmark F, Bluethmann W, Mehling J, Ambrose RO, Diftler M, Chu M, Necessary R (2005) Robonaut: the "short list" of technology hurdles. Computer 38:28–37. https://doi.org/10.1109/MC.2005.32
11. Ott Ch, Eiberger O, Friedl W, Bauml B, Hillenbrand U, Borst Ch, Albu-Schaffer A, Brunner B, Hirschmuller H, Kielhofer S, Konietschke R, Suppa M, Wimbock T, Zacharias F, Hirzinger G (2006) A humanoid two-arm system for dexterous manipulation. 2006 6th IEEE-RAS international conference on humanoid robots. university of genova, Genova, Italy, IEEE, pp 276–283
12. Comau's AMICO robot takes a starring role at the science museum in London. https://www.comau.com/en/media/news/2017/02/amico_sciencemuseum_london. Accessed 30 Apr 2020

13. ABB's collaborative robot -yumi—industrial robots from ABB robotics. https://new.abb.com/products/robotics/industrial-robots/irb-14000-yumi. Accessed 30 Apr 2020
14. Dual-arm SDA20D robot for assembly and handling| 20.0 kg. https://www.motoman.com/en-us/products/robots/industrial/assembly/sda/sda20d. Accessed 30 Apr 2020
15. Adroit® manipulator arm archives. In: HDT global. http://www.hdtglobal.com/series/adroit-manipulator-arm/. Accessed 30 Apr 2020
16. Sawyer collaborative robots for industrial automation. https://www.rethinkrobotics.com/sawyer. Accessed 30 Apr 2020
17. Biggs G, MacDonald B A survey of robot programming systems. 10
18. Tsarouchi P, Makris S, Chryssolouris G (2016) Human—robot interaction review and challenges on task planning and programming. Int J Comput Integr Manuf 29:916–931. https://doi.org/10.1080/0951192X.2015.1130251
19. Tsarouchi P, Makris S, Michalos G, Stefos M, Fourtakas K, Kaltsoukalas K, Kontrovrakis D, Chryssolouris G (2014) Robotized assembly process using dual arm robot. Procedia CIRP 23:47–52. https://doi.org/10.1016/j.procir.2014.10.078
20. Krüger J, Lien TK, Verl A (2009) Cooperation of human and machines in assembly lines. CIRP Ann 58:628–646. https://doi.org/10.1016/j.cirp.2009.09.009
21. Arai T, Kato R, Fujita M (2010) Assessment of operator stress induced by robot collaboration in assembly. CIRP Ann 59:5–8. https://doi.org/10.1016/j.cirp.2010.03.043
22. Singer S, Akin D (2011) A survey of quantitative team performance metrics for human-robot collaboration. In: 41st international conference on environmental systems. American Institute of Aeronautics and Astronautics, Portland, Oregon
23. Wilcox R, Shah J (2012) Optimization of multi-agent workflow for human-robot collaboration in assembly manufacturing. In: Infotech@Aerospace 2012. American Institute of Aeronautics and Astronautics, Garden Grove, California

Chapter 6
An Approach for Validating the Behavior of Autonomous Robots in a Virtual Environment

6.1 Introduction

The reduction of the product lifecycle as well as the increased number of new models and variants have forced the modern production systems to have sufficient responsiveness in order to be able to adapt their behavior according to the changes of today's global manufacturing environment [1]. One of the main challenges of the manufacturing companies nowadays is the demand for faster and more secure ramp-up process. Due to the aforementioned changes, the manufacturing companies need to launch an increasing number of ramp-up processes in order to satisfy the market demand variation. For achieving this, manufacturers should be ready to quickly and safely adjust their production systems which may include the production stoppage for an extended period of time. The complexity and diversity of the different line components, in terms of control systems and communication channels, requires a great amount of time for onsite setup, testing and validation of the assembly equipment. This leads to increased production time and cost.

For preventing the above increase in manufacturing cost every time a new ramp-up is needed, digital simulation of the production process has been introduced in modern manufacturing systems. Information technology systems have been over the past years an evolutionary technology, forwarding the concepts of digital manufacturing. These systems are based on the digital factory/manufacturing concept, according to which production data management systems and simulation technologies are jointly used for optimizing manufacturing before starting the production and supporting the ramp-up phases [2].

Virtual commissioning (VC) has taken virtual simulation one step further by including more validation capabilities by means of considering the mechatronic behavior of the resources. Commissioning defines the process by which an equipment or a factory is tested to validate its functions according to the prior specifications. Commissioning, usually the last step in the engineering process, can take up to 15–20% of the total delivery time of an automation system project. Unfortunately, nearly

S. Makris, *Cooperating Robots for Flexible Manufacturing*,
Springer Series in Advanced Manufacturing,
https://doi.org/10.1007/978-3-030-51591-1_6

the 2/3 of the time spent in commissioning is spent on fixing software errors, since the control software usually goes through proper integration testing only after all the hardware has been procured and assembled [3]. Virtual commissioning provides a solution for moving a significant portion of commissioning tasks to an earlier phase of the project, away from the critical path. In virtual commissioning, a simulation model of the system is created to replace the real factory. The virtual factory is then connected to the real control system, so that the simulation can be used simultaneously with the procurement and assembly to verify the design and test the control system. This allows for quicker detection of possible errors. As a result, VC allows for the reduction in the commissioning time, the advance of the start of the production time as well as the reduction in the shutdown times [4]. Virtual commissioning can facilitate the work of the production designers enabling decision making support and facilitating the production line setup in terms of resources, tools and tasks that are carried out by each one of them.

Various VC approaches have been presented over the years. All of them focused on validating PLC and other Hierarchical based ICT. Nowadays, there are several other ICT approaches in manufacturing which don't follow an hierarchical model. Hierarchical control approach utilizes a master-slave approach with the master (usually a PLC controller) coordinating and synchronizing the whole process. This approach, which is the most commonly used for robot synchronization, requires a lot of time for the setup, especially if there are more than two robots with several precedence constraints. Furthermore, the programmer should know which memory slots of the plc are available, in order to use them for synchronizing. A risk for mistakenly updating a wrong memory slot and signaling the wrong robot to continue is possible. For eliminating such risks and for facilitating the robot programming, nowadays new control architectures are already introduced, such as Service Oriented Approach (SoA). In addition, new tools such as ROS enable such SoA approach but their validation is still lacking. No VC tools have been developed for SoA control.

The objective of this chapter is to present a VC approach for validating a service-oriented architecture based control system in a virtual environment. The main focus is to introduce a VC application in manufacturing conditions where other control and synchronization methods where selected instead of the traditional PLC approach. Furthermore, the way the presented approach can be customized in order to be ready to be used with a service-oriented architecture will be described. Finally, the benefits from using this method as well as any possible problems or shortcomings will be discussed with respect to manufacturing cost and time saving. To provide a better insight on the real-life implication of the applying VC approach, a case study involving a vehicle floor assembly cell is presented. The case study is applied in an automotive manufacturing production line and it uses cooperating robots, mobile robots and flexible tools.

Virtual Commissioning

One of the first Virtual Commissioning approaches was presented in [5] and was referred to as "soft-commissioning", allowing the coupling of simulation models to real world entities and enabling the analyst to pre-commission and test a system's behavior, before it was built in reality. This approach, however, did not consider the entire life cycle of a technical system including requirements engineering and "classical" simulation analysis. In [6] another VC approach in an industrial robotic cell involving cooperating robots is presented. Two major Virtual Commissioning approaches have been identified. The Software in the Loop (SIL) method uses the same control software as the real world, installs it in the simulation controllers and through a network connection, it enables the communication between the mechatronic objects and the software emulating controllers. It lets the user adapt and test the system model with real-time simulation. SIL is a cost-effective method for evaluating a complex, mission-critical software system before it is used in the real world. Nevertheless, the exact reproduction of the control behavior is difficult due to the lack of control simulation software packages [7]. The second method, known as Hardware in the Loop (HIL), involves the simulation of production peripheral equipment in real time, connected to the real control hardware via fieldbus protocol. Under this setup, commissioning and testing of complex control and automation scenarios, under laboratory conditions, can be carried out for different plant levels (field, line or plant) [8].

6.1.1 Virtual Commissioning and Simulation Tools

Several commercial software packages that can be used for Virtual Commissioning have been developed over the last years. The Tecnomatix Process Simulate for Robotics and Plant Simulation are products that allow the engineers to debug and simulate PLC codes that control material handling systems including conveyors, transfer lines and overhead hangers as well as automation equipment, tooling and safety devices [9]. Delmia by Dassault Systemes allows the virtual prototyping of PLC control systems for cells, machines and production lines which uses Object Linking and Embedding for Process Control (OPC) communication for the coupling of the real control system with the simulated resource [10]. ESI's SimultionX package lets the user create, reproduce and check error scenarios in virtual prototypes [11]. ROS Gazebo is a set of packages that provide the necessary interfaces to simulate a robot. Gazebo also uses a physical engine for illumination, gravity, inertia and also integrates motion planning capabilities. The user can evaluate and test the robot in difficult or dangerous scenarios without any harm to the actual robot [12].

For the purposes of this chapter, although the robot control software was developed in ROS framework, we opted for one of the commercially available software simulation tools, namely the Siemens' Process Simulate over Gazebo for Virtual Commissioning due to the following reasons:

- Although Gazebo is integrated with ROS, it does not support out of the box a wide range of industrial robots. On the other hand, commercial tools offered a number of ready to use controllers that can be imported and facilitate the control of the simulated resources.

- Process Simulate includes a Software development kit namely SDK that can be used in order to expand the already existing functionalities according to the needs of the manufacturer. High flexibility of the SDK allows to VC integration with any type of resources and tools.

6.2 Virtual Resources Modeling

A service-oriented based framework is proposed in this paper in order to configure, control and coordinate dynamically robotic operations. Furthermore, ontology manufacturing principles were used for managing online and offline shop-floor information.

For the realization of this approach, several components were developed. First of all, distinct software modules were used for the control of the manufacturing resources and for the communication between each other. Each resource has its own interface and a software module that is running in parallel in order to provide control, sensing and communication capabilities.

Furthermore, a data model was developed in order to organize the elements of data and standardizes how they relate to one another and to properties of the real-world entities. This will facilitate the manipulation of the real-world resources and the definition of their workload in the specific cell. The logical data structure of a database management system (DBMS), whether hierarchical, network, or relational, cannot totally satisfy the requirements for a conceptual definition of data because it is limited in scope and biased toward the implementation strategy employed by the DBMS. Therefore, the need to define data from a conceptual view has led us to semantic data modelling techniques. For this reason, an ontological representation of the cell has been created. The semantics is often captured by ontology. The reason for integrating ontology with service oriented architecture based systems is that the explicit expression of service semantics will enable them to operate differently when the knowledge base changes, apart from enabling the integration of new agents to actually proceed without reprogramming the existing ones.

The above described components were connected to the Process Simulate environment in order to virtually perform the control and the communication of the simulated robots and perform their action for validation purposes.

An overview of the presented approach can be seen in Fig. 6.1. In the next section, each one of the components is described in more detail.

6.2.1 Services of Robotic Resources

In manufacturing, several attempts have been made recently by specialist robotics suppliers for the development of standard robot software platforms with software development kits (SDKs) in order to simplify integration and robot development.

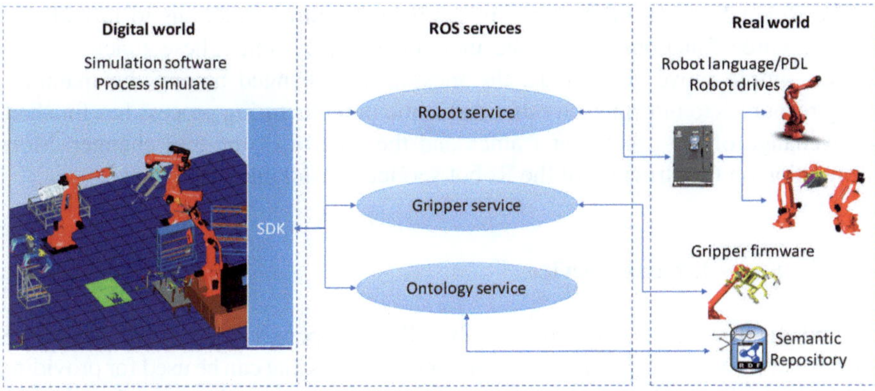

Fig. 6.1 Architecture for connectivity of software and hardware components

Robot Operating System (ROS) is one of the above attempts. ROS is an open framework for robot software development. It is a collection of tools, libraries, and conventions that aim to simplify the task of creating complex and robust robot behavior across a wide variety of robotic platforms.

Each resource or service has its own software modules to perform control and sensing of the manufacturing hardware as well as networking capabilities to exchange data with the other resources or services over a ROS communication channel. Following this scheme, the developed architecture allows the distributed control of the manufacturing system as well as increased autonomy for the manufacturing resources. Finally, each manufacturing resource will be also equipped with decision making software modules which will be able to perform dynamic rescheduling of the manufacturing system when required. Specifically for this approach, the resources consist of Robots and End effectors.

Each of these resources has its own software module to perform control and sensing of the manufacturing hardware as well as networking capabilities to exchange data with the other resources or services over a ROS communication channel.

6.2.1.1 Robot Service

The robot has an interface of itself and is able to communicate with the other resources using the server/client or the publisher/subscriber model [12]. The first allows synchronous communication between the resources. That means that a resource that will have the role of the client can request something from the resource that has the role of the server and then wait for the response. The publisher/subscriber model is used for asynchronous topic-based communication. All the resources subscribe to a topic and everyone can publish messages in it. The messages will be read by the subscribers and then will be parsed and translated into something useful by the resources who are interested.

The Robot Service interface includes all the required functions to provide the robot resource functionalities at the manufacturing system. These functionalities include simple move commands, the messages exchanged among the manufacturing resources to inform each other that some manufacturing process has finished, the exchange of the robot user frames and the control of the tool changer. More information on the structure of the Robot service is available in Chap. 2.

6.2.1.2 End—Effector Service

The end effector which is interchangeably called Gripper service, implements a ROS interface of its own which exposes a number of services that can be used for providing a number of functionalities (open clamps, close clamps, weld, reconfigure, etc.). These services can be used either by the end effector itself or by the other resources that need to control the end effector. For example the robot on which the end effector is attached will need to notify the gripper to close the clamps in order to grasp a part or to open them in order to release it. More information on the structure of the Gripper service is available in Chap. 2.

6.2.2 Data Model

To achieve the aforementioned functionalities a data model should encapsulate all the information needed to allow the resources and services to be autonomous and perform the decision making on their own. This information will need to depict the complete shop floor status both in terms of physical elements and in terms of operations. The overall UML schematic of the data model can be seen in Fig. 6.2 where all the data model classes are represented. Moreover, the connections between the classes are shown, stating either association or composition of the connected classes.

6.2.3 Ontology

The Ontology Service will be used to manage the online and offline shop floor information. An Ontology Repository will be created where all the facility and workload information, along with their interrelationship, will be stored. Each resource will be able to query the repository, using SPARQL language, in order to get the information that they need.

Building on top of the ROS, Ontology Service integration and communication software was developed. The Ontology Service is used for data management and storing services to the resources and services. A semantic repository is also utilized

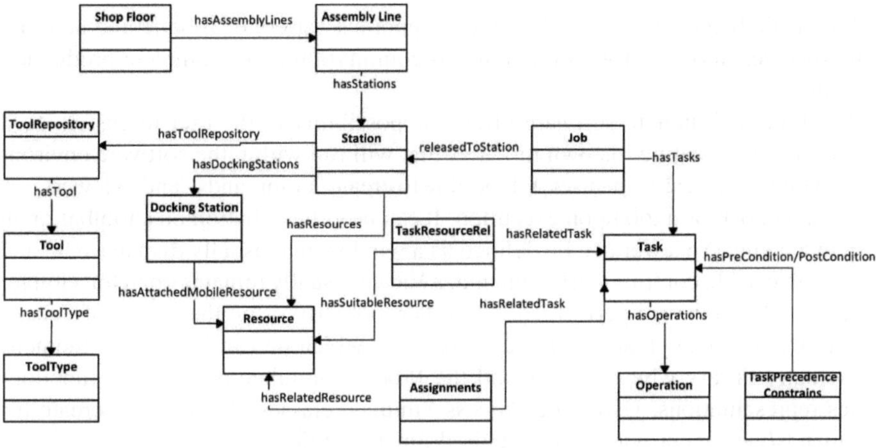

Fig. 6.2 Data model

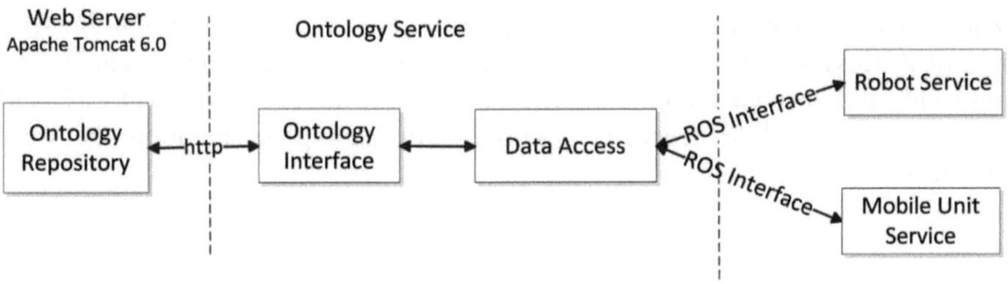

Fig. 6.3 Ontology service architecture

for these functionalities referred as the "Ontology Repository" software module. In Fig. 6.3 below the basic structure of the software architecture is shown.

6.2.4 Simulation Tool

The solution chosen for validating the above architecture was the virtual environment of the Process Simulate software. Process Simulate enables the validation of assembly plans virtually, from concept through to the start of production to help the mitigation of any possible risks.

Process Simulate provides the answer to the increased complexity issues through the design and validation of manufacturing processes in a 3D dynamic environment that is fully integrated within a data-managed environment, where manufacturing engineers can re-use, author, validate and optimize manufacturing process sequences with realistic behavior. Process Simulate supports a variety of robotic and automation processes allowing for simulation and commissioning of complete production systems.

The Process Simulate software offers the possibility to the user to customize it to his needs by creating his own modules that will run within the software environment. More specifically, the user will be able to create "commands" and "viewers". A command performs a job upon execution. It can be either a button on a toolbar or an item in a menu. A viewer can be defined as a window that usually displays data and state of objects. In contrast to a Command, a Viewer usually remains open for a longer duration, while jobs are performed with other tools. By using the Microsoft Visual Studio, the user can create a C# class library in which he can import the Siemens Tecnomatix library which contains all the Process Simulate objects and functionalities representations. Using the Process Simulate classes, the user can create his commands/viewers providing the required functionalities.

The library files that will be created will then be registered in the Process Simulate using a separate Tecnomatix tool. Then, the implemented command/viewer can be added and used like any other existing button in the PS. For achieving the communication between the already developed ROS framework and the virtual environment of Process Simulate, both a command and a viewer were implemented. The command implemented the communication between the two parts (ROS PC and Process Simulate), while the viewer was used for displaying the operations that were performed.

6.3 Illustrative Virtual Validation Example

6.3.1 Actual Assembly Cell

The case study was created in order to prove the feasibility of an autonomous and flexible production line which will be able to reconfigure dynamically according to the current situation. The case study took place at an automotive manufacturer premises and was applied onto a robotic cell that loads and welds the parts of a passenger car floor. The cell is consisted of two workstations: the loading station and the welding station. An overview of the actual assembly cell can be seen at Fig. 6.4.

The components that were installed at the assembly cell and used for the case study were the following:

- A Mobile Unit carrying a robotic arm and is called Robot1.
- A Welding Robot, called Robot2.
- A handling robot namely Robot3.
- A Flexible Gripper, able to reconfigure itself according to the geometry of the parts to be grasped.
- A Dexterous Gripper, strong and agile enough in order to be able to handle both bigger and smaller parts.

Fig. 6.4 Actual assembly cell

- Various automotive parts (9 single parts and 2 subassemblies) placed on especially designed racks.
- 1 mobile fixture especially designed for moving the Tunnel subassembly from the loading area to the welding area.

The operation of the cell can be summarized as follows:

- Robot1 attaches the dexterous gripper that is located on the tool stand
- Robot1 approaches the parts and dexterous gripper changes configuration and acquires each one of them.
- Robot1 releases the parts to the mobile fixture.
- The mobile fixture is moved by the human to the welding station.
- Robot2 performs some tack welding on the fixture.
- Robot3 with flexible gripper grabs the fixture.
- Robot2 and Robot3 cooperate. Robot3 holds the part and Robot2 performs some welding tasks.
- Robot3 breaks down. It notifies all the other resources. Rescheduling takes place. Robot1 will continue performing its tasks.
- Robot1 releases the gripper on the tool stand. The mobile unit undocks, approaches the welding station and docks.
- Robot1 and Robot3 perform the gripper exchange procedure. Robot1 attaches the flexible gripper.

- Robot1 and Robot2 continue their tasks (handling and welding).
- Robot1 moves to the rack and flexible gripper releases the part.

For the realization of the above scenario, service-oriented architecture was used for the control, communication and cooperation of the robots. The reason that this approach was chosen over the hierarchical control approach is that it increases the flexibility and facilitates the setup time by making it simpler to synchronize the robotic tasks and easier to reprogram without having to stop the production line in order to make a new PLC code for the new data. Furthermore, each resource has its own interface and is completely autonomous. There is no master—slave relationship between any of the entities of the manufacturing system. Finally, this autonomy permits easier and quicker rescheduling in case of any unexpected event. Whenever any robot fails, it broadcasts a message informing of the failure and one of the available resources performs the rescheduling in order to reassign the remaining tasks to another suitable resource.

6.3.2 Virtual Assembly Cell

For achieving the Virtual Commissioning approach described in this paper, another PC was used where the Process Simulate software was installed. An accurate representation of the real cell in terms of layout, resources and equipment was created as a new PS study.

The PS SDK was used for providing to the software some additional functionality. Microsoft Visual Studio and C# programming language was used for implementing the required functionalities. The outcome was a new Process Simulate command which implemented the following:

- TCP socket connection between the Process Simulate and the resources' services. The PS side opens a socket and waits for incoming connection from the ROS services.
- Definition of the communication protocol between the two sides (i.e. communication initialization, commands formatting, response after successful r unsuccessful execution of operations, communication termination etc.).
- Reading the incoming operations and translating them in order to control the simulated resources. More specifically.
  - Every time a "MOVE" operation is sent to the PS, the program will be aware of which resource sent this operation, it recognizes the position and the orientation and creates inside the PS study a Generic Robotic Operation which moves the robotic arm accordingly.
  - When a "NAVIGATE" operation is sent, an Object Flow Operation is created that moves the mobile unit to the desired location.

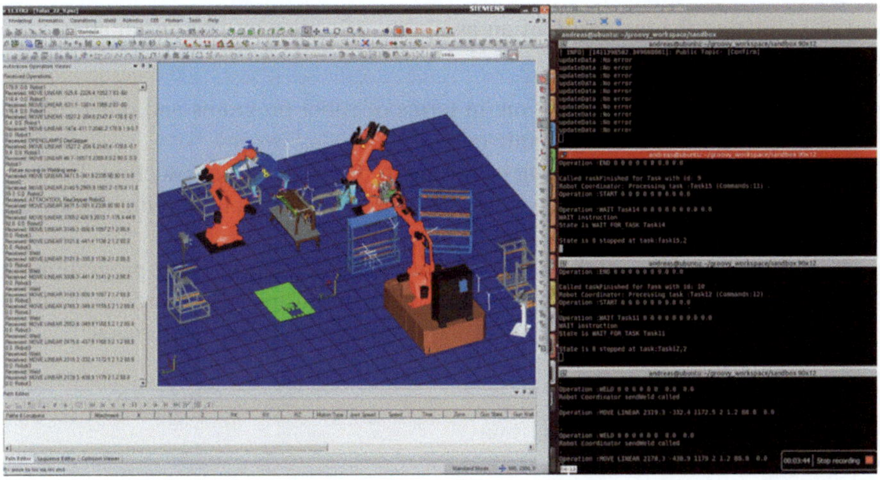

Fig. 6.5 Virtual assembly cell—robot services

- When an “ATTACH TOOL DexGripper/FlexGripper” operation is received, the PS is instructed to attach the specific tool to the resource that sent the operation.
- If a gripper operation is sent (“OPEN CLAMPS” or “CLOSE CLAMPS”), the PS attaches the part for which it is configured or de-attaches the attached part.

- Display of the operations sent by the resources to the PS by using a PS viewer that also connects with the services that are running in the Master PC.

Figure 6.5 presents the virtual cell in the left side, along with the resources’ services that are running in the right-hand side terminals and sending the operations to the PS in order to be executed.

The robot supplier, which in this case is Comau, provides a PS plugin which can be imported in the PS in order to emulate the real controller. Nevertheless, for having more flexibility in terms of robot, gripper and mobile unit operations, the implemented Process Simulate command played the role of the controller of all the resources inside the cell. The high-level architecture of the virtual validation compared to the execution is depicted in Fig. 6.1.

6.4 Discussion

This chapter examined the latest advances in VC technologies and presented the complete workflow of applying a VC method to an industrial assembly cell the control of which was conducted by a novel service oriented approach instead of the traditional PLC. The results confirm that Virtual Commissioning provides a reliable way of validating the operation of an assembly cell prior to each installation. The main benefits of VC involve:

- Reduction of the installation time in terms of robot programs and other related software installation at the robotic cell.
- Reduction of the manufacturing costs due to the set-up time reduction. Furthermore, the human resources costs for debugging and troubleshooting during the ramp-up process.
- High reconfigurability of the VC process itself. With relatively little programming effort, the solution presented in this study could be customized in order to be used in other types of robots in any agent-based or service-oriented approach.

References

1. Chryssolouris G (2006) Manufacturing systems: theory and practice. Springer
2. Bär T (2008) Flexibility demands on automotive production and their effects on virtual production planning. In: Proceedings of the 2nd CIRP conference on assembly technologies and systems. Ontario, Canada, pp 16–28
3. Liu Z, Suchold N, Diedrich C (2012) Virtual commissioning of automated systems. In: Kongoli F (ed) Automation. InTech
4. Drath R, Weber P, Mauser N (2008) An evolutionary approach for the industrial introduction of virtual commissioning. 2008 IEEE international conference on emerging technologies and factory automation. IEEE, Hamburg, Germany, pp 5–8
5. Farrington PA, Nembhard HB, Sturrock DT, Evans GW Interface driven domain-independent modeling architecture for "soft-commissioning" and "reality in the loop"
6. Makris S, Michalos G, Chryssolouris G (2012) Virtual commissioning of an assembly cell with cooperating robots. Adv Decis Sci 2012:1–11. https://doi.org/10.1155/2012/428060
7. Reinhart G, Wünsch G (2007) Economic application of virtual commissioning to mechatronic production systems. Prod Eng 1:371–379. https://doi.org/10.1007/s11740-007-0066-0
8. Kain S, Schiller F, Frank T (2010) Monitoring and diagnostics of hybrid automation systems based on synchronous simulation. 2010 8th IEEE international conference on industrial informatics. IEEE, Osaka, Japan, pp 260–265
9. Tecnomatix. In: Siemens digit. ind. softw. https://www.plm.automation.siemens.com/global/en/products/tecnomatix/. Accessed 23 Dec 2019
10. DELMIA Global operations—dassault systèmes®. https://www.3ds.com/products-services/delmia/. Accessed 23 Dec 2019
11. SimulationX: Software for virtual commissioning | ESI ITI. https://www.simulationx.com/simulation-software/beginners/virtual-commissioning.html. Accessed 23 Dec 2019
12. Yoonseok Pyo DL Hancheol Cho, Leon Jung (2017) ROS Robot Programming (English). ROBOTIS

Chapter 7
Physically Interacting Cooperating Robots for Material Transfer

7.1 Introduction

One of the greatest advantages of using robotic structures in production lines is that of a robot being used in many different processes by only changing its tools [1]. For example, a robot may have a welding gun during the welding process and then replace it with a gripper to be used in a gripping process. This research refers to an assembly process, which two robots cooperate in so as to exchange a gripper. The tool exchanging is based on a pegs-in-holes insertion. The aim of this research is to develop a method that calculates all the required quantities about a pegs-in-holes model and be used to fully align and splice the pegs with the holes. The above method is feasible to be used in simulation programs thus enabling them to control the exchanging process of the tool in simulation mode and not immediately in real time. Contact forces are developed in a variety of assembly process such as using a dual arm robot for cooperative assembly [2].

The peg-in-hole assembly problem has been analysed extensively. Hybrid controller based on sequential hybrid communication of the processes has been implemented in previous research to deal with the peg-in-hole assembly problem [3]. For the same reason, Lagrange based mathematical approaches have also been applied [4]. Chamfer-less peg-in-hole passive assembly has also been implemented in previous research to study the inaccuracy of the problem [5]. The insertion problem has been studied performing a force analysis based on screw theory [6, 7]. The peg-in-hole assembly problem has also been analysed through visual feedback, too [8]. The compensation of pegs' orientation has also been investigated utilizing a passive compliant center device [9]. Another method suggested the configuration space would make the system analysis visible and that the assembly strategy could be easily designed in the two subspaces [10]. The calculation of the critical angles of the peg's declination and the critical depth of insertion into the hole for the assembly of the peg-hole type parts has been studied in a previous research [11]. Dynamic analyses have also been performed utilizing generalized inequality equations in order to

S. Makris, *Cooperating Robots for Flexible Manufacturing*,
Springer Series in Advanced Manufacturing,
https://doi.org/10.1007/978-3-030-51591-1_7

describe the problem [12–14]. Furthermore, this assembly problem can be solved with the use of sensors that correct the lateral and angular errors [15]. Finally, lots of research have used the simulation programs in order to attend a variety of process before the real time execution of them. Besides, the simulation programs give the opportunity to optimize the process or parts of it [16–19].

The difficulty of finding geometric features and calculations of contact forces in a 3-dimensional pegs-in-holes model results in insertions which are limited to the analysis of 2-dimensional pegs in holes models. Besides, most of this research does not refer to multiple pegs in holes, but they are models with one or two pegs and holes. In this work, we are dealing with a 3-dimensional pegs-in-holes model. The aim of this method is to analyze the contact area between the pegs and the holes in a random contact position. Then, the necessary geometrical sizes of the contact area will be defined. Finally, with the use of the contact forces as well as the forces and torques on the robot flange will be calculated.

7.2 Approach

In the context of this work, a robot that can be used in more than one production processes cooperates with another one having a gripper as its tool. This gripper contains a tool exchanger, which allows the attaching and detaching of the gripper between the two robots (Fig. 7.1).

The robot which has not attached the gripper has a flange with pegs and the goal is that the pegs be inserted into the holes of the tool exchanger. The assembly process is

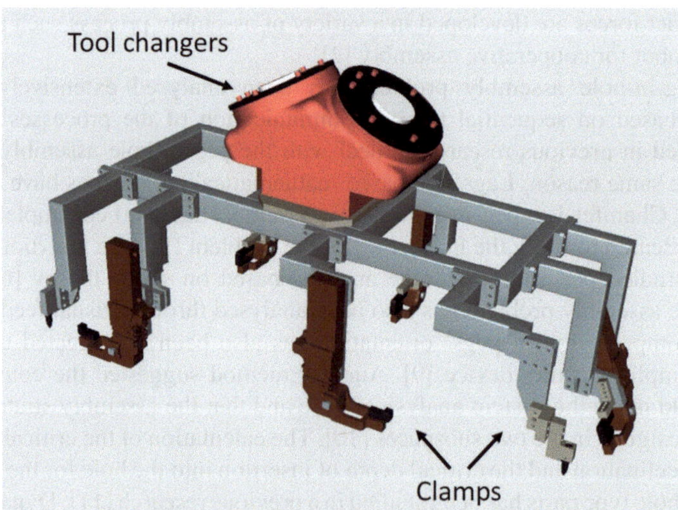

Fig. 7.1 Gripper concept with multiple coupling points

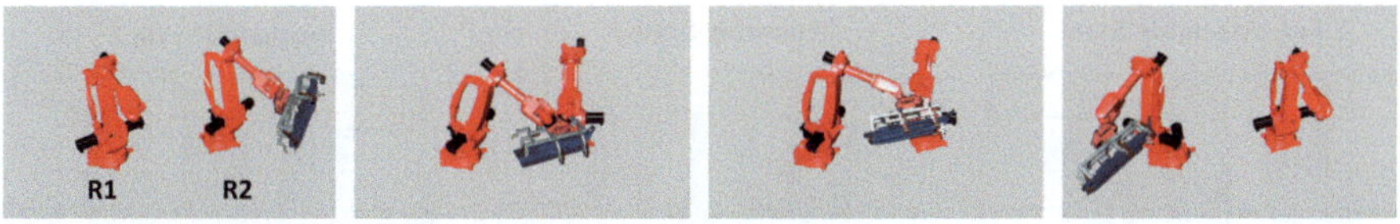

Fig. 7.2 Exchange of parts and gripper between two robots

depicted in Fig. 7.2. If the pegs and the holes are not correctly inserted, contact forces are generated at the pegs and consequently, there are forces and torques applied on the robot flange. This is harmful for the robots and their controllers because when a force is applied to robot 1, an equal force with different orientation is applied from robot 1 to robot 2. This force interaction between the two robots creates for each robot displacements "confusing" the controllers which try always to bring the robots to their "correct", programmed position.

For the analysis of this 3-dimensional triple pegs-in-hole method, there are two flanges used. The first flange contains three pegs and the second one three holes. The radius of the pegs and the holes is the same. The pegs and the holes have the same height. The two flanges and their geometrical elements have the same dimensions (Fig. 7.3).

There are three assembly states of each peg as they are depicted in Fig. 7.4.

Moreover, there are eight cases for State 1 and eight cases for State 2 as it is depicted in Fig. 7.5.

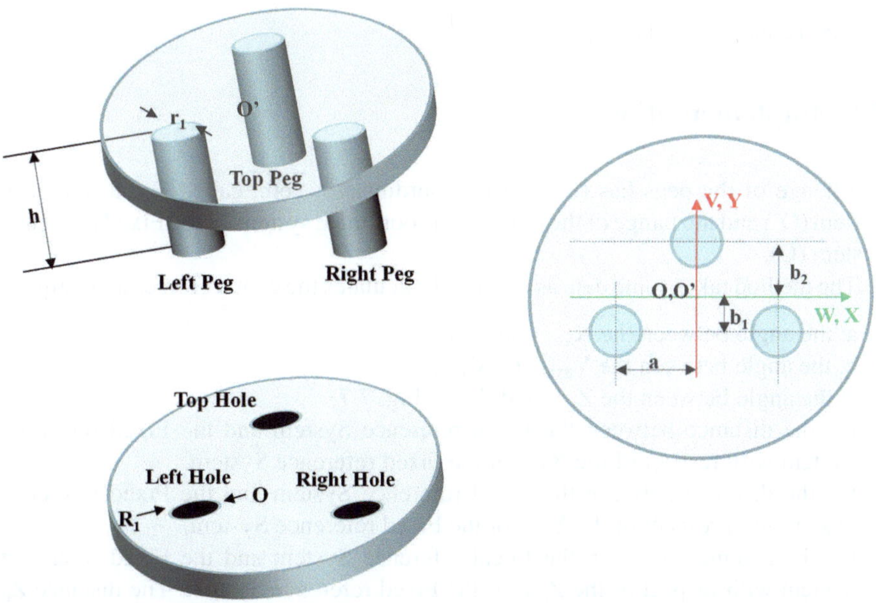

Fig. 7.3 Geometry of the flanges

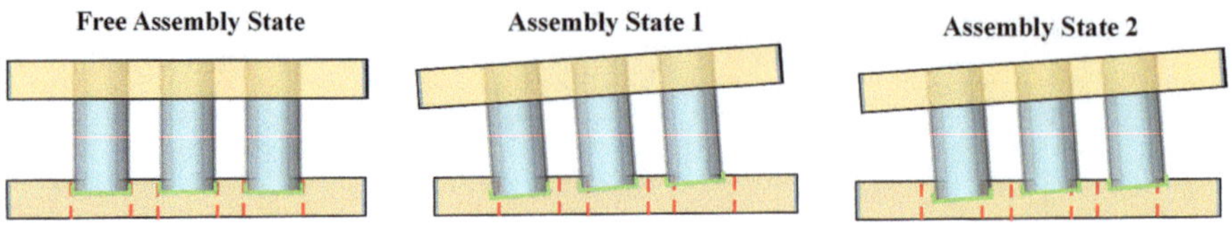

Fig. 7.4 Assembly states of each peg

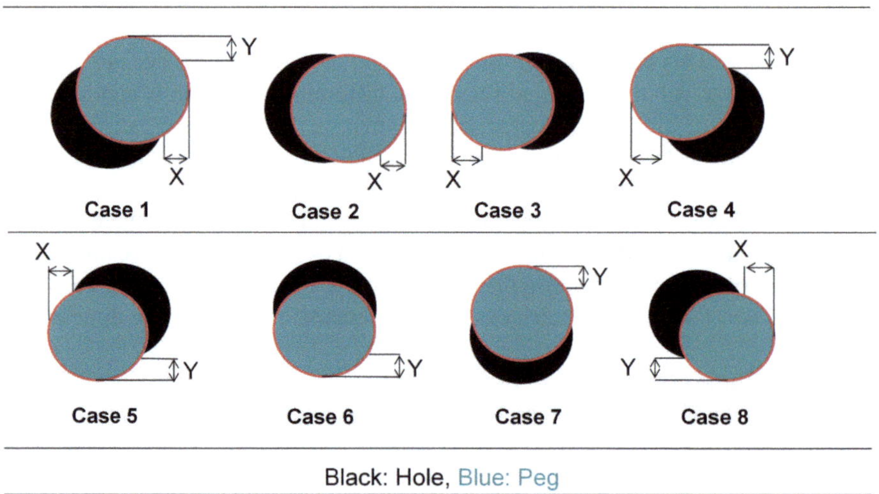

Fig. 7.5 Position Cases of each peg

7.3 Implementation

The flange of the pegs has (Fig. 7.6) a coordinate system, called Local reference System (O') and the flange of the holes has a coordinate system called Fixed reference System (O).

The method takes some data as input and calculates the contact forces and torques:

- a: the angle between the X_{axis} and U_{axis}.
- b: the angle between the Y_{axis} and V_{axis}.
- c: the angle between the Z_{axis} and W_{axis} (Fig. 7.7).
- P_x: the distance between the Local reference System and the Fixed reference System with respect of the X_{axis} of the Fixed reference System.
- P_y: the distance between the Local reference System and the Fixed reference System with respect of the Y_{axis} of the Fixed reference System.
- P_z: the distance between the Local reference System and the Fixed reference System with respect of the Z_{axis} of the Fixed reference System. The distance Z_d for this description is smaller than the height of the pegs in order to have collisions.

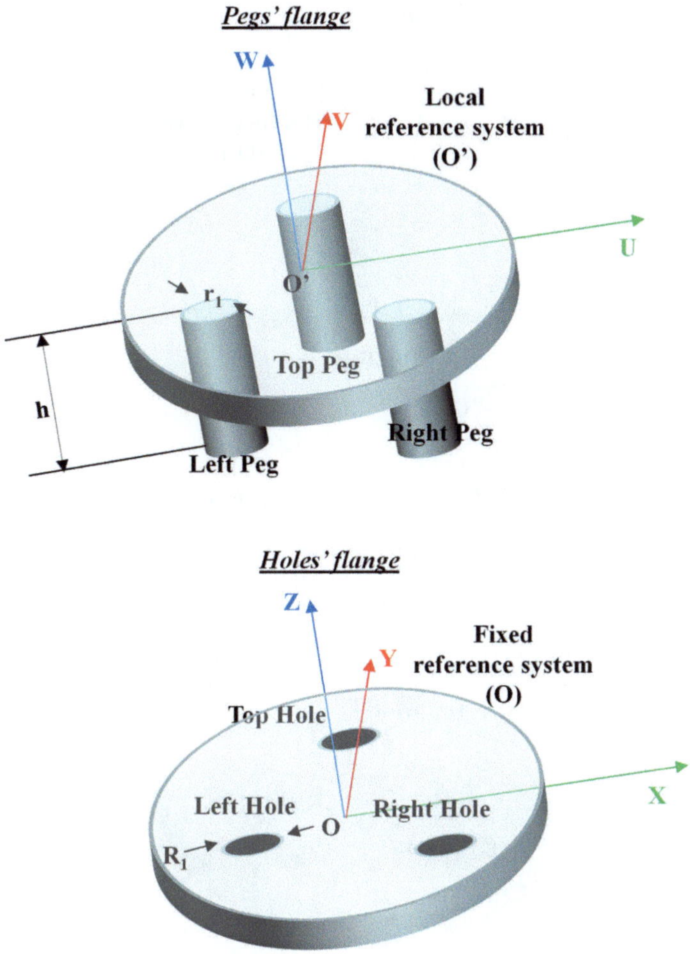

Fig. 7.6 Reference System of the flanges

Fig. 7.7 Points with known coordinates of the flanges

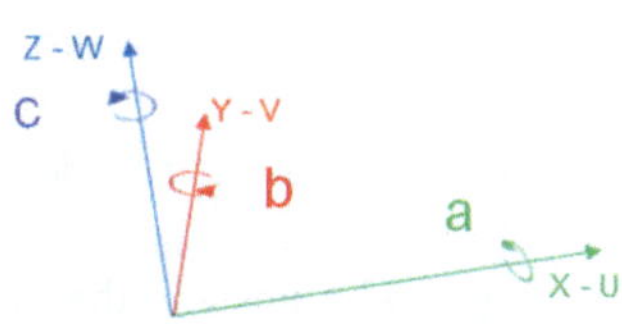

The contact forces and torques can be calculated in the following sequence:

1. The contact forces and torques depend on the distances X_{force}, Y_{force}, P_z and on the coordinates of the contact points.
2. In order to calculate the distances X_{force}, Y_{force} and to define the coordinates of the contact points, the collision States should be defined.

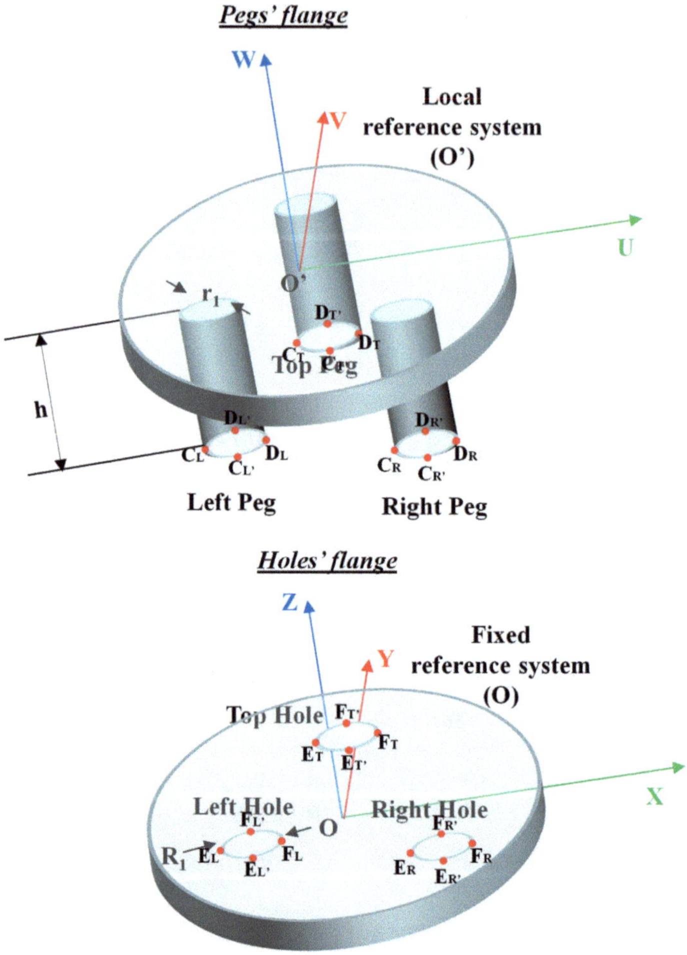

Fig. 7.8 Points with known coordinates of the flanges

3. The collision States of the pegs depend on the Case of the peg and on the angles a and b.
4. The Cases of the peg depend on the distances X_{case} and Y_{case}.
5. The distances X_{case} and Y_{case} depend on the coordinates of the limit points of the pegs and holes and the distances P_x, P_y, P_z and the angles a, b and c.

The coordinates of the limit points are depicted in Fig. 7.8.

The coordinates of these points are known from the flanges' geometry. In order for the Case of each peg to be defined, all the limit points with respect to the Fixed reference system have to be translated. The translation matrix is:

$$T=\begin{pmatrix} c\gamma\cdot c\beta & -s\gamma\cdot c\alpha + c\gamma\cdot s\beta\cdot s\alpha & s\gamma\cdot s\alpha + c\gamma\cdot s\beta\cdot c\alpha & P_x \\ s\gamma\cdot c\beta & c\gamma\cdot c\alpha + s\gamma\cdot s\beta\cdot s\alpha & -c\gamma\cdot s\alpha + s\gamma\cdot s\beta\cdot c\alpha & P_y \\ -s\beta & c\beta\cdot s\alpha & c\beta\cdot c\alpha & P_z \\ 0 & 0 & 0 & 1 \end{pmatrix}$$

Rotation Matrix

$$R=\begin{pmatrix} n_x & s_x & a_x \\ n_y & s_y & a_y \\ n_z & s_z & a_z \end{pmatrix}$$

Where, ca is an abbreviation of cos(a) and sa is of sin(a).
Now, the distances X_{case} and Y_{case} can be calculated as:

- X_{case}: This distance can be calculated with the subtraction of the distance on the X-axis of the D point and the distance on the X-axis of the F point with respect to the Fixed reference System.
- Y_{case}: This distance can be calculated with the subtraction of the distance on the Y-axis of the D' point and the distance on the Y-axis of the F' point with respect to the Fixed reference System.

Following combinations are to be considered which are depicted in Table 7.1.
If the Case of the peg is known, the State of the pegs comes from the following combinations which are depicted in Tables 7.2 and 7.3.

Table 7.1 Contact Cases with respect of Xcase and Ycase distances

Y_{case}/X_{case}	$X_{case} > 0$	$X_{case} < 0$	$X_{case} = 0$
$Y_{case} < 0$	Contact Case 8	Contact Case 5	Contact Case 6
$Y_{case} = 0$	Contact Case 2	Contact Case 3	No contact
$Y_{case} > 0$	Contact Case 1	Contact Case 4	Contact Case 7

Table 7.2 Assembly States with respect of the angles α and β

	C. Case 1	C. Case 4	C. Case 5	C. Case 8
$(\alpha > 0)$ & $(\beta > 0)$	A. State 1	A. State 2	A. State 2	A. State 1
$(\alpha > 0)$ & $(\beta < 0)$	A. State 2	A. State 1	A. State 1	A. State 2
$(\alpha < 0)$ & $(\beta > 0)$	A. State 1	A. State 2	A. State 2	A. State 1
$(\alpha < 0)$ & $(\beta < 0)$	A. State 2	A. State 1	A. State 1	A. State 2

Table 7.3 Assembly States with respect of the angles α and β

	C. Case 2	C. Case 3	C. Case 6	C. Case 7
$(\beta > 0)$	A. State 1	A. State 2	A. State 2	A. State 1
$(\beta < 0)$	A. State 2	A. State 1	A. State 1	A. State 2

If the State of the peg is known, the distances X_{force} and Y_{force} can be calculated as:

- State
  - X_{force}: This distance can be calculated with the subtraction of the distance on the X-axis of the D point and the distance on the X-axis of the F point with respect to the Local reference System.
  - Y_{force}: This distance can be calculated with the subtraction of the distance on the Y-axis of the D' point and the distance on the Y-axis of the F' point with respect to the Local reference System.
- State 2
  - X_{force}: This distance can be calculated with the subtraction of the distance on the X-axis of the D point and the distance on the X-axis of the F point with respect to the Fixed reference System.
  - Y_{force}: This distance can be calculated with the subtraction of the distance on the Y-axis of the D' point and the distance on the Y-axis of the F' point with respect to the Fixed reference System.

The coordinates of the contact points depend on the Case and the Collision State. If the case of the pegs-in-holes model is case-1, the coordinates of the "worst" contact point can be calculated as:

The distances *m, n* and *o* as well as the angle *d,* required for the definition of the coordinates, can be calculated by the following equations: (Fig. 7.9)

$$m = R_{peg} - \left(X_{force}^2 + Y_{force}^2\right) \tag{7.1}$$

$$d = \tan^{-1}\left(Y_{force}/X_{force}\right) \tag{7.2}$$

$$n = m \cdot \cos(d) \tag{7.3}$$

$$o = m \cdot \sin(d) \tag{7.4}$$

where the distances X_{force} and Y_{force} can be calculated as:

And the coordinates of each state are:

- <u>*State 1*</u>
  If the coordinates of the center of the peg are (a, b, c), the coordinates of the contact point are (a + n, b + o, w_{hole}).
- <u>*State 2*</u>
  If the coordinates of the center of the peg are (a, b, c), the coordinates of the contact point are (a + n, b + o, h_{peg}).

The forces which are generated by the contact can be calculated as:

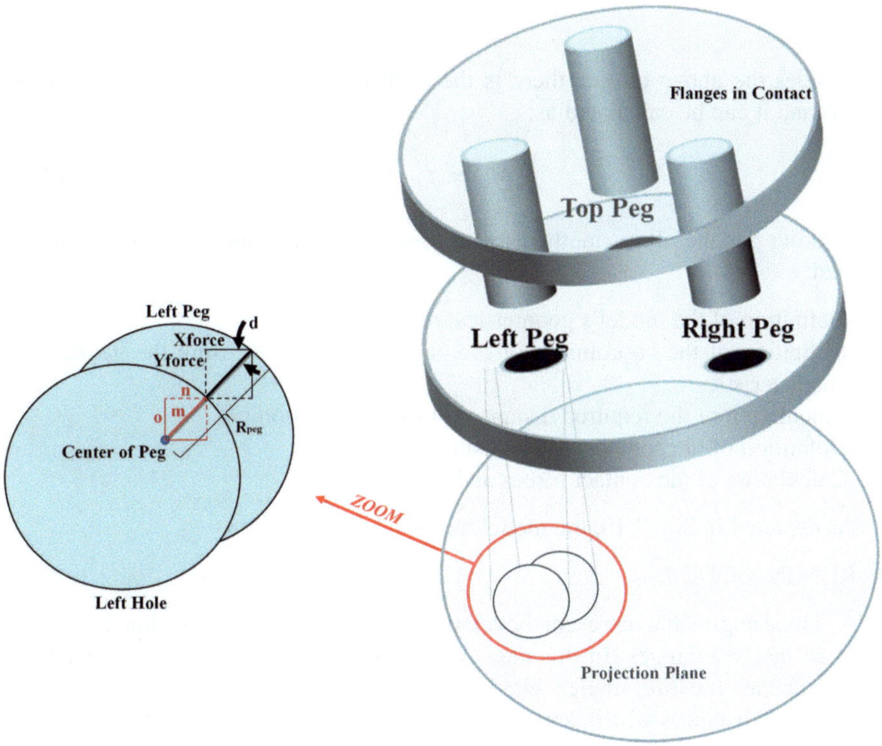

Fig. 7.9 Definition of "worst" contact point coordinates

$$F_x = k_x \cdot X_{force} \tag{7.5}$$

$$F_y = k_y \cdot X_{force} \tag{7.6}$$

$$F_z = k_z \cdot Z_{force} \tag{7.7}$$

$$F = \sqrt{F_x^2 + F_y^2 + F_z^2} \tag{7.8}$$

Except from these forces, there is the friction force, which is exerted on the Z-axis and can be calculated as:

$$f_q = \mu \cdot F \tag{7.9}$$

The coefficient of friction μ and the coefficients of stiffness k_x, k_y and k_z depend on the material of the pegs and holes as well as on the robot's stiffness.

The contact torques can be calculated as:

$$M = F \cdot m \tag{7.10}$$

Besides the above torque, there is the friction torque, which is exerted on the Z-axis and it can be calculated as:

$$M_q = f_q \cdot m \tag{7.11}$$

In order for the above method to be tried out a simulation example has been created:

1. Definition of the model's geometrical characteristics.
2. Definition of the coordinates of the pegs and the holes using the geometrical characteristics.
3. Calculation of the required distance for the contact points.
4. Definition of the contact case and states.
5. Calculation of the contact forces and torques.

As depicted in Fig. 7.10, the model has as inputs:

- RUN IN MODE 1
  - Three angles that represent the rotations with respect to the coordinates systems of the two flanges. In this case, we set the flange of the holes as fixed and calculate the three angles with respect to it.
  - Three distances which depict those of the two flanges' coordinates systems. In this case, we set the flange of the holes as fixed and calculate the three distances with respect to it.
- RUN IN MODE 2

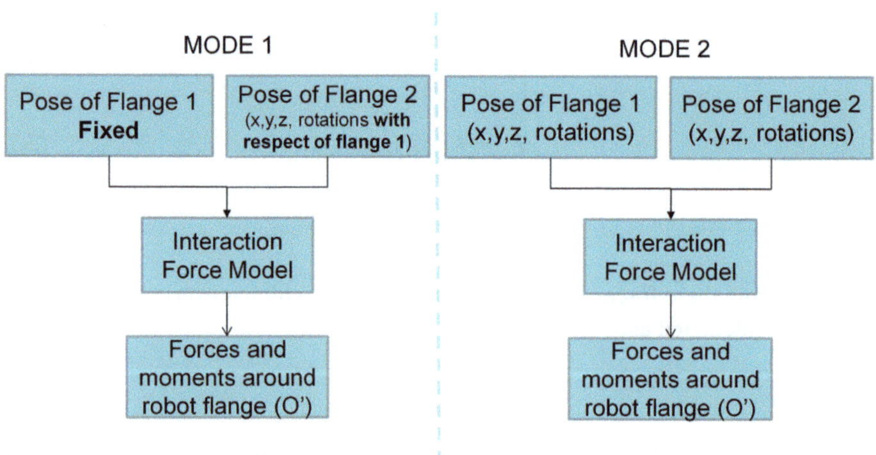

Fig. 7.10 Execution Modes of algorithm

– Three angles and three distances for each flange. In this case, we set the angles and the distances of each flange at random, but the same coordinate system for the two flanges.

When the code is executed has as outputs:

- The state and the case of each peg.
- The distances X and Y, required for the calculation of the contact forces.
- The coordinates of the contact points required for the calculation of the contact torques.
- The contact forces and torques of the worst points.
- The figure of the moved pegs-in-hole model with respect to the fixed one.

7.4 Industrial Example

A case study was setup in a simulation mode in order for the exchange gripper assembly process to be demonstrated. The scenario involved two fixed robots and a gripper with a tool exchange (Fig. 7.11).

Firstly, the robot R1 uses the gripper. When this process is finished, it passes the gripper to the robot R2 in order for another process with the gripper to be continued (Fig. 7.12).

As described in Sects. 7.2 and 7.3, the tool exchanging takes place on the gripper. The tool exchanger is the flange of the holes, while the robot contains the flange of the pegs as we can see in Fig. 7.13.

In order for the gripper exchanging to take place, the pegs have to be fully aligned and spliced with the holes. In case that they are not aligned, contact forces are created. To simplify the above tool exchange assembly problem the simple model with only two flanges is created and depicted in Fig. 7.14.

In Fig. 7.14, the two flanges are fully aligned. In order for the calculation method to be checked, the flange with the pegs is moved and rotated with respect to the other

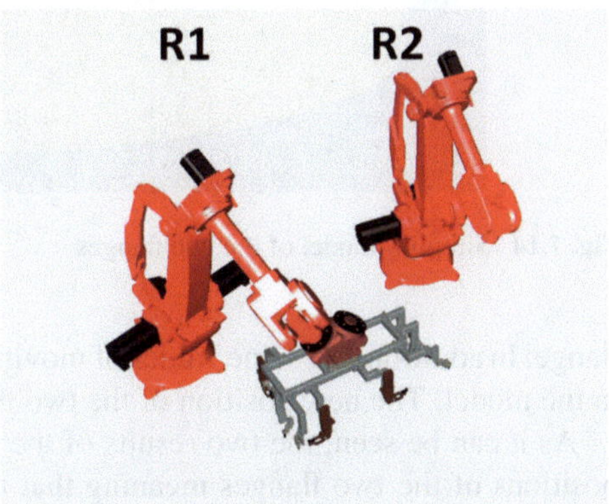

Fig. 7.11 Simulation scenario with two fixed robot and one gripper

Fig. 7.12 Tool exchange process

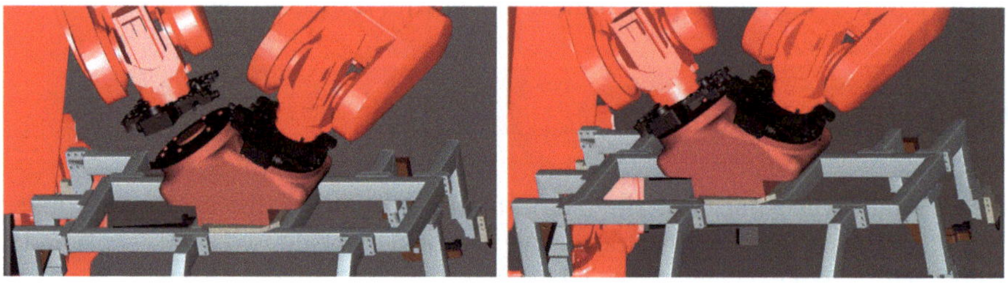

Fig. 7.13 Tool exchanger flanges

Fig. 7.14 Simplify model of the two flanges

flange. In addition, the same values of moving and rotating the flange are set as inputs in the model. The new position of the two flanges is depicted in Fig. 7.15.

As it can be seen, the two results of the translation and rotation depict the same positions of the two flanges meaning that the definition of the assembly Case and State with the model is successful. This in turn, suggests that the calculation of

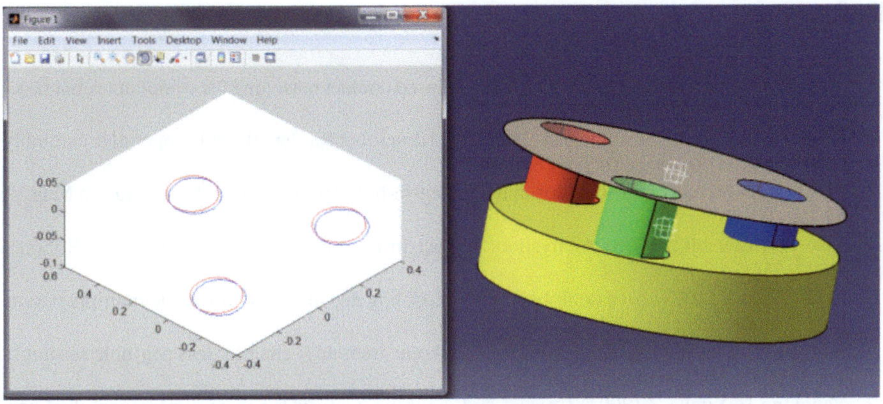

Fig. 7.15 Results of model simulation

the distances X, Y and Z for the determination of the contact forces is correct. As described in Sect. 7.3, the contact forces can now be calculated by multiplying the distances X, Y and Z with the corresponding stiffness coefficients k_x, k_y and k_z.

7.5 Discussion

This chapter considers the concept of fixtureless assembly process using industrial robots and describes a method for the control and compensation of the contact forces and torques, which are developed in the process. More specifically, the tool exchanging process in which the fixtureless assembly is based on, has been described. A pegs-in-holes model is used for the calculation of the contact forces and torques. The coordinates of the two flanges of the robot (pegs) and of the tool changer (holes) as well as the level of the alignment define the model's Assembly State and Contact Case. Using this information, the coordinates of the critical contact points can be extracted and the contact forces and torques can be calculated. By adopting such technologies, over the traditional control techniques, one may avoid the use of sensors for the exchange tooling. The method provides information about:

- The position of each peg with respect to the corresponding insertion hole.
- The coordinates of the worst contact points of this collision for the points with the largest developing forces and torques.
- The calculation of the above developing forces and torques in the entire model.

The illustrated method is not limited to the geometry of the model, but it is flexible to adapt to any 3-dimensional pegs in holes assembly model. By adopting such technologies over the traditional control techniques, one could avoid the use of sensors for the exchange tooling. Future research should focus on the improvement of the calculation accuracy.

References

1. Maraghy A, Payandeh S (1988) Knowledge-based contact reasoning for compliant robot tasks. Int J Adv Manufact Technol 3:61–80
2. Krüger J, Schreck G, Surdilovic D (2011) Dual arm robot for flexible and cooperative assembly. CIRP Annals–Manufact Technol 60:5–8
3. Li Y (1997) Hybrid control approach to the peg-in-hole problem. IEEE Robot Autom Mag, pp 52–60
4. Liao HT, Leu MC (1998) Analysis of impact in robotic peg-in-hole assembly. J Robotica 6:347–356
5. Fei Y, Zhao X (2003) An assembly process modeling and analysis for robotic multiple peg-in-hole. J Intell Robot Syst 36:175–189
6. Haskiya W, Maycock K, Knight J (1999) Robotic assembly: chamfer less peg-hole assembly. J Robotica 17:621–634
7. Yanqiong F, Xifang Z (2004) Contact and jamming analysis for three dimensional dual peg-in-hole mechanism. Mech Mach Theory 39:477–499
8. Pauli J, Schmidt A, Sommer G (2001) Servoing mechanisms for peg-in-hole assembly operations. Robot Vision, pp 157–166
9. Cheng CC, Chen GS (2002) A multiple RCC device for polygonal peg insertion. JSME International Journal, Series C, p 45
10. Su J, Qiao H, Ou Z, Zhang Y (2012) Sensor-less insertion strategy for an eccentric peg in a hole of the crankshaft and bearing assembly. Assembly Autom 32:86–99
11. Usubamatov R, Leong K (2011) Analyses of peg-hole jamming in automatic assembly machines. Assembly Autom 3:358–362
12. Zohoor H, Shahinpoor M (2003) Dynamic analysis of peg-in-hole insertion for manufacturing automation. J Manufact Syst 10:99–108
13. Yanchun X, Yuehong Y, Zhaoneng C (2006) Dynamic analysis for peg-in-hole assembly with contact deformation. Int J Adv Manufact Technol 30:118–128
14. Shirinzadeh B, Zhong Y, Tilakaratna PDW, Tian Y, Dalvand M (2010) A hybrid contact state analysis methodology for robotic-based adjustment of cylindrical pair. Inte J Adv Manufact Technol 52:329–342
15. Jain RK, Majumdera S, Dutta A (2013) SCARA based peg-in-hole assembly using compliant IPMC
16. Fleischer J, Munzinger C, Trondle M (2008) Simulation and optimization of complete mechanical behavior of machine tools. Prod Eng 2:85–90
17. Bernard A, Delplace JC, Perry N, Gabriel S (2003) Integration of CAD and rapid manufacturing for sand casting optimization. Rapid Prototyping J 9:327–333
18. Abele E, Wörn A, Fleischer J, Wieser J, Martin P, Klöpper R (2007) Mechanical module interfaces for reconfigurable machine tools. Prod Eng 2:421–428
19. Reinhart G, Weissenberger M (1999) Multibody simulation of machine tools as mechatronic systems for optimization of motion dynamics in the design process. Int Conf Adv Intell Mechatron

Chapter 8
Generating Motion of Cooperating Robots—The Dual Arm Case

8.1 Introduction

During the last decades, there is a tremendous need for flexibility of industrial manipulators in order to fit the industrial and market goals [1]. In this direction, the case of motion planning in industrial robots has been promoted rapidly. Motion planning is a special case of general planning, which is concerned with the problem of figuring out how the robots should move to get from one point to another, and how to perform a desired task [2]. The main problem of designing different paths with industrial robots is the differences in factories' environments as well as the big effort needed from specialized programmers. The main and widely known method of programming a new path in industrial robots is implemented using the teach pendant. The programmer has to record each point of the desired path separately and moving the robot manually. In the end, the recorded points are connected and form the desired trajectory. Although, this method seems simple, it underlies many risks as the programmer has to be exquisitely accurate and cognizant of the robot's constraints and dynamics. In this way, some method which based on the intuitive programming of industrial manipulators have be developed [3].

The last few years, lots of research worked on the construction of approximate models using sampling-based motion planning algorithms. The motion planner algorithm can be separated in two main categories, the roadmap-based algorithm and the tree-based algorithm [4]. As far as the roadmap-based algorithm, an approach separates the planner in the phase of the learning where the random configurations of the robot are generated and in the phase of the query [5]. Another approach describes strategies for node generation and multi-stage connection of an obstacle probabilistic roadmap method [6]. In the other side, the basic idea of the tree-based algorithm is the using of an initial sample and the newly produced samples are connected with it. This procedure is continuously, and the final result is a creation of a random tree [7]. Another approach uses and repairs the random trees when the dynamic environment is changed. The algorithm takes under consideration the data of the environment

S. Makris, *Cooperating Robots for Flexible Manufacturing*,
Springer Series in Advanced Manufacturing,
https://doi.org/10.1007/978-3-030-51591-1_8

and removes the invalid paths of the random tree [8]. A path planning algorithm includes an on-line stage and an off-line stage for the generation of a collision free path [9]. In addition, a parameterized search algorithm which based on the random tree construction has be developed. Parameters such as the grid resolution and the smoothness of the path can be defined by the user [10].

Almost all of the above approaches have been used for the motion planning of single arm manipulators. In contrast, this study refers to the developing of a parameterized motion planner for dual arm robots. The using of industrial dual arm robot system in the production line enables the performance of operations that are carried out by humans which offers a number of advantages [11]. The proposed search algorithm creates the random tree of all the possible configurations of a 13 degrees of freedom mechanism and selects the best one each time until to derive the desired path. It is worth to mention that this motion planner generates two paths, one of each arm, while simultaneously evaluates the 13 DOF configurations of the robot using multiple criteria.

Lots of approaches have been developed for the path planning of multi arm robots. An explicit treatment of singular configurations of the robot have been used for the path generation which requiring the robot reconfiguration [12]. A bidirectional RTP-connect algorithm has been used in other cases for the concurrent path planning of robot with two arms [13]. Another approach uses the multi-objective optimization domain and implements the proposed co-evolutionary algorithm for the simultaneously optimization of the trajectory length, velocity distribution, rotate angle and the number of collision object [14]. Using compact roadmaps and super-graph which constructed by the roadmaps, the path planning of multi arm systems can be derived [15]. In other cases, motion planning of parallel kinematics is used as an input in order to complete operations fast and accurate enough under the usage of a drilling system [16].

Also, a possible planning method is proposed in multiple robot systems, where different robotic mechanisms adapt to the same environment and accomplish different tasks taking in consideration possible elastic and field forces [17]. Moreover, in such environments, a point to point path planning technique is introduced in order to minimize the assembly time of cooperating robots [18]. One different planning method is experimented using a robot with a laser scanner in order to achieve both accuracy and optimization of the time elapsed during the whole movement process [19].

On the terms of safety, dual arm robotic systems are studied in order to find possible ways of interaction between the human and the robot itself [20]. Working under this subject, path optimization is investigated in order to find the safest and fastest trajectory for the operators inside an industrial environment [21]. Last but not least, a planning method is used in order to assembly sequential tasks in manufacturing systems considering complexity [22].

8.2 Related Work

The most well-known method for generating the motion of robots is the robot programming. The field of robot programming is divided into manual programming, automatic programming and software architectures as shown in Fig. 8.1.

Manual systems (Fig. 8.2a) require the user/programmer to directly enter the desired behavior of the robot, usually using a graphical or text-based programming language. In automatic programming systems (Fig. 8.2b) the user/programmer has little or no direct control over the robot code. These include learning systems, Programming by Demonstration and Instructive Systems. Software architectures are important to all programming systems, as they provide the underlying support, such as communication, as well as access to the robots themselves.

The concept of automated robot programming implies many complex modules like user-friendly human interfaces (spoken commands, gestures etc.) but one of the most difficult parts in automatic robot programming found to be the execution and control of automatically generated motion sequences including automated generation of collision-free paths [23]. Automatic motion planning for industrial robots brings

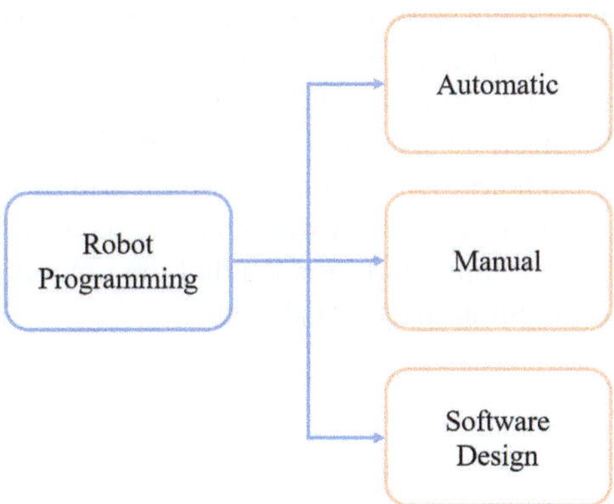

Fig. 8.1 Robot programming approaches

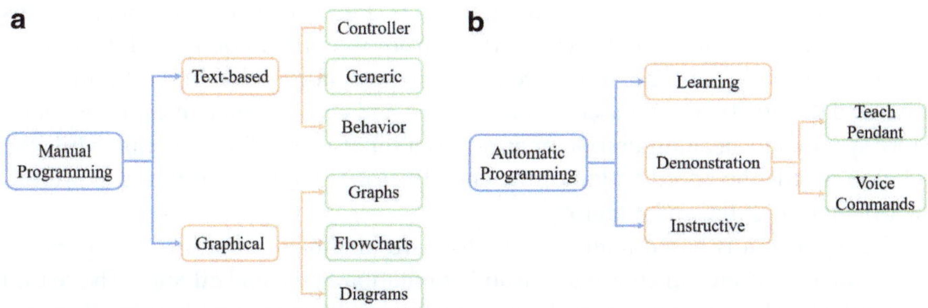

Fig. 8.2 **a** Manual programming systems, **b** Automatic programming systems

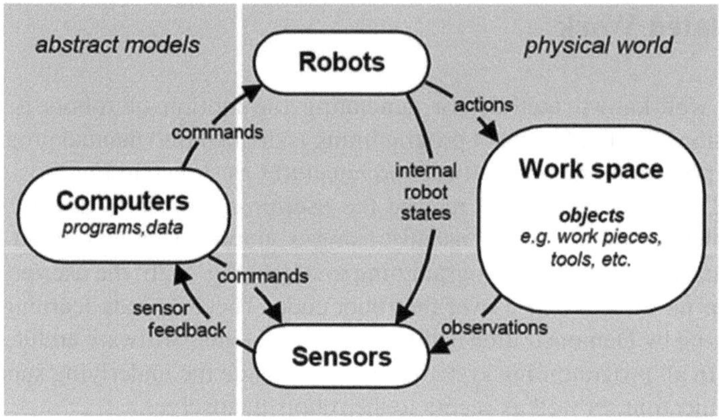

Fig. 8.3 General robot programming paradigm

new capabilities in robot programming where the goal is to specify a task in high-level language and have the robot automatically generate the trajectories by considering all the constraints (Fig. 8.3).

In this chapter, an intelligent search algorithm for the motion planning of dual arm robots is described. The random tree with the alternatives configurations of the robot is generated while each of them are evaluated using a variety of criteria. The best configurations of each step are selected and constitute the intermediate configurations of the desired paths of each arm. The resolution of the grid as well as the smooth of the paths depend on a set of parameters which are defined by the user. In the next section, the searching algorithm and more detailed the random tree generation and the evaluation of the configurations are analyzed.

8.3 Approach

An intelligent search algorithm is proposed with aim the definition of the intermediate Dual Arm robot's configurations which leads the robot's end effectors from the initial positions to the desired positions and orientations (Fig. 8.4). All these configurations constitute the path of each arm of the robot from the initial configurations to final positions for each one. The difference between the configuration of each arm of the robot and the position of it based on the forward and inverse kinematics of the robot. More specifically, a configuration of an arm is a set of values for each joint while the position of it is the coordination of the end effector from the base frame as well as the orientation of the end effector.

The algorithm is based mainly in the forward kinematics where a configuration of an arm is translated in coordination and orientation of the end effector. The reason is that the final configuration of the arm can easily be estimated but this does not

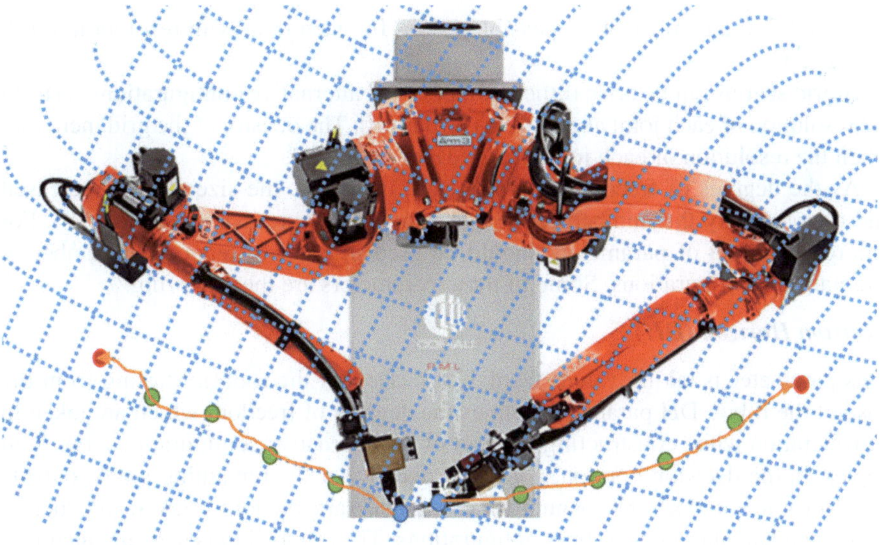

Fig. 8.4 Grid based path search

assume that it concurs with the target position without taking in consideration of the motion of the torso.

The paths which were constituted from the intermediate configurations of each arm are not unique as there are lots of alternatives intermediate configurations which lead the arms to each target position. The search algorithm which concludes the intermediate configurations of each arm as well as the intermediate configuration of torso, based on the gradually approach of all the alternatives configurations and evaluates them. The gradually approach of the configurations is limited from a grid which based on a set of parameters and aims to reduce the computational time. Grid resolution as well as the values of the parameters are changeable and depends on the estimation level. As far as the evaluation of the alternative configurations, multiple criteria are used in order for the different requirements to be fulfilled. This method provides an inexperienced robot programmer with flexibility to generate automatically a path for a Dual Arm robot that would fulfil the desired criteria without having to record intermediate points to the goal position.

Grid Search of the Alternative Configuration

As alternative configuration, it defined a set of n joint angles where n the degrees of freedom of the robot. The number of the alternative configurations is given from the following equation:

$$N = (2k + 1)^n \tag{8.1}$$

where, n is the degrees of freedom.

2k + 1 is the number of all possible values for each joint with resolution dΘ (k = 1, 2 …).

A grid search can be arise if the number of the alternative configurations (N) and the resolution of each joint angle (dΘn) be defined. The density of the grid increased when the resolution of each joint angle (dΘn) decreased.

As the degrees of freedom of a robot are increased, the size of grid increased and consequently the number of the alternative configurations increased, too. For this reason, a set of parameters should be used in order to reduce the number of alternatives configurations. Some of these parameters are the following:

Decision Horizon (DH)

This parameter is taking values from 1 to n (DoF of the robot). Starting from the base of the robot, DH parameter defines the degrees of freedom which are taken in consideration while constructing the grid of the alternative configurations. For joint angles, in the decision horizon, a grid is created. For the remaining joints, outside the decision horizon, only a number of samples are randomly taken in order to have complete alternative robot configurations. The robot's joints are separated into those that mainly affect the robot's movement in the workspace (position of the end effector) and those that mainly affect the orientation of the end effector. When only the target position has to be reached and the orientation of the end effector is ignored this parameter could be reduced for better performance and less computational time.

Maximum Number of Alternatives (MNA)

A maximum number of alternatives from the grid in the decision horizon are randomly selected for evaluation. If MNA > N then automatically the parameter MNA = N.

Sample Rate (SR)

A sample rate is defined as the number of samples taken from the joints, outside the decision horizon, in order to form the robot's complete alternative configurations. When the orientation of the end effector is considered, SR parameter should be increased in order to generate more alternative configurations which affect the orientation of the end effector.

For a Dual Arm industrial robot with two arms (6 DOF each arm) and one torso (1 DOF) the total degrees of freedom are n = 13 (Fig. 8.5). If the total number of nodes in grid is k = 1 and the resolution is dΘ = 10° for each degree of freedom, the number of alternatives can be calculated using the $N = 2k + 1n$ Eq. 8.1:

N = 313 = 1,594,323 alternative configurations of Dual Arm robot

Applying the first reduction parameters, DH, the number of the alternative configurations will be reduced. Indeed, by setting DH = 7 which means that only the degree of freedom of torso as well as the first three degrees of freedom of each arm taken into consideration, the number of the alternatives configuration according the is $N = 2k + 1n$ Eq. 8.1:

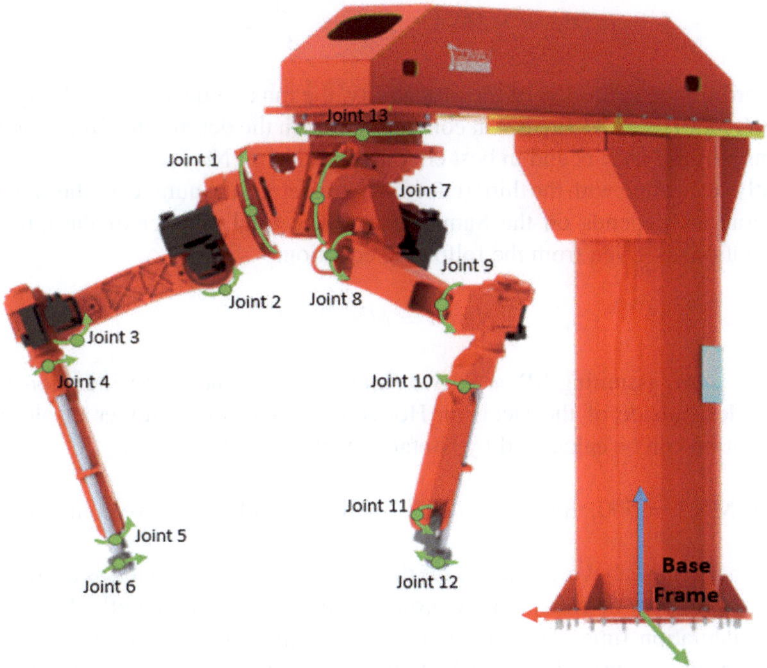

Fig. 8.5 COMAU Dual Arm, 13 DOF, Industrial Manipulator

$$\mathrm{N(DH = 7) = 37 = 2187\ alternative\ configurations\ of\ Dual\ Arm\ robot}$$

As it observed, the number of the alternative configurations have been reduced about 1,592,136 which leads at a large desired reduction of the calculation time.

According with the second reduction parameters, Maximum Number of Alternatives, the probability of getting the alternative configuration closer to desired position is given by the following equation:

$$p(DH, MNA) = \frac{MNA}{N} \tag{8.2}$$

Applying the $N = 2k + 1n$ Eq. 8.1 to $p(DH, MNA) = \frac{MNA}{N}$ Eq. 8.2 arise the final equation:

$$p(DH, MNA) = \frac{MNA}{(2k+1)^{DH}} \tag{8.3}$$

In this case, assuming MNA = 500, the probability to get the best alternative configuration of the robot in DH = 7 with k = 1 and dΘ = 10o can be calculated by Eq. 8.3:

$$p(DH = 7, MNA = 500) = \frac{500}{3^7} = 22.8\%$$

According the definition of the probability, the max value of a probability is p = 1. $p(DH, MNA) = \frac{MNA}{N}$ Eq. 8.2 in combination with the definition of the probability explains why MNA < N and in best conditions MNA = N.

Lastly, according with the third reduction parameters, the number of the alternative configurations depends on the Sample Rate. The total number of the alternative configurations is given from the following equation:

$$Ntotal = MNA * SR \tag{8.4}$$

In this case, assuming SR = 2, two samples taken under consideration for the joint angles outside of the Decision Horizon. So, the total number of alternative configuration can be calculated by Ntotal = MNA * SR Eq. 8.4 as:

$$\text{Ntotal(MNA} = 500,\ \text{SR} = 2) = \text{MNA} * \text{SR} = 1000 \text{ alternative configurations}$$

As it mentioned above, the aim of the definition of these parameters is the reduction of the total number of the alternative configurations and consequently the reduction of the calculation time. The calculation time is the time which the search algorithm running to approach and evaluate the alternative configurations until the two arms reach their target positions. In this section, the gradually approaching of the alternative configurations is analyzed.

Evaluation of the Alternative Configurations

The next step, after the selection of the total number of alternative configurations, is the evaluation of them. There is a variety of criteria which can be used for the evaluation. In this study, two criteria taken under consideration, those of the distance due to translation and distance due to rotation. The search algorithm evaluates the alternative configuration with aim to select the best one each time. The sequence of these configurations, translated to end effector's positions through the direct kinematic calculator, consists one path of each arm of the robot which fulfil the evaluation criteria. Each criteria characterized by a weighted factor. The final utility of each alternative is calculated as the weighted sum of the distance due to translation and to orientation (Table 8.1).

$$\text{Ui} = \text{Wt}\ ||\text{Xi} - \text{X}|| + \text{Wr f(qi, q)} \tag{8.5}$$

where,

Xi – X is the Euclidean distance of the end effector from the target position.
f(qi, q) is the distance duo to rotation of the end effector from the target orientation.
Wt, Wr are the weight factors of each criteria (Wt + Wr = 1).

Table 8.1 Evaluation of the alternatives according to the distance criteria

Alternative configurations	Normalized criteria		Utility value
	Distance due to translation	Distance due to rotation	$U_i = W_1 - C_{i1} + W_2 C_{i2}$ (*where* W_1 *and* W_2 *the criteria weights*)
Alternative 1	C_{11}	C_{12}	U_1
Alternative 2	C_{21}	C_{22}	U_2
Alternative 3	C_{31}	C_{32}	U_3
⋮	⋮	⋮	⋮
Alternative m = MNA * SR	C_{m1}	C_{m2}	U_m

For the calculation of the distance duo to rotation, the metric of metric of the distance is the Min of the Difference of Quaternions.

$$f(qi, q) = \min\{||qi - q||, ||qi + q||\} \tag{8.6}$$

where, || || denotes the Euclidean norm (or 2-norm) and q the orientation of the end effector, expressed in quaternions. The metric gives values in the range $[0, \sqrt{2}]$.

Industrial Manipulator Motion Generation

The proposed algorithm has as final goal the generation of a path for an industrial manipulator which leads it from the initial configuration to the desired position and orientation. This path consists of sequential, intermediate configurations which have selected from a variety of alternatives configuration after an evaluation procedure. More detailed, the User gives as input in the algorithm the target positions of each arm as well as the grid parameters, the search parameters and the evaluation criteria. The algorithm uses the above input and searching about the best path. This procedure is presented at Fig. 8.6.

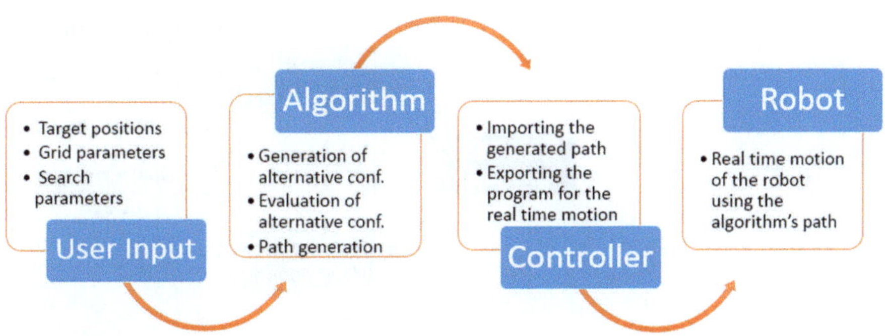

Fig. 8.6 Robot motion generation

When the algorithm estimates the best path of each arm, it passes the paths at robot controller which translates the path in motion programming. The final step is the real time motion of the robot following the generated by algorithm path.

In order to overcome out of range errors of the robot in real time motion, it is needed the definition of the robotic structure in the algorithm. Some parameters which should be defined are the limits of each joint, the length of each link, the working envelope of the robot and the type of each joint. The algorithm should take under consideration the above restrictions when it searches the intermediate alternative configurations of the paths.

8.4 Industrial Example

An industrial pilot case has been used to evaluate the efficiency of the proposed algorithm for automatic motion generation. The pilot case involves the pre-assembly of a dashboard traverse for the automotive industry. The robot used for the assembly is a COMAU Dual arm robot. Automatic motion planning has been integrated in a complete intuitive solution for dual arm robot programming for assembly operations. The assembly steps are described using different hierarchical levels which help the programmer to use high level commands during the programming and execution phases (Fig. 8.7).

In the initial step of the proposed method the whole assembly process is described using the hierarchical levels. Then a wizard is responsible to guide the programmer in order to define all the information required for each of the assembly operations. When a "move" operation has to be defined, for example "Approach" a specific position, the programmer is able to use intuitive human–robot interfaces to move the robot in the desired potion and record it. If the workspace is free of obstacles and

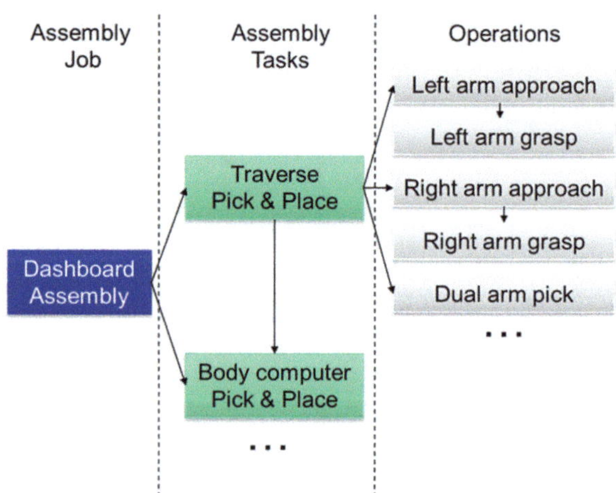

Fig. 8.7 Hierarchical levels for assembly operations

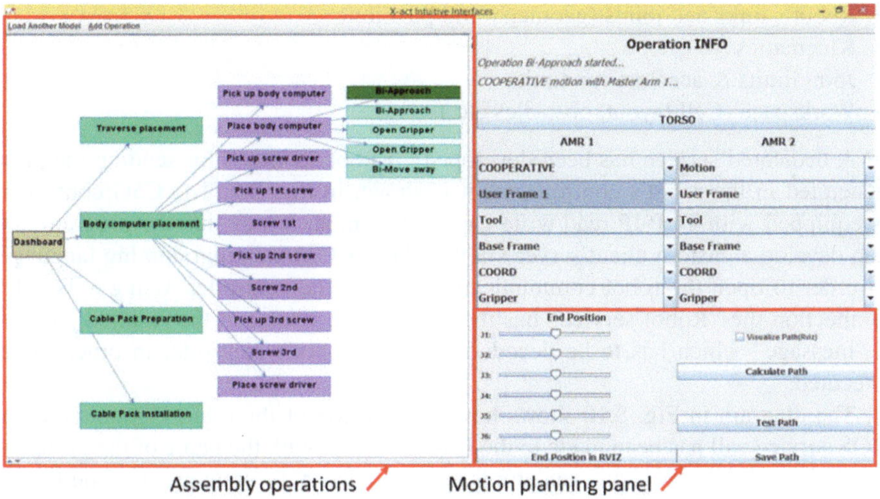

Fig. 8.8 Programming interface

the desired goal position is easily approached with a linear movement of the robot the programmer has just to command the robot (voice, gestures) to go in the desired position and then to define the motion type to "Linear". When the desired end position is not easily accessible and the risk of collision exists, the programmer has the option of the automatic motion planning. When the motion planning module is enabled a virtual environment of the robotic cell is starting up, where the programmer has just to define the desired end position of the robot and a free collision path is automatically generated and visualized. If the programmer is satisfied from the resulted path, the intermediate points that lead to the end position are automatically recorded and the path can be executed.

The interface for the automatic motion planning module has been integrated in a user-friendly programming interface (Fig. 8.8) which has been developed using Java. Motion planner has been considered as an alternative motion type to the built-in options of an industrial robot manipulator.

The integration has been done into the ROS environment [24]. The proposed motion generation method has been implemented using C++ programming language in combination with Robotic Operating System (ROS) software as well as Moveit! and RViz for visualization. The benefit of selecting Moveit! is that as a core part of ROS it offers seamless integration with the interaction modules which have been developed for the programming platform. The basic services include the robot position monitoring, desired end position definition which are presented in detail in the following paragraphs of this section.

The virtual environment that is used is described using the Unified Robot Description Format (URDF). URDF is one kind of XML file used for robot modeling and visualization, including the following information:

- Robot's links and joints (geometry and position)
- Kinematics
- Joint limits & acceleration limits
- Workspace (Cell layout, obstacles etc.).

A dedicated service has been developed ("Robot service") for sending the path generated in the robot's controller as it is shown in Fig. 8.9. The C5G controller is equipped with TCP-IP read/write capabilities in such way that the programmer can develop a custom client/server via the PDL2 COMAU programming language in order to open the robot communication outside the controller. Using a TCP/IP connection the "Robot service" has the responsibility to send the resulted path as a "message" which has to be decoded from the robot's controller in order to be executed.

The diagram in Fig. 8.10 shows how the position of the robot is monitored. A ROS service call has been made to the Robot Service with the name of the group as input (for a dual arm robot, each arm is considered as a different group). The Robot service forms a TCP/IP connection with the actual robot, retrieves the joint angles and sends them back to the motion planning module (proposed algorithm integrated with ROS-Moveit).

The diagram in Fig. 8.11 shows how the motion planning service is called through the Human Service GUI. When the button "Calculate path" is pressed, a service call is made with inputs: start joint angles, group name, goal joint names and visualization option. The MoveIt component integrated with the proposed motion generation algorithm, will execute the plan and return an array of points. The path is passed to

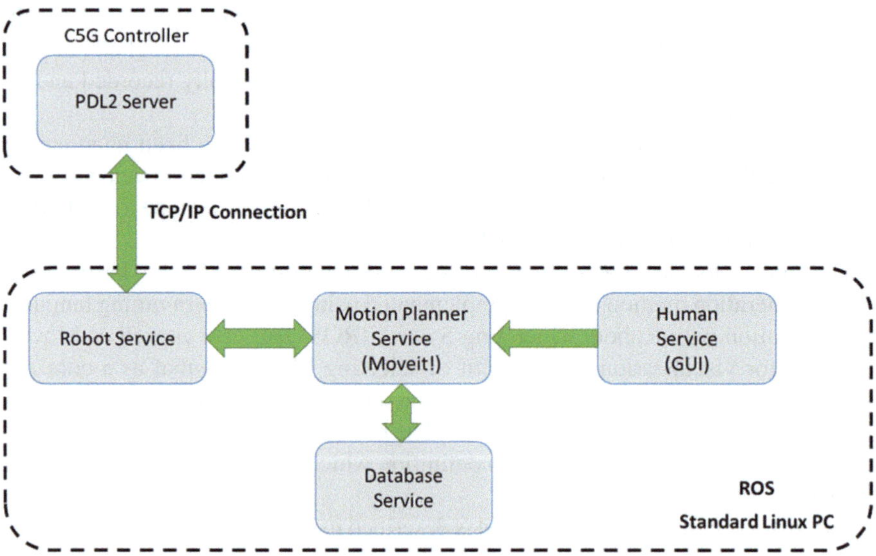

Fig. 8.9 Overall architecture

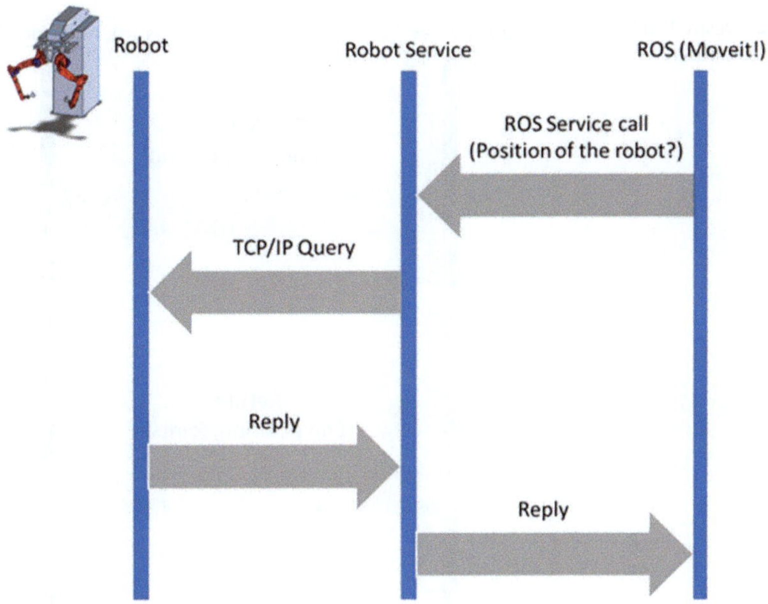

Fig. 8.10 Robot position monitoring service

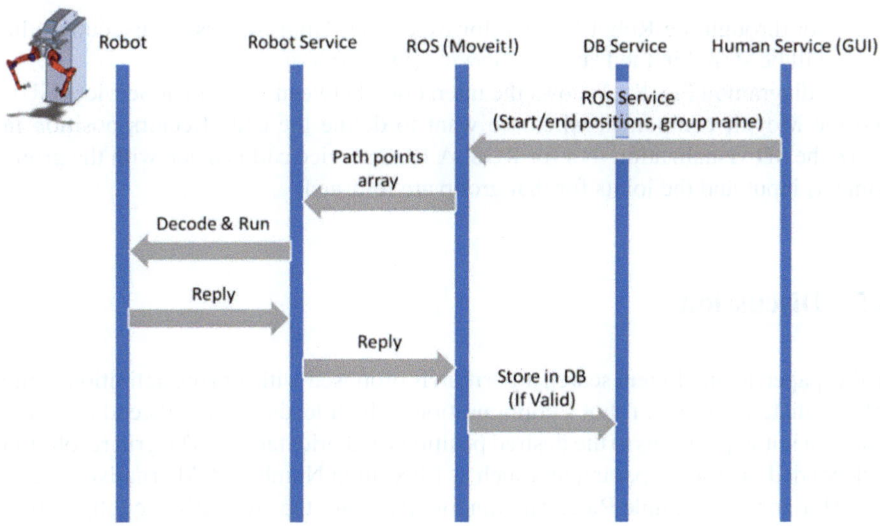

Fig. 8.11 Path calculation service

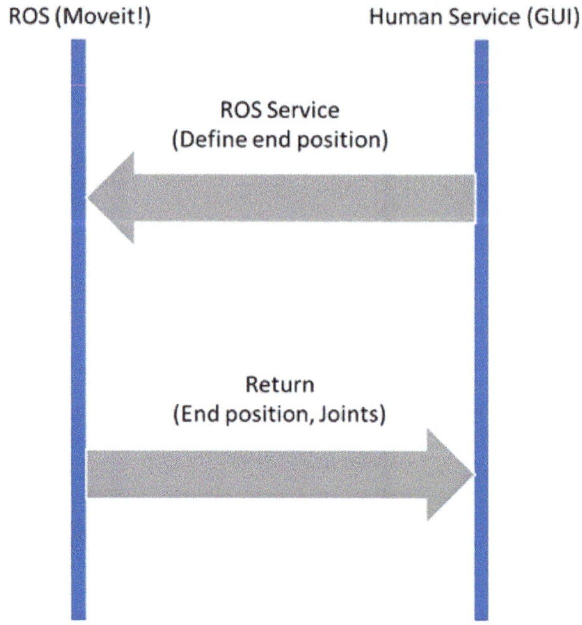

Fig. 8.12 Desired end position definition

the robot through the Robot Service for execution. Upon successful execution the array will be stored in the Database Service (DBService).

The diagram in Fig. 8.12 shows the interaction between the HumanService (GUI) and the MoveIt component when we want to define the end effectors position in RVIZ the 3D visualization tool for ROS. A ROS service call is made with the group name as input and the joints for that group are returned.

8.5 Discussion

In this paper, an intelligent search algorithm is proposed with aim the definition of the intermediate Dual Arm robot's configurations which leads the robot's end effectors from the initial positions to the desired positions and orientations. The grid resolution is depended on a set of parameters such as Maximum Number of Alternatives, Decision Horizon and Sample Rate. The random tree with the alternatives configuration is generated taking under consideration the above parameters. Each configuration is evaluated using two main criteria, the criteria of minimum distances and the criteria of minimum rotations. An acceptable error can be defined by the user and be used in order to reduce the computational time of the algorithm. The search algorithm based on a 13 DOF mechanism while the final aim is the generation of one individual path of each arm of the robot. Both target positions of each end effector are used for the evaluation of the alternatives. Thus, the importance of automatic motion planning keeps rising over the resent years with research being mainly focused on the case of collision avoidance.

Automatic path planning on site reduces the overall time of programming. The points to be recorded are reduced compared to conventional programming by demonstration methods. This is more obvious when complex cell layouts are taken in consideration where the free collision paths are not easily defined manually. In contrast to the path planning tools offered in some robot simulation packages for offline programming, in the proposed solution free collision paths are calculated while the programmer is on site. In this way the programmer is able to test the resulted path not only in the virtual environment but directly in the shop floor, making all the required modifications which result in a feasible path before the export of the final robot program. The proposed method enables the use of advanced programming tools like automatic path planning for programming on site. This becomes more essential for SME's with lower budget where advanced offline programming simulation packages are not affordable. Using the advanced programming platform which is proposed, the system integrator is not required for the robot programming, allowing even unexperienced programmers to handle with demanding assembly scenarios. Especially for SME's is of high importance to be more independent for programming/re-programming robotic cell using only internal resources without the need of specialized and expensive external system integrators.

References

1. Chryssolouris G (2006) Manufacturing systems: theory and practice, 2nd edn. Springer, New York
2. Mourtzis D, Alexopoulos K, Chryssolouris G (2012) Flexibility consideration in the design of manufacturing systems: an industrial case study. CIRP J Manuf Sci Technol 5:276–283. https://doi.org/10.1016/j.cirpj.2012.10.001
3. Makris S, Tsarouchi P, Surdilovic D, Krüger J (2014) Intuitive dual arm robot programming for assembly operations. CIRP Ann 63:13–16. https://doi.org/10.1016/j.cirp.2014.03.017
4. Tsianos KI, Sucan IA, Kavraki LE (2007) Sampling-based robot motion planning: towards realistic applications. Comput Sci Rev 1:2–11. https://doi.org/10.1016/j.cosrev.2007.08.002
5. Kavraki LE, Svestka P, Latombe J-C, Overmars MH (1996) Probabilistic roadmaps for path planning in high-dimensional configuration spaces. IEEE Trans Robot Automat 12:566–580. https://doi.org/10.1109/70.508439
6. Amato NM, Bayazit OB, Dale LK, Jones C, Vallejo D (1998) OBPRM: an obstacle-based PRM for 3D workspaces
7. Lavalle SM (1998) Rapidly-exploring random trees: a new tool for path planning
8. Ferguson D, Kalra N, Stentz A (2006) Replanning with RRTs. In: Proceedings 2006 IEEE international conference on robotics and automation. ICRA 2006. IEEE, Orlando, FL, USA, pp 1243–1248
9. Wu XJ, Tang J, Li Q, Heng KH (2009) Development of a configuration space motion planner for robot in dynamic environment. Robot Comput Integr Manuf 25:13–31. https://doi.org/10.1016/j.rcim.2007.04.004
10. Kaltsoukalas K, Makris S, Chryssolouris G (2015) On generating the motion of industrial robot manipulators. Robot Comput Integr Manuf 32:65–71. https://doi.org/10.1016/j.rcim.2014.10.002

11. Tsarouchi P, Makris S, Michalos G, Stefos M, Fourtakas K, Kaltsoukalas K, Kontrovrakis D, Chryssolouris G (2014) Robotized assembly process using dual arm robot. Procedia CIRP 23:47–52. https://doi.org/10.1016/j.procir.2014.10.078
12. Gharbi M, Cortes J, Simeon T (2008) A sampling-based path planner for dual-arm manipulation. In: 2008 IEEE/ASME international conference on advanced intelligent mechatronics. IEEE, Xian, China, pp 383–388
13. Lim S-J, Han C-S (2014) Operational space path planning of the dual-arm robot for the assembly task. Int J Precis Eng Manuf 15:2071–2076. https://doi.org/10.1007/s12541-014-0565-9
14. Ćurković P, Jerbić B (2010) Dual-arm robot motion planning based on cooperative coevolution. In: Camarinha-Matos LM, Pereira P, Ribeiro L (eds) Emerging trends in technological innovation. Springer, Berlin, Heidelberg, pp 169–178
15. Gharbi M, Cortes J, Simeon T (2009) Roadmap composition for multi-arm systems path planning. In: 2009 IEEE/RSJ international conference on intelligent robots and systems. IEEE, St. Louis, MO, pp 2471–2476
16. Li Z, Katz R (2005) A reconfigurable parallel kinematic drilling machine and its motion planning. Int J Comput Integr Manuf 18:610–614. https://doi.org/10.1080/095119205000 69218
17. Seo DJ, Ko NY, Simmons RG (2009) An elastic force based collision avoidance method and its application to motion coordination of multiple robots. Int J Comput Integr Manuf 22:784–798. https://doi.org/10.1080/09511920902741083
18. Bonert M, Shu LH, Benhabib B (2000) Motion planning for multi-robot assembly systems. Int J Comput Integr Manuf 13:301–310. https://doi.org/10.1080/095119200407660
19. Hatwig J, Minnerup P, Zaeh MF, Reinhart G (2012) An automated path planning system for a robot with a laser scanner for remote laser cutting and welding. In: 2012 IEEE international conference on mechatronics and automation. IEEE, Chengdu, China, pp 1323–1328
20. Vick A, Surdilovic D, Kruger J (2013) Safe physical human-robot interaction with industrial dual-arm robots. In: 9th international workshop on robot motion and control. IEEE, Kuslin, Poland, pp 264–269
21. Shahrokhi M, Bernard A, Fadel G (2011) An approach to optimise an avatar trajectory in a virtual workplace. Int J Comput Integr Manuf 24:95–105. https://doi.org/10.1080/0951192X.2010.531290
22. Zhu X, Hu SJ, Koren Y, Huang N (2012) A complexity model for sequence planning in mixed-model assembly lines. J Manuf Syst 31:121–130. https://doi.org/10.1016/j.jmsy.2011.07.006
23. Wahl FM, Thomas U (2002) Robot programming-from simple moves to complex robot tasks
24. Quigley M, Conley K, Gerkey B, Faust J, Foote T, Leibs J, Wheeler R, Ng A (2009) ROS: an open-source robot operating system

Chapter 9
Physics Based Modeling and Simulation of Robot Arms

9.1 Introduction

There are lots of reasons which explain the increasing use of robotic structures in production lines. The ability of the robots to perform a variety of different tasks such as machining processes [1] or human–robot cooperative tasks [15, 22, 23] with aim to increase the productivity and the quality of the manufacturing process [5, 8] is the main explanation of this phenomenon. In this way, lots of researchers deal with how to improve the robot's behavior in order to be more flexible and accurate. Every robot is characterized by two main factors, the repeatability and the accuracy. Repeatability is defined as the ability of the robot to execute the same repetition of the same task [16]. Usually, the level of the repeatability of a robot is acceptable and it does not affect at the execution of different tasks. On the other hand, the accuracy of the robot causes lots of problems at the execution of a task while lots of time restricts the ability of a robot to execute it. The inaccurate behavior of a robot is caused by geometric and non-geometric factors which analyzed below [4]. In Fig. 9.1, a comparison between the ideal versus real posture and trajectory of a robot is presented.

One way to control this inaccurate behavior of a robot is to select, estimate and control the parameters which affect to this behavior. This approach is well known as parameter estimation or parameter identification and constitutes an active area of researchers for decades. There are two major approaches in terms of parameter estimation for robotic manipulators, the numerical estimation and the mathematical estimation of parameters.

The numerical approach is based on estimation algorithms in order to identify the robot parameters. This method can be implemented using stochastic methods as Monde Carlo method which allows to calculate the uncertainty in the identification of each single robot parameter [20]. Moreover, numerical algorithms like Rapid Prototyping Algorithm (RPA) in combination with traditional methods (Newton, Levenberg–Marquardt) or with stochastic methods are used in order to manipulate the data configurations for robot calibration [14].

S. Makris, *Cooperating Robots for Flexible Manufacturing*,
Springer Series in Advanced Manufacturing,
https://doi.org/10.1007/978-3-030-51591-1_9

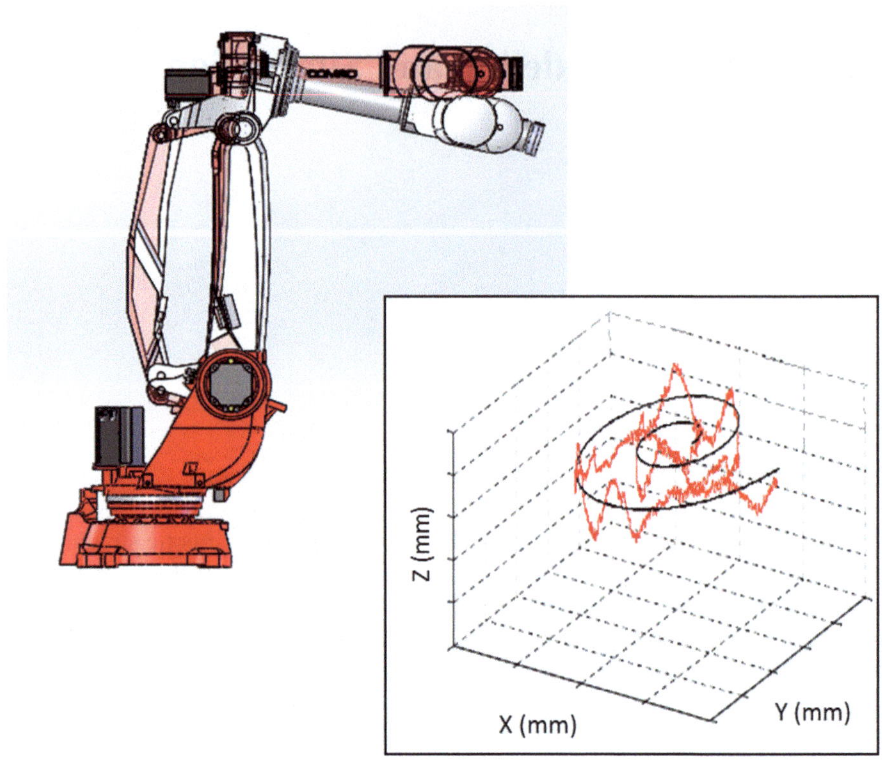

Fig. 9.1 Ideal versus real posture and trajectory of robot

Mathematical methods are based on solving systems of equations in order to identify the parameter's values. Through these methods, mathematical expressions and characteristics of the system are used for identification procedure. Mathematical methods consist the most conventional methods in robot identification able to be implemented with a variety of different ways. The usage of sensors is often applied in these methods as can be determine the real data of the system such as the positions of the joints, the velocity and the acceleration of each motor. Circle Point Method combines the sensors data with robot's structure description with aim to execute an identification procedure for robot kinematic calibration [21]. Within this context, robot internal joint sensors and recurrent neural network (RNN) approach was applied to solve robot calibration problem [25]. Other studies define a linear differential model of independent parameter errors and use the least-squares method to calculate the calibration model [4].

In this approach, a numerical estimation method in combination with physical-based simulation models is used for the estimation of robot parameters. The main reason is that the estimated parameters of the robot has a nonlinear behavior. This means that the designing of a mathematical method is more complex to solve a nonlinear system. In addition, there is possibility to fail to solve a system like this.

Last but not least, the numerical estimation is more flexible and could be used in a variety of robot models in contrast with a mathematical estimation method.

The most difficult phase of a parameter estimation method is the selection of the robot's parameters which will be estimated. In this procedure, the selection of the robot parameters based on the calibration procedure. As calibration procedure characterized the process by which it is possible to improve the static accuracy of industrial robots. The calibration processes are classified in three levels based on the type of accuracy error (Fig. 9.2) and are presented below [9, 17, 19]:

- First level calibration or joint calibration: The purpose at this level is to ensure that the joint sensor value and correspond to the real joint displacement. The mathematical model which is required for this calibration is the mathematical equation between the output of the joint sensor and the actual joint displacement. Comparing the model output values with the robot real time measurements, it can be determined the deviation between real and theoretical values. The correction in this level can be completed either by a software approach or using a hardware implementation [12, 24].
- Second level calibration: This level covers the entire robot model calibration. Having as goal to improve the accuracy of kinematic model of the manipulator and the relationship between the joint transducers and the actual joint displacement. Numerous approaches have been developed for the kinematic model of a robot manipulator, however the most know is Denavit and Hartenberg. The identification of the parameters in a robot kinematic model is a problem that has been addressed by a number of researchers using a variety of models and identification algorithms [3, 25].
- Third level calibration: This calibration, known as non-kinematic calibration, presents the errors in positioning of the end effector of the robot that are due to

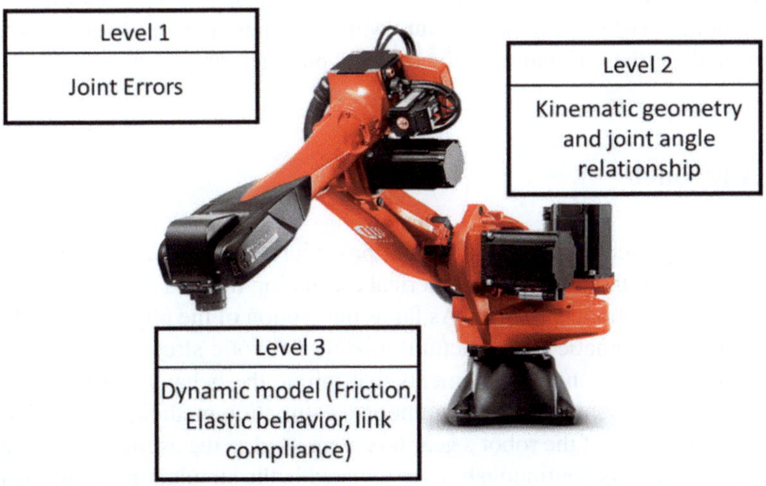

Fig. 9.2 Levels of robot calibration

factors that do not connect with joint displacements and the kinematic structure of the robot. Factors such as joint compliance, friction, link compliance, elastic behavior and other elements affect the robot accuracy and the estimation of their parameters is needed [11, 18].

In this approach, the parameters which are selected for the estimation procedure belong to the third level of the calibration procedure and more specific are the elasticity and the damper coefficients of each gearbox of the robot. The challenging part of this selection is the non-linear behavior of these parameters which lead to a high complexity mathematical description.

Simulation is an integral part of identification, playing a strategic role in many fields of robotic. In order to satisfy the increasing demands for simulation a number of software tools are available. MATLAB\Simulink are the most known simulation programs as these are used in a variety of fields [6, 26]. Simulink is integrated with MATLAB enabling the incorporation with MATLAB algorithms into models and export simulation results. OpenModelica is an open-source simulation program which based on Modelica language for simulating mechanical, electrical, fluid and other systems [7, 10]. Dymola is a multi-engineering modeling and simulation package which uses Modelica language in order to create models [7, 13]. Physical-based simulation models have been used in a variety of application such as for predictive maintenance [2].

While this topic has been researched in the past, it seems that there is a gap in the connection of numerical estimation methodologies with physical-based simulation models able to interact with the real machines for data exchanging and eventually, improving the accuracy. The current study is focused into the increasing of a robot accuracy using an identification procedure based on physical-based simulation model in combination with a continuously feedback of the model with real time data. This combination enables the concept of digital twin via the online data gathering, model simulation and feedback to the real robot model. OpenModelica software is used for the simulation of the robotic structure while a numerical identification method is implemented for the estimation of the selected parameters.

9.2 Approach

This chapter presents an identification procedure which based on simulation modelling in combination with a numerical estimation method with aim to improve the accuracy of a robotic structure. As far as the section of the physical-based model simulation, the kinematic and structural model of robotic structure were modelled using a set of components and elements that mimic the behavior of the real robot. As far as the implementation of the numerical estimation method, the elasticity and damping coefficients of the robot's gear box were used as the estimation parameters. The above parameters continuously were updated in the simulation environment and the output data were compared with the real motion of the robot. As motion of a robot, is refers at a predefined trajectory of the end effector of the robot. When the simulated trajectory of the robot, which is the output of the physical-based model simulation, and the real one was very close, then the estimation procedure stop, and the values of the elasticity and damper coefficients were exported. In this way, the simulation model is able to predict the position errors of each robot's joint. Knowing the error of each joint position, it is feasible to control the robot's motor aiming to drive them at the desired position without accuracy errors.

9.2.1 Physical-Based Simulation Modeling

The simulation modeling has as aim to correspond the real behavior of the robot in a simulation environment. The simulation model structured using OpenModelica, an object-oriented language that is used for component-oriented modelling. The structure of the robot model was composed using components of the OpenModelica libraries. Every component is characterized by a number of parameters which correspond to a specific robot characteristic. The robot model consists of input, processing and output components. An indicative physical-based model of a robotic structure is presented in Fig. 9.3.

More detailed, the physical-based simulation model of a robot consists of the below components.

9.2.1.1 Mechanical Structure

The main concept for the simulation of the mechanical structure of robot is the using of two subcomponents, the joint component and the link component. The link components represent the rigid bodies of a robot while the joint components represent the rotation of a rigid body around one axis. The combination of the above subcomponents leads to the simulation of the kinematic model of the robot included the mass properties of it. Parameters such as the length and the mass of the links, the center of mass of each link and the inertia of it, the rotation axis of each link etc. are defined using these subcomponents. Last but not least, the orientation of the base frame of the robot is defined in this model. If the robot is equipped with a tool, it should be defined as an extra link located at the end effector of the robot. As far as the connection among these subcomponents, the links are connected each other using the joints as connectors. Finally, joints are connected to flanges which use the output data of each joint as input at the next component of the model. In Fig. 9.4, a model of the robot's mechanical structure is presented.

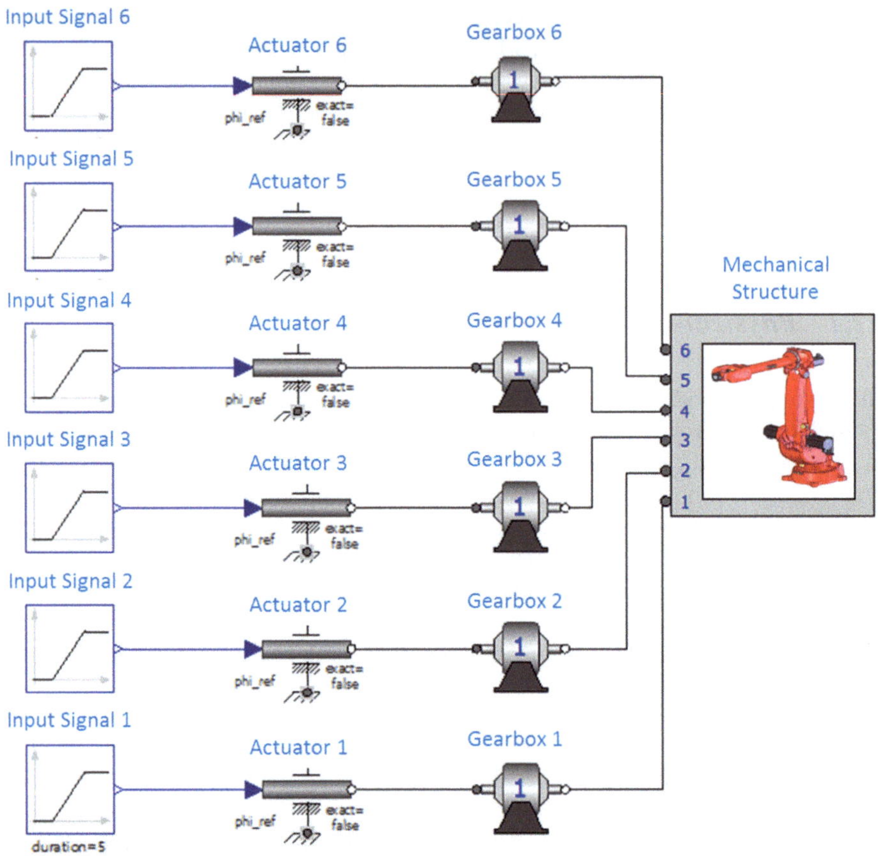

Fig. 9.3 Simulation model of the robotic structure

9.2.1.2 Actuators

This component enables the movement of the gearbox flange according to the reference input signal of the next component. The role of this component is to translate the input angle signal to an output torque signal. This component replaces the motor component with aim to simplify the whole model and to focus at the estimation of the dynamic parameters of the robot.

9.2.1.3 Gearbox

The gearbox is another component of the robot simulation model which consists of three subcomponents, the gear component, the elasticity component and the friction component (Fig. 9.5). This is the most important component of the whole model because here are defined the estimation parameters of the presented identification

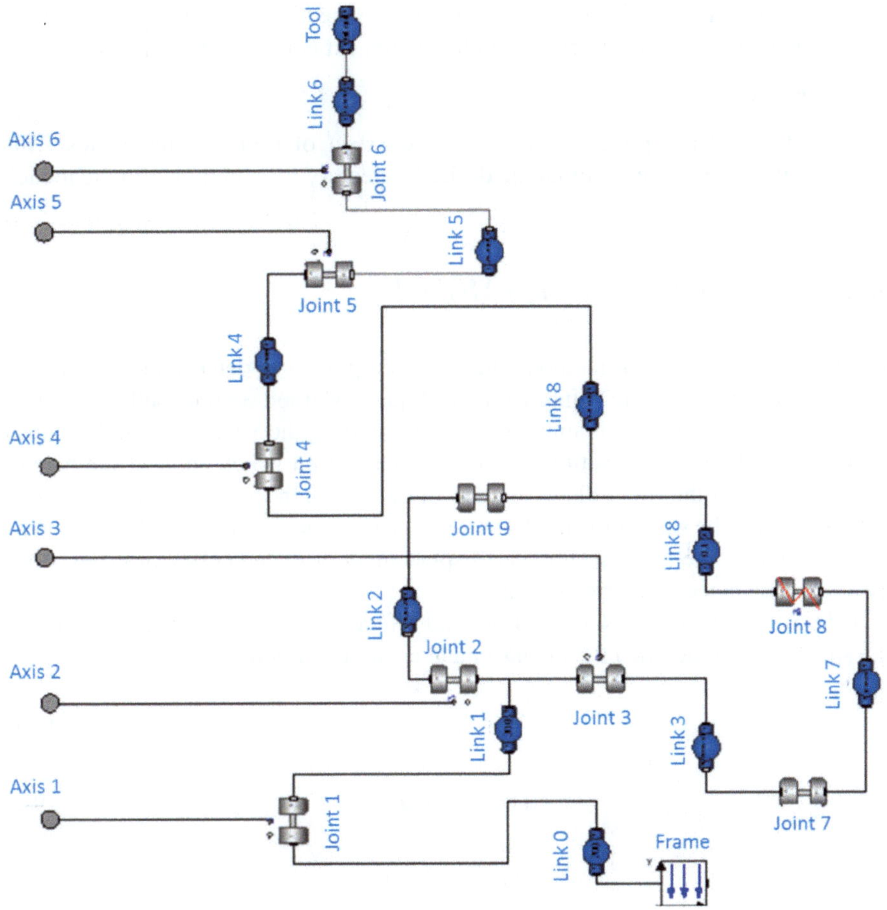

Fig. 9.4 Simulation model of the mechanical structure of the robot

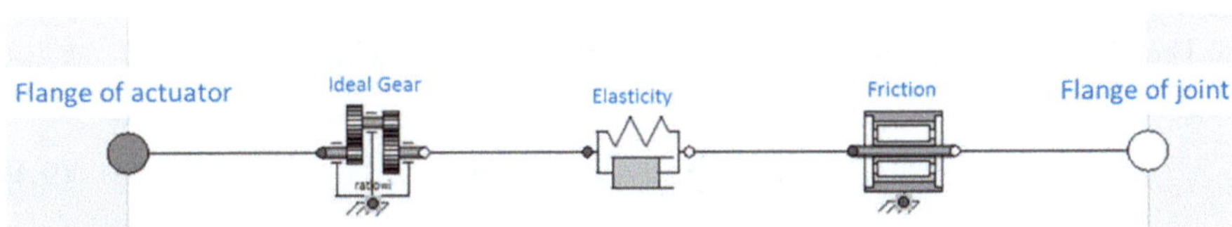

Fig. 9.5 Gearbox of the robot simulation model

procedure. More specifically, the elasticity component includes the data about the elastic and damping behavior of the model. The calculation of these parameters values is the object of this study as they play a key issue for improving robot accuracy. As far as the gear subcomponent, it represents an ideal gear in which the drive ratio of each axis has be defined. The friction subcomponent represents the Coulombic friction of the model. The role of this component is the connection of the flange of the joint subcomponents with the actuators which are the next component of the model.

Input signals

The input signal component consists of a data set of robot's joint angles. This component used in order to input the desired trajectory for simulation in the model.

9.2.2 *Numerical Estimation Method*

The numerical estimation method which was employed to determine the dynamic parameters is based on the Nonlinear Least Square Method. In this method are used *m* number of sets of experimental data so that finding the values of *n* independent parameters. The model function is a nonlinear function which connect the parameters values with the robot displacement. In this research the role of the parameter function places the simulation model as it has as input the parameter and as output the displacement of the end effector. The displacement of the end effector is described by a 3×1 displacement error vector.

If D_r is the real displacement of the end effector and D_c the calculated error of the simulation model then the displacement error vector is the ΔD.

$$D_r = \left[x_r y_r z_r\right]^T \tag{9.1}$$

$$D_c = \left[x_c y_c z_c\right]^T \tag{9.2}$$

$$\Delta D = [\, D_r - D_c] \tag{9.3}$$

ΔD is the deviation between target and real trajectory. The estimation of the dynamic parameters is executed minimizing the sum of square of displacement error vector ΔD.

The objective function of the model is defined as following:

$$E = \sum_{i=0}^{m} [\Delta D]^T [\Delta D] \tag{9.4}$$

The initial values of elastic—damper parameters were set based on the robot data. The initial parameters values are very critical for the whole procedure as they define the duration of the estimation method, the closer to the real values is the initial parameters values the shorter time will be demanded. The parameter values are changed by an *iteration step*. This step has specific value so that if it is applied to the parameter values of the model neither cause not change to the displacement error vector or cause important differences to it. The estimation method was executed iteratively until the displacement error is close to the determined termination condition. The displacement error in Cartesian system is described by the root mean square (RMS) of the displacement error.

$$RMS = \sqrt{\frac{1}{m}\sum\nolimits_{i=1}^{m}(\mathrm{D}_r - \mathrm{D}_c)_i^2} \tag{9.5}$$

9.2.3 *Identification Procedure*

The identification procedure based on a closed loop estimation of the parameters with aim to improve the accuracy of the robot. The simulation results and the real data are compared continuously until the position errors to be reduced in a desired level. While the position errors remain in high level, the estimation parameters of the model were updated using the above numerical estimation method. The identification procedure is depicted in Fig. 9.6.

The identification procedure consists of the following steps.

9.2.3.1 Programming of the Robot to Execute a Trajectory

The first step of the identification procedure is the designing of a trajectory and the programming of the robot to execute it. The designing of the trajectory can be designed in a random way or based on a trajectory generation theory. In this study, the trajectory designed using the least squares method. The main advantage of this method is that it eliminates the execution time of the program.

9.2.3.2 Execution of the Trajectory in Real Time

After the designing of the desired trajectory and the robot programming, the real time execution of it is the next step. As it is expected, the robot did not follow the trajectory because of its inaccurate behavior. A simple graphical method is used for the data collection of the real trajectory which robot executed. Using this graphical method, the data of the execution are exported by 3d designing software. The containing of these data are the values of translation and orientation of the end effector while the execution of the trajectory.

9.2.3.3 Execution of the Trajectory in Simulation Environment

Now, the same trajectory used as input in the simulation model which run and execute the trajectory in a simulation environment. The results of the simulation contain

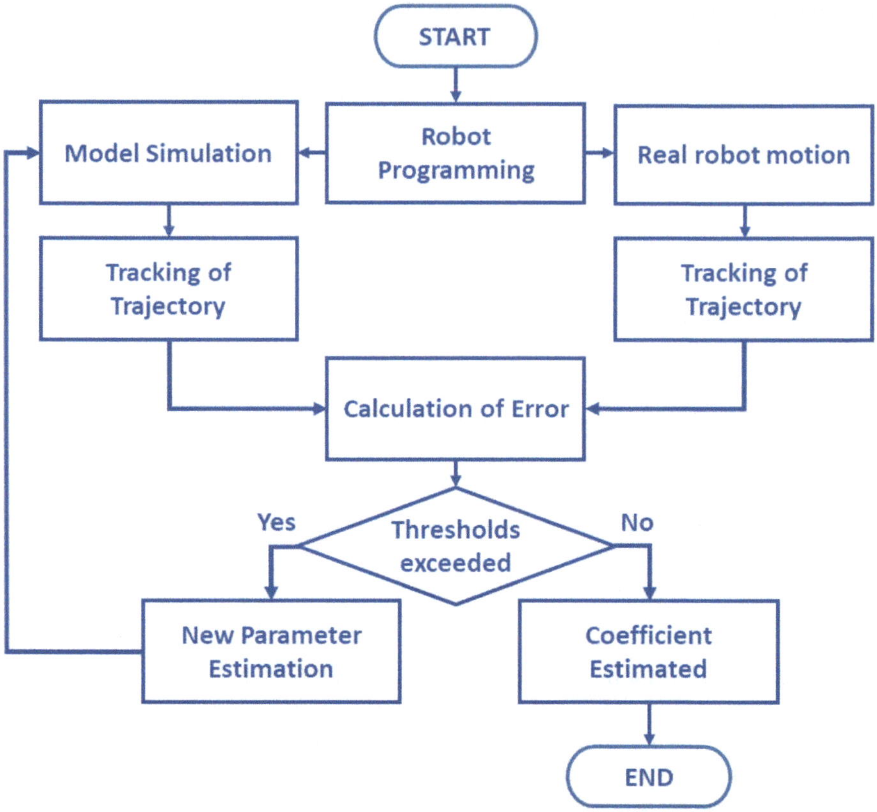

Fig. 9.6 Identification procedure

the values of translation and orientation of the end effector while the simulated execution of the trajectory, too. Due to the elastic behavior of the robot which have been modeled, the trajectory which the simulation model execute is different from the reference one.

9.2.3.4 Comparison of the Output Data of Each Execution

The data of the real execution and the data of the simulated execution are compared and the position errors are calculated. This comparison consists of six comparison of values, three for the translation of each axis and three for the orientation of each axis and the corresponding position errors are calculated. If the all errors are smaller than a desired level, it means that the estimation parameters have been estimated successful and the simulation model of the robot has the same behavior with the real one. In different case, the procedure goes ahead to the next step.

9.2.3.5 Estimation of the Parameters

The estimation of the dynamic parameters of the model is based on the aforementioned Nonlinear Least Square Method. Using this method the value ranges for parameters were restricted significantly. An iterative process was executing using the estimated parameters with aim to evaluate behavior of the simulated model. The dynamic parameters for which the calculated errors were minimum was saved for the next execution of the program. This process was executed as many times as the number of the robot experiment having as output an excel file which involved the dynamic parameters of the robot. When the parameters have been estimated successfully, it is feasible to know the real position of the robot for each trajectory without to run the identification procedure again. In this way, it is feasible to control the robot to compensate the positions errors which lead to an accurate behavior of the robotic structure.

9.3 Industrial Examples

9.3.1 Single Robot Operation—Implementation

In this section, the software which was implemented in order to execute the presented approach are described in detail. The main procedure that this method demands is the modelling of the robot model, the designing of the target trajectory and the implementation of the numerical method using a programming language. The structure of the robot model was composed using components of the OpenModelica libraries. The target trajectory was designed using the PDL programming language. The numerical estimation method that was implemented was the least squares method and it was applied using Python programming language. The software which were used selected with aim to reduce the calculation time and the programming effort. Moreover, the communication architecture including the services and the way of communication is presented.

The modelling procedure was developed in the OMEdit environment, a Modelica connection editor for OpenModelica. OpenModelica is an open-source Modelica-based modeling and simulation environment intended for industrial usage. Modelica is a non-proprietary, object-oriented, equation based language to conveniently model complex physical systems. This software allows to user to create models that describe the behaviour of real world systems by two ways. The first way is to use components from free Modelica Standard Library and the second is to create their own components. Combining these components can be created large and complex systems. So, no particular variable needs to be solved manually, as the Modelica tools has been created with aim to solve them automatically [10].

As this method described above, the least squares estimation method eliminates the time of estimation procedure calculating the sum of square of displacement error vector. The displacement error in Cartesian system is described by the root mean square (RMS) of the displacement error. The estimation method was executed iteratively until the displacement error is close to the determined termination condition.

For the robot programming, the corresponding programming language was used. PDL or Process Description Language is a robot programming language for programming autonomous agents (robots) in a dynamical way. Using the PDL programming language, the paths of the robot motion can be designed and executed. A path is a number of coordinates or joint sets which followed by the robotic structure and leads from an initial position to the target one.

Python programming language was used for the programming of the whole procedure as well as the implementation of the least squares estimation method. This language includes a rich standard library, providing easy interface with OpenModelica—OMPython [10] and other programs.

The target trajectory was designed in a 3D software and used as input in the robot controller. Using the PDL language, this data translated to a robot execution program. The real trajectory was monitored and the needed data was extracted using a 3D design software. On the other hand, the same initial trajectory was used as input of the simulation model in Modelica. The monitored data of the real execution of the trajectory as well as the results of the robot simulation in Modelica was imported in a Python script in which the least squares estimation method was implemented. Initial values of the estimation parameters are defined in Python code and a repeat procedure is started until the final estimation of the parameters.

9.3.2 Experiments and Results

In this section, the experiments which took place with aim to evaluate the method are presented. The experimental setup consists of a fixed robot equipped and a fixture. The robot is equipped with a graphic recording mechanism of its path on a millimeter sheet which is on the fixture as it presented in Fig. 9.7. In order to increase the accuracy of estimation, some experiments taken place using an extra metallic part approx. 75 kg adapted to the last flange of the robot. In this way, a second Modelica robot model was created including the modelling of the metallic part as an extra feature. The center of mass as well as the moments of inertia of the metal weight are calculated using a 3D design software and all the information imported in Modelica robot model.

Initially, the robot motion is programmed using PDL programming. The target trajectory consists of ten robot configurations and was executing in a vertical to robot plane. Each robot configuration is a set of six values which correspond to the robot's joint values. The contribution of every robot's joint rotation lead to the real trajectory. Using the graphic recording mechanism, the real robot trajectory is recorded on the

Fig. 9.7 Experimental setup

millimeter sheet. In the next step, an image of the millimeter sheet is imported in a 3D design software where the needed data of the trajectory are extracted and saved in a temp file (Fig. 9.8).

The same robot configurations were imported in the simulation model of the robot as input. After the simulation of the robot motion, the simulated trajectory was generated as output (Fig. 9.9). The corresponding needed data of the trajectory are extracted and saved in a temp file, too. As needed data they mentioned to the coordinates of each configuration of the robot both in real and simulated trajectory in the vertical XY plane.

In the next step, all the data was imported in a Python code in which the identification procedure was implemented. Random initial value of the estimation parameters were defined. In this code, the following functions were implemented continuously:

- Function for reading and importing the data of the robot both in real and simulated trajectory.
- Comparison of the real and simulated position data and calculation of the corresponding position error.

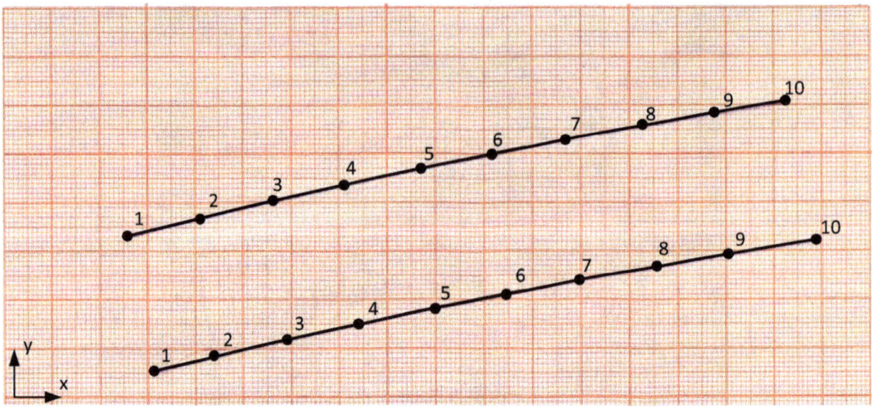

Fig. 9.8 Real trajectory of the robot

Fig. 9.9 Simulation trajectory of the simulated model

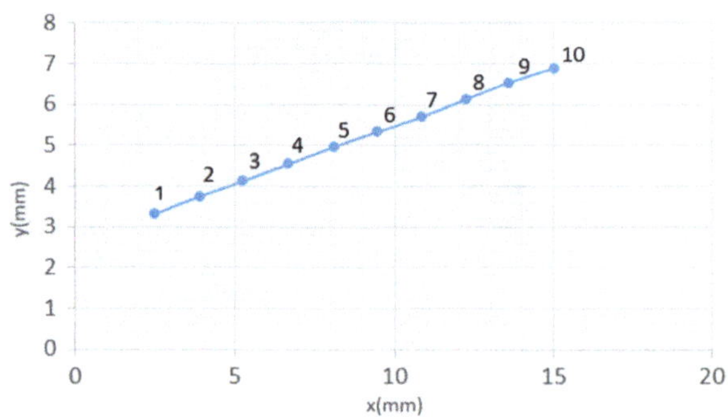

- If the error were unacceptable, the values of the dynamic parameters of the simulated model in Modelica were updated using the Nonlinear Least Square Method and the procedure was repeated.
- If the position error was acceptable, the estimation of the parameters was completed and the simulated robot model had almost the same behavior with the real one.

Results

In this section of the chapter, the results of a number of experiments are presented. The experiments taken place for different trajectories of the robotic structure. At first level, the estimation of the parameters only for the first axis of the robot taken place. The estimated values of the spring and damper coefficients are presented in the Table 9.1.

It is useful to mention that the set of the estimated values is not unique. Lots of set of elastic-dumper coefficient values can be exported. The selection of the final values based on the combination of these set through the execution of the experiments. More detailed, all the set of the values are saved in a database after the execution of each

Table 9.1 One axis model parameter's values

Robot axis	Spring coefficient	Dumper coefficient
First	*1565 N/m*	*3477 N*s/m*

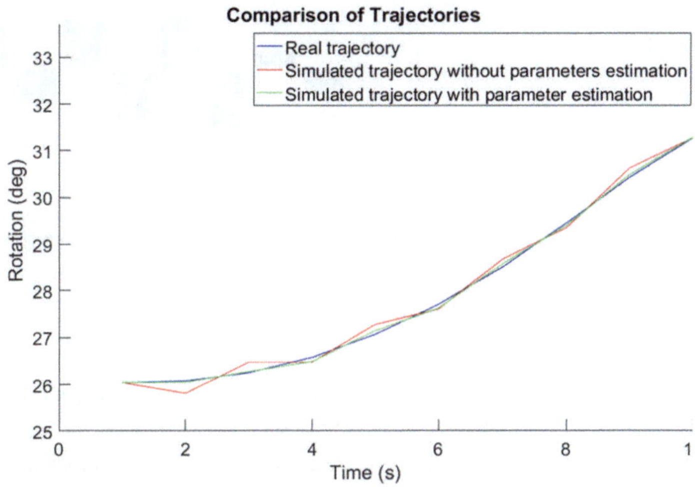

Fig. 9.10 Comparison of trajectories—one axis model

experiment. When a set of experiments with different trajectories are completed, the set of values which is most common to all the experiments is selected.

In the Fig. 9.10, the real trajectory of the robot, the initial trajectory of the simulation model and the trajectory of the simulation model using the estimated parameter's values are depicted.

The corresponding errors between the real trajectory and the simulated one before and after the simulation are depicted in Fig. 9.11.

At second level, the estimation of the parameters for the full robot model, included six axis, taken place. It useful to refer that the parameter's values of the last three axis of the robot (axis 4, axis 5 and axis 6) were assumed equal to the theoretical due to there are responsible about the orientation of the end effector. The estimated values of the spring and damper coefficients are presented in the Table 9.2.

In Fig. 9.12, the real trajectory of the robot, the initial trajectory of the simulation model and the trajectory of the simulation model using the estimated parameter's values are depicted.

9.3.3 Cooperating Robot Concept—Implementation

In this work, a method for reducing the position errors during robot cooperation is proposed and tested. Models simulating the behavior of the robots used were

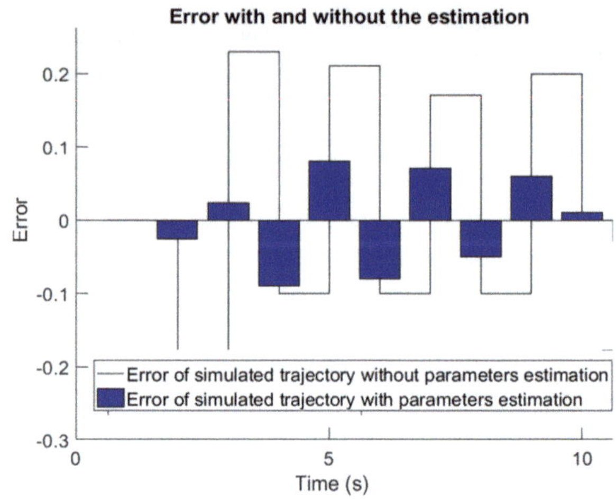

Fig. 9.11 Comparison of errors with and without the parameter's estimation

Table 9.2 Full robot model parameter's values

Robot axis	Spring coefficient	Dumper coefficient
First	2761 N/m	1377 N*s/m
Second	3542 N/m	1856 N*s/m
Third	3678 N/m	981 N*s/m

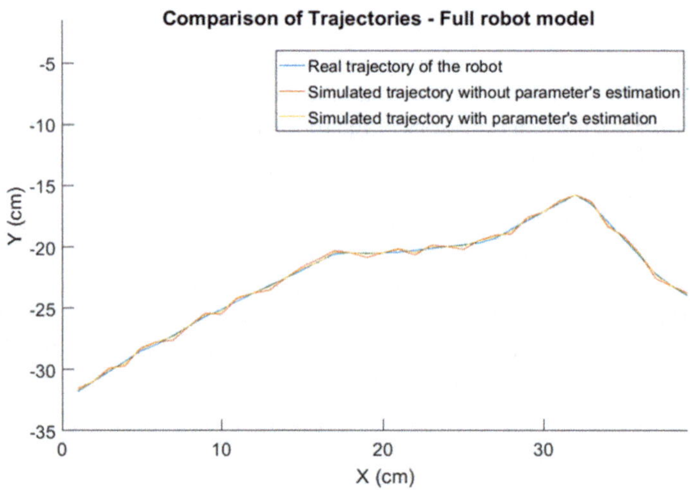

Fig. 9.12 Comparison of trajectories—full robot model

developed using the Open Modelica object oriented programming language. The development of the models is based on the elastic lumped parameter model. The parameters were provided by the robot manufacturer. In the next paragraph we will present a brief overview of the problem. Based on the simulation results the robots are commanded to common point. The distance between the two end effectors is measured before and after the correction. The results along with some proposal for further improvement are presented in the last section (Fig. 9.13).

Two industrial robots were used in this experiment, namely a COMAU NJ-130 and a COMAU NJ-370. The aim was to develop a method that would be used for moving the robot flanges to the same Cartesian position. A common user frame was set. Defining the common User frame (UFrame) allows two robots with different base frame to reach the same positions without making conversions between intermediate frames of references. The user frame was set using the POS_FRAME function provided by the manufacturer and includes the following steps: Move robot to Origin->Move to X->Move to XY->Repeat for second robot. The resulting frame can be seen in Fig. 9.14. A common user frame with the following parameters is created:

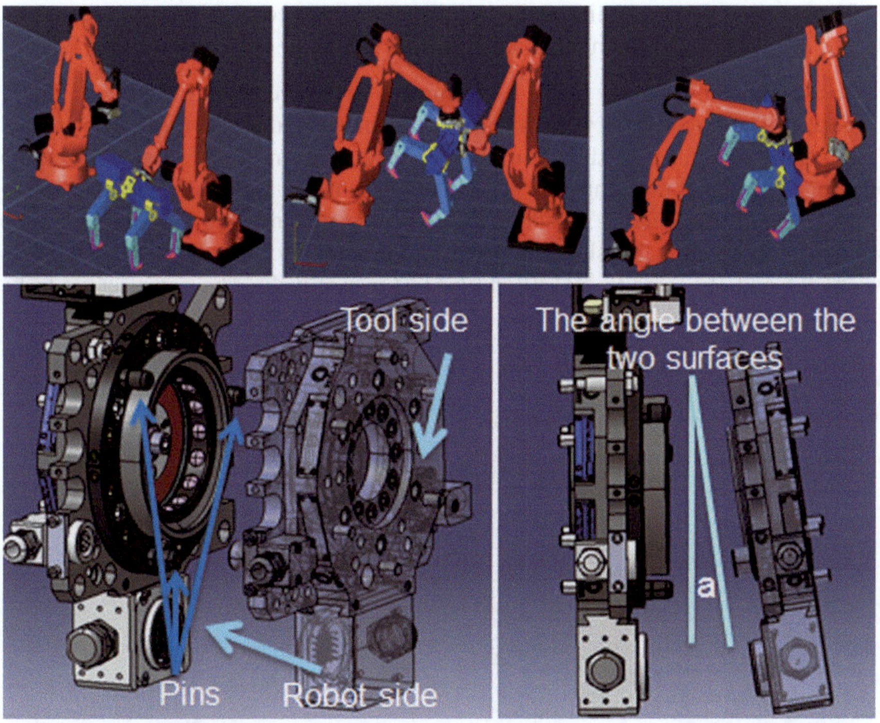

Fig. 9.13 Robot to robot gripper transfer

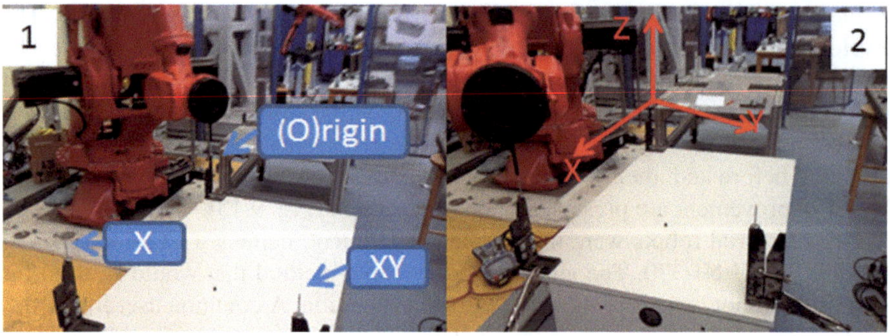

Fig. 9.14 Setting up the common UFrame

- Origin is the origin of the new user frame.
- X-axis is defined by the points O and x.
- The xy-plane is parallel to a plane defined by the points O, x, and xy. xy is on the positive half of the xy-plane.
- The y-axis is on the xy-plane and is perpendicular to the x-axis.
- The z-axis is perpendicular to the xy-plane and intersects both the x- and y axes.
- The positive direction on the z-axis is found using the right hand rule.

To evaluate the accuracy of the user frame two tests were conducted.

Test 1: The two robots were manually moved until the two tool center points (TCP) came in contact as shown in Fig. 3. Since a common UFrame was used the position recorded at the controller should be the same for both robots (Fig. 9.15).

However as presented in Fig. 9.16, the controller readings were different for each robot. The following table shows the positions recorded for three random test points where R1 and R2 refer to NJ130 and NJ370 respectively. The error seems to be

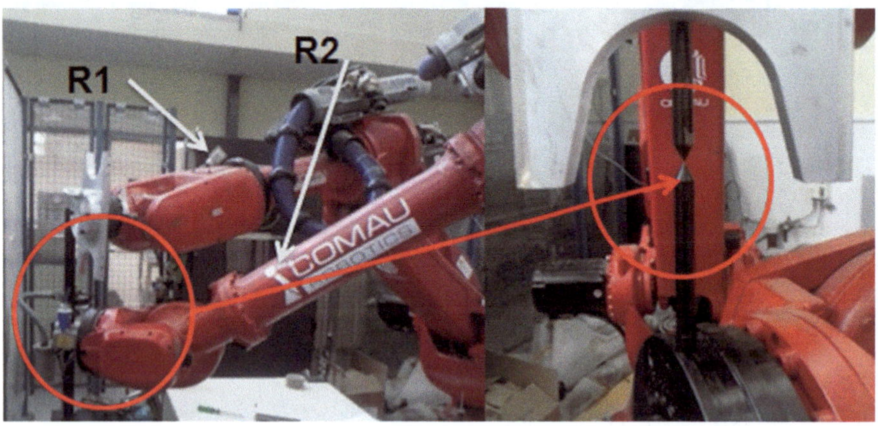

Fig. 9.15 The two robots in a "Stretched" configuration manually commanded to the same point

Status		R1	R2	Error(mm)
Not Streched	X(mm)	419,2	422,389	3,189
	Y(mm)	-667,074	-666,327	0,747
	Z(mm)	762,086	760,527	-1,559
Semi Streched	X(mm)	499,156	497,156	-2
	Y(mm)	424,334	425,816	1,482
	Z(mm)	582,056	577,663	-4,393
Fully Streched	X(mm)	427,9	421,753	-6,147
	Y(mm)	1131,509	1133,016	1,507
	Z(mm)	246,074	239,873	-6,201

Fig. 9.16 The values recorded at each robot controller when the TCPs were at the same point. Ideally both robot controllers should record the same values

related with the distance of the point from the base of the robots i.e. how stretched the robot configuration is.

Test 2: Both robots were commanded to move to the same point using a motion command with the same Cartesian values for both robots. When the motion completed the distance between TCP1 and TCP2 was measured. Normally the two TCP should reach exactly the same position. However the distance between them varied from 1–7 mm. For example in Fig. 9.17 while the two robots were commanded to the same point a distance of about 5 mm.

The method proposed in this research aimed to reduce the error during robot cooperation and can be split in to three phases. An overview of phases 2 and 3 can be seen in Fig. 9.17. In every case the compliance of the robot due to the elastic behavior of the robot is taken into consideration (Fig. 9.18).

Phase1: Models simulating the behavior of the robots were developed and tested.

Fig. 9.17 Typical examples of the cooperation results using the conventional method

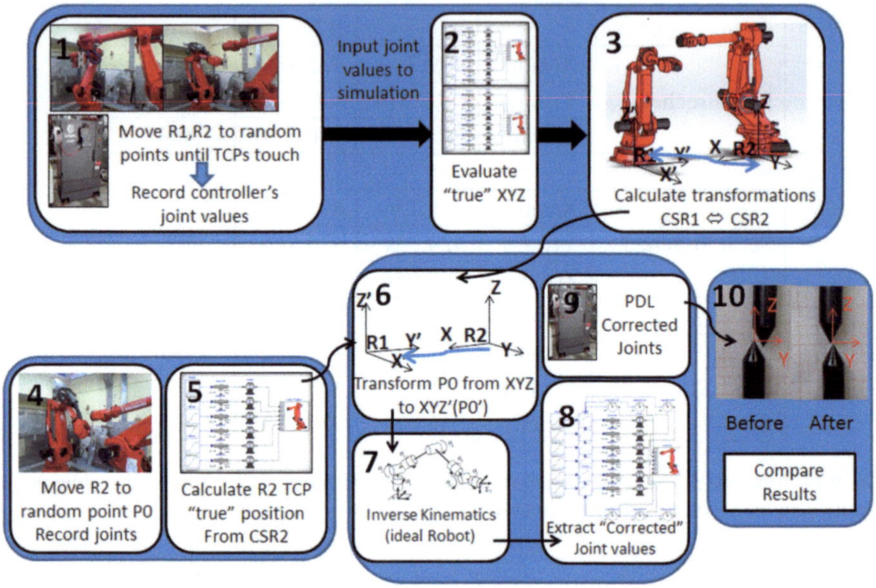

Fig. 9.18 Proposed method overview

Phase2: The common user frame is calculated manually.

(1) The two robots were manually commanded to random positions until Tool Center Point 1(TCP1) and Tool Center Point 2 (TCP2) contacted.
(2) For the joint angles recorded at the controller, the actual position of each robot to its base was calculated using the robot models developed.
(3) The above data were used to calculate the 3D transformation-rotation matrix between the coordinate system of R1 (CS1) and that of R2 (CS2). Positions for both robots can be expressed with CS1 as the common user frame.

Phase 3

(1) R2 is commanded to the desired point.
(2) For the joint values recorded at the controller and using the simulation model the actual position of R2 from its base is calculated (P2).
(3) Using the transformation matrix, P2 is expressed according to CS1(P2′).
(4) The joint values of R1 to reach P2′ are calculated using inverse kinematics.
(5) Using the simulation model R1 joint values are recalculated taking into consideration the compliance of the robot.
(6) R1 is commanded in joint space using the corrected joint values extracted from previous step.

The modeling of each component along with the full robot model will be described in the following sections.

9.3.4 Mechanical Structure Model

This is the model that describes how the links are connected to each other. The mechanical Structure model is made by rigid bodies connected to each other with revolute joints. Each revolute joint allows rotation of the connected body only around the plane of rotation of the joint. All elasticity effects such as the elasticity of the links and of the joints are represented separately in the gear models. Figure 9.19 presents the coordinate system used for every joint along with the joint-link allocation while Fig. 9.20 (left) shows how the kinematic chain loop was assembled and modeled. The model had to be further modified in the case of NJ370 to model the gravity compensator attached between links 2 and 4. The gravity compensator and the respective model is highlighted in Fig. 9.20 (right). The model can be considered as a concentrated load joint model since the total load of each link is concentrated in a single point at the center of gravity of the link. However the Coriolis and inertia effects are taken into consideration so the model can simulate the static to an extent

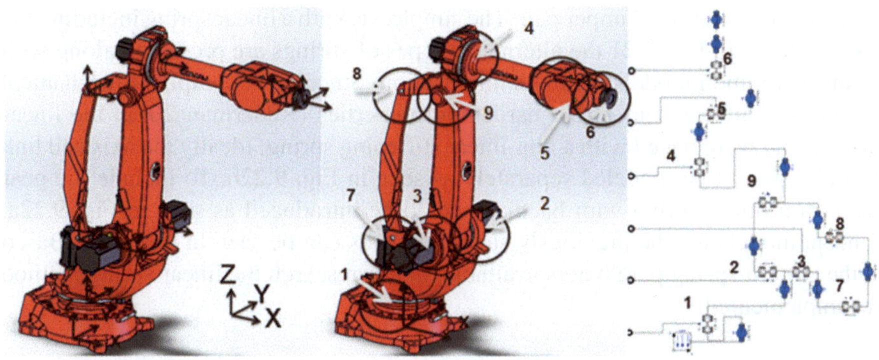

Fig. 9.19 Coordinate systems and joint-link allocation in mechanical structure model

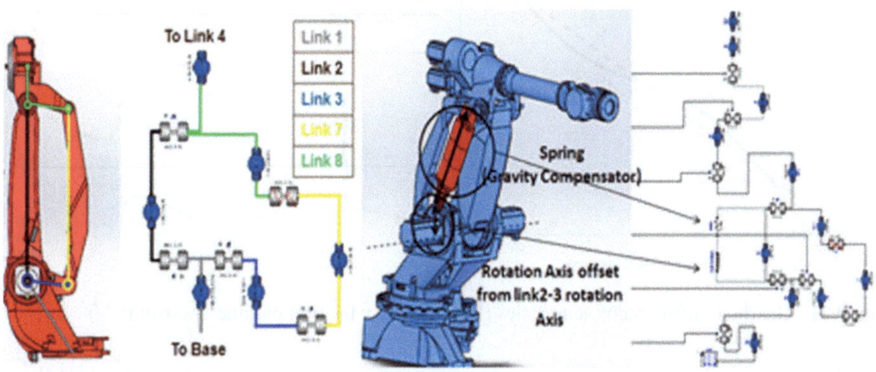

Fig. 9.20 Assembly of the kinematic chain loop model and modeling of NJ370 gravity compensator

the dynamic behavior of the robot. The model contains parameters based on the standard Denavit–Hartenberg parameters but recalculated to fit to the coordinate system used to model the robot. The parameters include the mass and the inertia tensor calculated at the center of mass of the links, the axis of rotation of each joint, the kinematic parameters of the links and the center of mass of each link according to the coordinate system of the model. To include the link deformation while not increasing complexity the flexible joint model was used and the link elasticity was simulated by replacing the axis torsional spring by a torsional spring that includes both link and axis elasticity.

9.3.5 *Gearbox Model*

Joint elasticity

According to the lumped parameter model the elasticity of the gears and the links is represented by a spring damper pair. The simplest is with a linear spring including the link elasticity. In Fig. 9.21 the alternative types of springs are presented along with the most common models for modeling friction. Since most compliant mechanical structures exhibit a kind of the hardening properties (Ruderman 2012) the linear spring could be replaced with a non-linear stiffening spring. Ideally the axis and link elasticity should be modeled separately as seen in Fig. 9.22b. To include the gear backlash a linear spring with backlash could be introduced as seen in Fig. 9.22a. Concept models for the previously stated methods can be seen in Fig. 9.22. Based on the current spring parameters available, in this research the linear spring method was implemented.

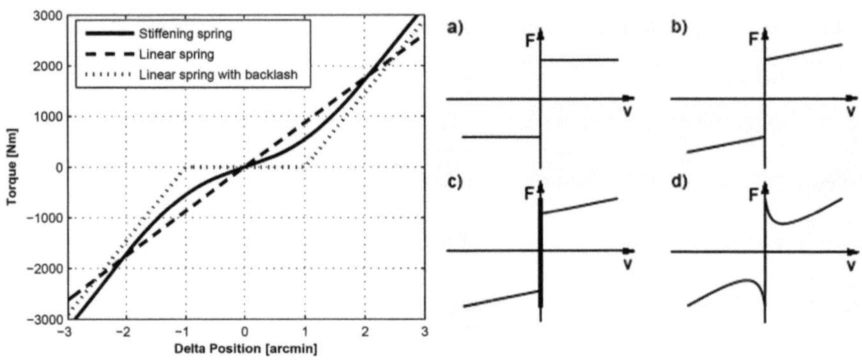

Fig. 9.21 Gearbox stiffness characteristics (left), gearbox friction parameters (right)

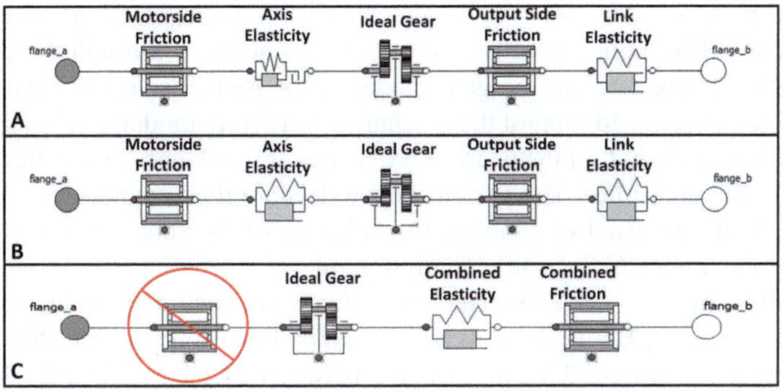

Fig. 9.22 **a** Concept model with backlash spring. **b** Concept model with stiffening spring. **c** The gear mode used in the research

9.3.5.1 Friction

As the axis rotates it is normal that friction forces are developed inside the gearbox. Two major types of friction, namely the Coulomb and Viscous friction are considered. In this work the classical constant Coulomb friction and the linear velocity-dependent viscous friction was used to model the frictional forces generated at the bearing supports and inside the gear mechanism during the coupling of the gears. The Stribeck effect should be involved to describe the nonlinear transition between the break-away friction at zero velocity and linear viscous friction in a more accurate way.

For the presented case study, due to the possibly high rotation speed of the axis, motor friction should include both coulomb and viscous parameters while the viscous friction of output side could be neglected as after the gear reduction rotation speeds are considerably lower. The modeling of the Striebeck Effect is a topic to be investigated in future work. Therefore, the developed models include Coulomb and Viscous parameters (type b from Fig. 9.21). The gear ratio is modeled by an ideal gear with a ratio equal to that of the joint gearbox.

9.4 Full Robot Models

The models described before were combined to create the full robot model of NJ130. However before we proceed with the description of the model, we should make some clarifications regarding the actuator and controller models. The actual robot joints are actuated by motors connected to the gear axis. These motors are neither perfect nor with zero response time. In addition all signals are processed by the robot's controller. In this model the motors and controllers were replaced with ideal positioning actuators with instant response time which certainly lead to reduced accuracy of the results. The input of the simulation is a time-table with the desired robot joints over time. The joint values are calculated by first extracting the desired

Cartesian Space position values and then calculating the inverse kinematics for the specific robot. The input signal is sent to the joint actuator through the controller. The actuator is connected to the gear model of the joint. Each gear model is then connected to the actuated joint of the mechanical structure model. A virtual sensor is placed at the output shaft of the gearbox axis, after any elasticity takes effect sending feedback to the controller. The various submodels are then connected as presented in Fig. 9.23 and are used to simulate the behavior of the robot or to calculate the corrected joint values for a certain configuration.

The following test was conducted to evaluate the accuracy of the cooperative motion of the robots. Both robots were commanded to a single point. We will name the point that R2 moved P2. The distance between TCP1 and TCP2 when using the automatically set common user frame was recorded. We will name the position of R1 as P1. The distance P1-P2 was measured. Based on the coordinate system transformation P2 was expressed according to the coordinate system located at the base of R1. We will refer to this point as P2′. Taking into consideration the compliance of both robots and using the simulation model the inverse kinematics for moving R1 to P2′ were extracted. The motion was completed through a PDL command, including the corrected joint values for R1. We will refer the position acquired through the corrected joint values as P1′. The new distance P1′-P2 was recorded and compared to P1-P2. The tests were conducted using unloaded robots. A total of 22 points were recorded within the common working area of the robot as shown in Fig. 9.24. To evaluate the effect of payload a 60 kg dead weigh was loaded on one robot and additional 8 points were recorded (Fig. 9.25).

In Fig. 9.26 a typical example of the TCPs distance before and after applying the correction is presented. As can be seen in Fig. 9.26 for the 0 kg scenario it can be

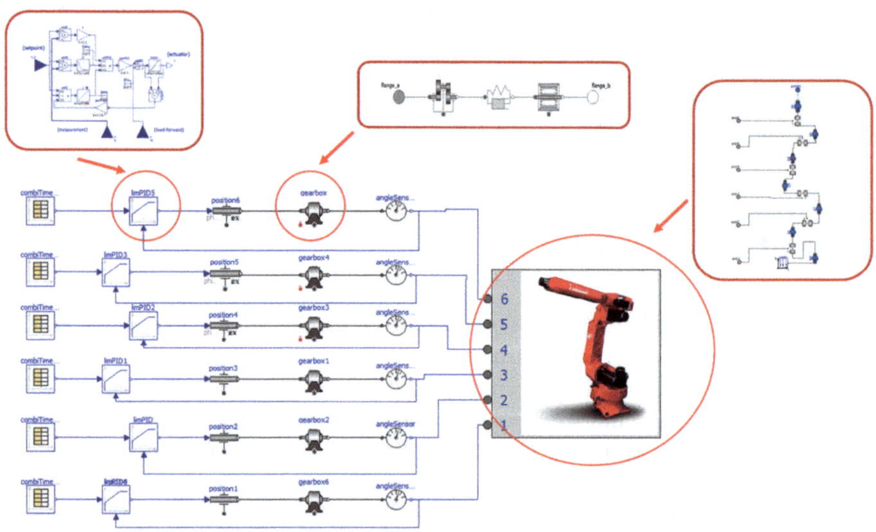

Fig. 9.23 Assembly of "full robot" model

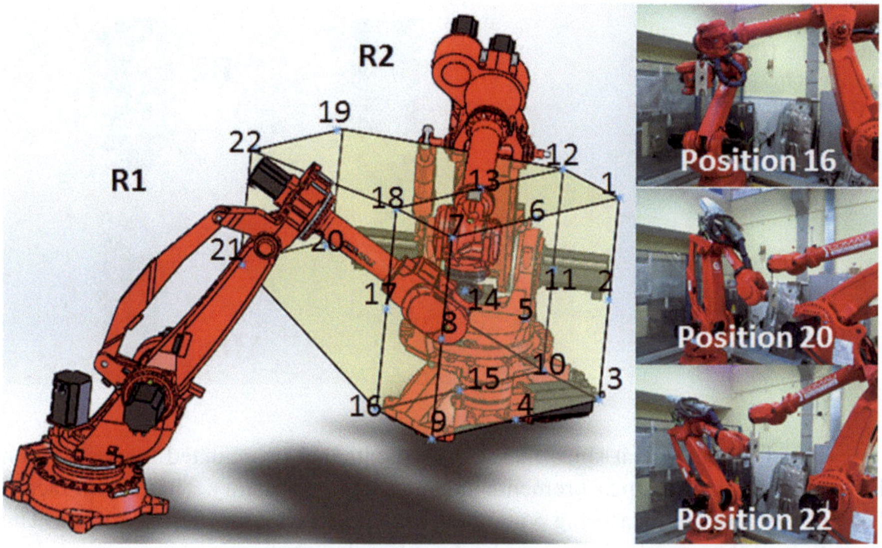

Fig. 9.24 The area of experimental measurements for the 0 kg payload scenario

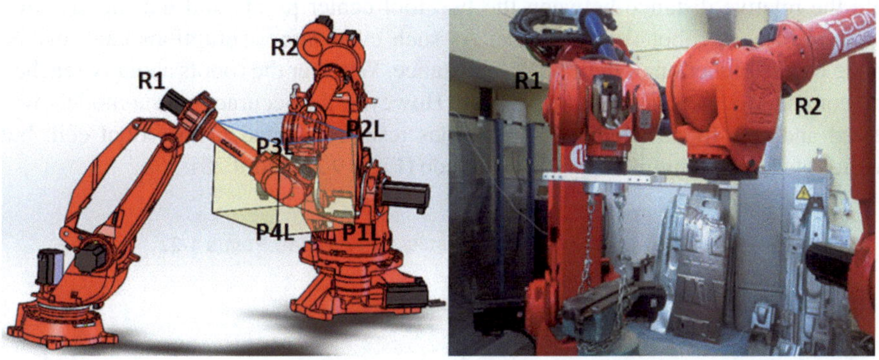

Fig. 9.25 The testing area for the 60 kg payload, picture of the loaded robots at a random position

observed that for most of the points the error was reduced. It should be noted that in a couple of occasions the distance between the tool center points actually increased. However the average relative error was reduced by nearly 39% or from 2.8 to 1.7 mm. The maximum error was reduced from 5.9 mm at point 22 to a maximum of 3.7 mm at point 19, almost 37%.

For the 60 kg scenario while an overall improvement is observed, it is quite small. Even if we do observe an average error reduction of 20% the points tested are quite few to come to a definitive conclusion. Also in this occasion checking the average is misleading since in 3 out of the 8 points tested an increase in the error was recorded.

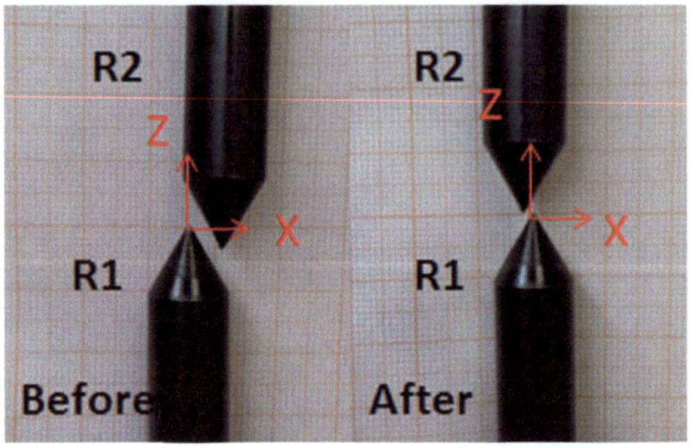

Fig. 9.26 The distance between the TCPs before and after the correction with the proposed method

It should be noted that since the measurements were conducted manually with the use of a caliper the measurements include some uncertainty. The human factor, the calibration of the measuring equipment may have contaminated the results. The magnitude of the uncertainty cannot be directly calculated.

However, even if we include a 10% measurement error the results still indicate improvement of the overall robot motion. Also, the measurements could only calculate the relative distance between the two tool center points and not the absolute position of each robot from its base. As such even safe assumptions can only be made for the reduction of their relative distance. Whether the robots actually reached the commanded point is hard to evaluate. However, the accuracy of the models was tested and was found adequate in previous tests conducted in the robot cell, but without further testing it is hard to conclude (Figs. 9.27 and 9.28).

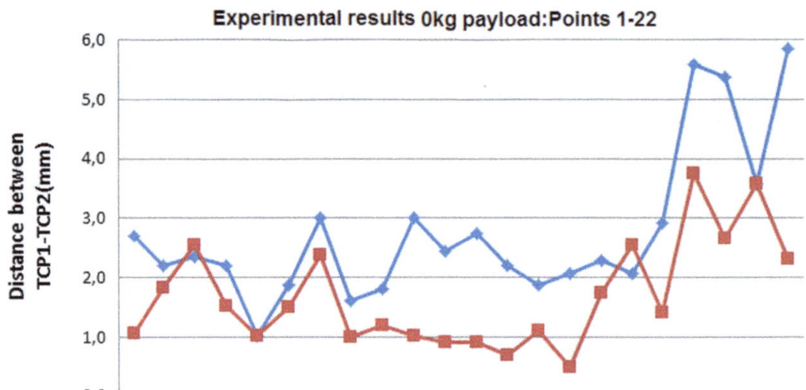

	1	2	3	4	5	6	7	8	9	10	11	12	13	14	15	16	17	18	19	20	21	22
Before correction	2,7	2,2	2,4	2,2	1,0	1,9	3,0	1,6	1,8	3,0	2,4	2,8	2,2	1,9	2,1	2,3	2,1	2,9	5,6	5,4	3,6	5,9
After correction	1,1	1,8	2,5	1,5	1,0	1,5	2,4	1,0	1,2	1,0	0,9	0,9	0,7	1,1	0,5	1,8	2,5	1,4	3,7	2,7	3,6	2,3

Fig. 9.27 Experimental results before and after the correction with 0 kg payload

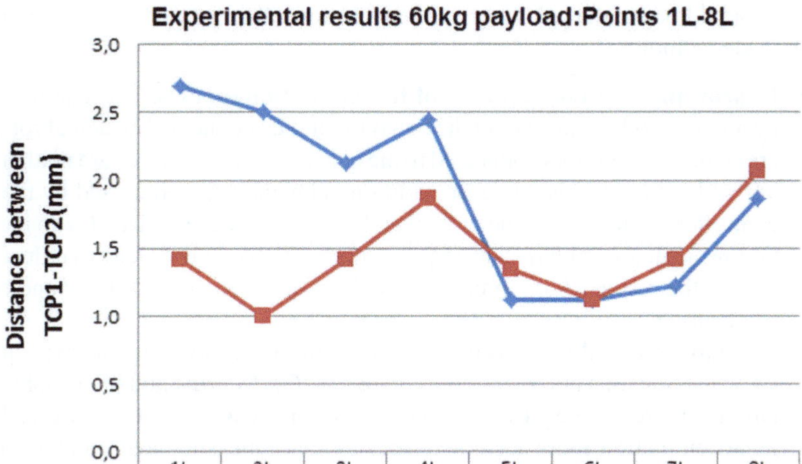

	1L	2L	3L	4L	5L	6L	7L	8L
Before correction	2,7	2,5	2,1	2,4	1,1	1,1	1,2	1,9
After correction	1,4	1,0	1,4	1,9	1,3	1,1	1,4	2,1

Fig. 9.28 Experimental results before and after the correction with 60 kg payload

9.5 Discussion

A method of robot identification based on simulation modelling was presented in this chapter. The path accuracy of the robot was evaluated before and after the estimation of the spring and dumper parameters with the proposed method in order to evaluate the efficiency of this method. The results of one axis scenario indicate an up to 85% reduction of inaccuracy while the full robot scenario indicate an up to 81%. The difference results of one axis model and the full robot model caused to the complexity of the full model in contrast with the simplified one. As future activity, this approach could be improved using more detailed simulation models as far as the elastic behaviour not only in the gear box of each axis but inside the motors of the robot. The estimation method could be also improved using intelligent algorithms to reduce the computation time.

A model based compensation approach was presented and tested for improving the robot cooperation of NJ130 and NJ370. The models were developed using the OMEdit environment and were based on the elastic lumped parameter model. The accuracy of the cooperation task was evaluated before and after the correction with the proposed method. The results for the 0 kg payload scenario the magnitude of the correction is up to several mm thus suggesting that the approach is worth some further testing. However, for the 60 kg scenario are inconclusive. Overall, the results indicate an up to 35% reduction of the distance between the two cooperating robots. The number of tested points does cover the area where most of the tasks are more likely to be completed when robots are cooperating, but further testing should be repeated for more points. However more tests where the same point is reached but with different pose should be conducted. Overall, there are several areas where improvements can be made including:

- **Improvement of the accuracy of the simulation models**: This can be achieved by adding models simulating the behavior of the actuators. The actual robot joints are actuated by motors connected to the gear axis. These motors are neither perfect nor with zero response time. In addition, all signals are processed by the robot's controller. In this model the motors and controllers were replaced with ideal positioning actuators with instant response time which certainly lead to reduced accuracy of the results. Additionally, more degrees of freedom for links and bearings can be used.
- **Use equipment able to record the absolute positioning accuracy**: Equipment such as laser trackers, theodolites, cameras, CMM arms and telescopic armbars have been used in the past for accurate position measurements. Using such equipment could allow more accurate Cartesian space measurement s and would allow us to measure the pose accuracy as well.
- **Repeat the tests to improve the reliability of the results**: To increase the reliability of the results more tests should be conducted increasing the number of test points and altering experiment parameters such as the robot payload and using different poses (alter Euler angles).

References

1. Abele E, Weigold M, Rothenbücher S (2007) Modeling and identification of an industrial robot for machining applications. CIRP Ann 56:387–390. https://doi.org/10.1016/j.cirp.2007.05.090
2. Aivaliotis P, Georgoulias K, Chryssolouris G (2017) A RUL calculation approach based on physical-based simulation models for predictive maintenance. 2017 international conference on engineering, technology and innovation (ICE/ITMC). IEEE, Funchal, pp 1243–1246
3. Barati M, Khoogar R, Nasirian M (2013) Estimation and calibration of robot link parameters. IEEE Int Conf Rob Autom 225–234
4. Caenen JL, Angue JC (1990) Identification of geometric and nongeometric parameters of robots. In: Proceedings., IEEE international conference on robotics and automation. IEEE Comput. Soc. Press, Cincinnati, OH, USA, pp 1032–1037
5. Chryssolouris G (2006) Manufacturing systems: theory and practice, 2nd edn. Springer, New York
6. Du Z, Iravani P, Sahinkaya MN (2014) A new approach to design optimal excitation trajectories for parameter estimation of robot dynamics. 2014 UKACC international conference on control (CONTROL). IEEE, Loughborough, UK, pp 389–394
7. Dwiputra R, Zakharov A, Chakirov R, Prassler E (2014) Modelica model for the youbot manipulator. pp 1205–1212
8. Edwards M (1984) Robots in industry: an overview. Appl Ergon 15:45–53. https://doi.org/10.1016/S0003-6870(84)90121-2
9. Elatta AY, Gen LP, Zhi FL, Daoyuan Y, Fei L (2004) An overview of robot calibration. Inf Technol J 3:74–78. https://doi.org/10.3923/itj.2004.74
10. Elmqvist H, Mattsson SE, Otter M (1999) Modelica-a language for physical system modeling, visualization and interaction. In: Proceedings of the 1999 IEEE international symposium on

computer aided control system design (Cat. No. 99TH8404). IEEE, Kohala Coast, HI, USA, pp 630–639
11. Joubair A, Bonev IA (2015) Non-kinematic calibration of a six-axis serial robot using planar constraints. Precis Eng 40:325–333. https://doi.org/10.1016/j.precisioneng.2014.12.002
12. Le QV, Ng AY (2009) Joint calibration of multiple sensors. 2009 IEEE/RSJ international conference on intelligent robots and systems. IEEE, St. Louis, MO, USA, pp 3651–3658
13. Loh CC, Traechtler A (2012) Cooperative transportation of aload using nonholonomic mobile robots. Procedia Eng 41:860–866. https://doi.org/10.1016/j.proeng.2012.07.255
14. Marie S, Courteille E, Maurine P (2013) Elasto-geometrical modeling and calibration of robot manipulators: application to machining and forming applications. Mech Mach Theory 69:13–43. https://doi.org/10.1016/j.mechmachtheory.2013.05.003
15. Michalos G, Makris S, Spiliotopoulos J, Misios I, Tsarouchi P, Chryssolouris G (2014) ROBO-PARTNER: seamless human-robot cooperation for intelligent, flexible and safe operations in the assembly factories of the future. Procedia CIRP 23:71–76. https://doi.org/10.1016/j.procir.2014.10.079
16. Mooring B, Pack J (1987) Aspects of robo repeatability. Robotica. 223–230
17. Nubiola A, Slamani M, Joubair A, Bonev IA (2014) Comparison of two calibration methods for a small industrial robot based on an optical CMM and a laser tracker. Robotica 32:447–466. https://doi.org/10.1017/S0263574713000714
18. Olabi A, Damak M, Bearee R, Gibaru O, Leleu S (2012) Improving the accuracy of industrial robots by offline compensation of joints errors. 2012 IEEE international conference on industrial technology. IEEE, Athens, pp 492–497
19. Roth Z, Mooring B, Ravani B (1987) An overview of robot calibration. IEEE J Rob Autom 3:377–385. https://doi.org/10.1109/JRA.1987.1087124
20. Santolaria J, Ginés M (2013) Uncertainty estimation in robot kinematic calibration. Rob Comput-Integr Manuf 29:370–384. https://doi.org/10.1016/j.rcim.2012.09.007
21. Santolaria J, Conte J, Pueo M, Javierre C (2014) ROTATION error modeling and identification for robot kinematic calibration by circle point method. Metrol Meas Syst 21:85–98. https://doi.org/10.2478/mms-2014-0009
22. Tsarouchi P, Spiliotopoulos J, Michalos G, Koukas S, Athanasatos A, Makris S, Chryssolouris G (2016) A decision making framework for human robot collaborative workplace generation. Procedia CIRP 44:228–232. https://doi.org/10.1016/j.procir.2016.02.103
23. Tsumugiwa T, Yokogawa R, Hara K (2001) Variable impedance control with regard to working process for man-machine cooperation-work system. In: Proceedings 2001 IEEE/RSJ international conference on intelligent robots and systems. Expanding the societal role of robotics in the the next millennium (Cat. No.01CH37180). IEEE, Maui, HI, USA, pp 1564–1569
24. Veryha Y, Kurek J (2003) No title found. J Intell Rob Syst 36:315–329. https://doi.org/10.1023/A:1023048802627
25. Xiao-Lin Z, Lewis JM (1995) A new method for autonomous robot calibration. In: Proceedings of 1995 IEEE international conference on robotics and automation. IEEE, Nagoya, Japan, pp 1790–1795
26. Zakharov A, Halasz S (1999) Genetic algorithms based identification method for a robot arm. In: ISIE '99. Proceedings of the IEEE international symposium on industrial electronics (Cat. No. 99TH8465). IEEE, Bled, Slovenia, pp 1014–1019

Chapter 10
Vision Guided Robots. Calibration and Motion Correction

10.1 Introduction

This chapter discusses on the use of vision-based systems for guiding robot arms in performing accurately assembly operations. Such systems are needed when there are errors in positioning parts to be processed by robots as well as tolerances and errors of robot arms themselves. In this case, a vision system is used to calculate the position of an object and guide the robot to the calculated position. The discussion on these aspects will be facilitated with the help of a real-world examples on welding parts of a vehicle body.

The need for using vision guided robot systems arises from the technical aspects of the welding operation that is used as well as the product characteristics which dictate the use of the specific joining technology. An indicative example is the case of the assembly of metal parts, such as the vehicle door components or the vehicle body sides. In this case, flanges (thin strips of metal) are created during the stamping process in order to provide an area where the welding of two parts can take place. As far as the product is concerned, flanges are undesirable since they signify the alteration of the product design and add more weight. However, the welding of the components may not be otherwise feasible since the contact area between the parts may be limited. For this purpose, the flanges of the parts to be welded, are designed to be adjacent and parallel to each other allowing accessibility to the welding equipment. The flanges in the CAD model of a door and the actual frame are shown in Fig. 10.1.

The width of a flange may vary depending on the part and the selected joining technology, with typical dimensions ranging from 5 to 10 mm. The thickness of the welding gun, the laser beam etc. are the main parameters that define the flange area dimensions. Due to a number of errors that can be traced in the process of stamping car door elements, the flanges' real positions do not always match their theoretical ones. Since the precision of the process is not always guaranteed, the flanges are designed wider than they need to be, in order to ensure that the welding spots/seams are always placed within the flange area and sufficiently away from the flange edge.

S. Makris, *Cooperating Robots for Flexible Manufacturing*,
Springer Series in Advanced Manufacturing,
https://doi.org/10.1007/978-3-030-51591-1_10

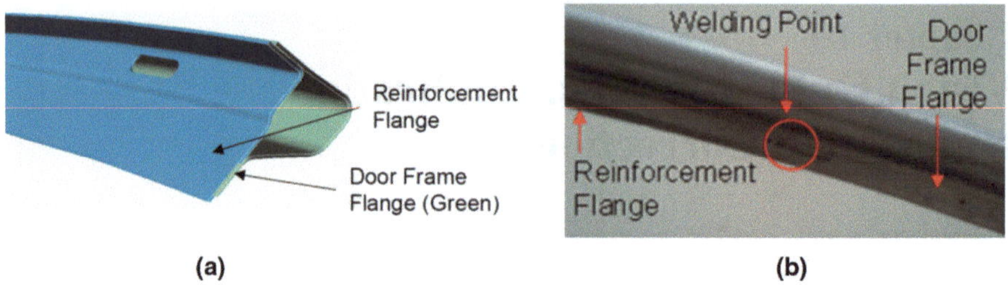

Fig. 10.1 Door frame and reinforcement—CAD and physical parts

For this purpose, vision systems can be used in order to identify the flange's position and edges and provide a feedback to the robot controller indicating the necessary offset to ensure that the welding spot will be performed inside the flange area. However simple such an application may sound, there are several limitations and challenges that are introduced by the assembly environment [6]. The challenges considered in this chapter are going to be explained with the help of Fig. 10.2, which shows a possible arrangement of an industrial robot and a part to be welded at position Y. In this case, the assumption is that the exact position of parts is not well known beforehand due to errors in positioning parts or parts geometrical errors. In the general case, it is assumed that the precise location of the physical parts is calculated with the help of an external measurement system, for example an external camera. Therefore, the target position for the robot to reach is position Y.

An industrial robot arm is not an ideal rigid body and due to its structural elasticity and tolerances [1] the actual motion of the tool center point namely point A, which is located at the tip of the welding gun used for welding, is position X. Therefore, there is a need for a method to compensate for the introduced inaccuracy. For this reason, the vision system can be used for compensation comprising of a camera and a coordinates reference system attached to its sensor. A difficulty to overcome is the

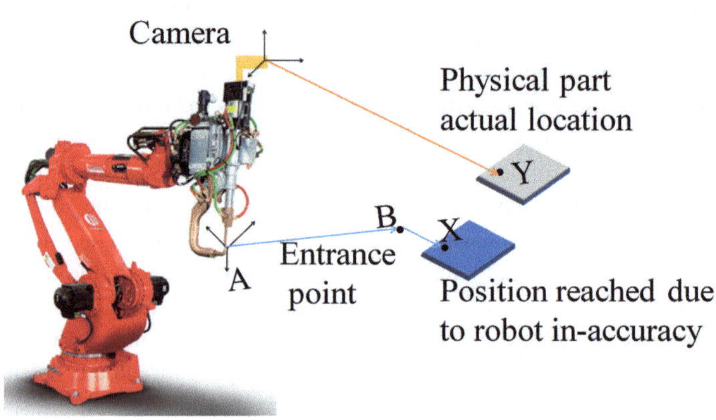

Fig. 10.2 Concept of correcting robot motion

calculation of the robot base with respect to the camera's frame of reference. This task which is called hand-eye-calibration and a method for performing it is discussed in this chapter. In order for the robot and camera to work together the information from the camera need to be transformed to the robot base coordinate system. After that, the robot controller can use this information to modify its program so that it better fits the actual environment conditions.

Considering the errors in locating the actual parts to be welded as well as the structural errors of the robot arm, this chapter discusses a method to guide industrial robot arms to be more precise in their operation.

10.2 Calculating 3D Coordinates Using Stereo Vision

The method is based on the pinhole camera model as shown in Fig. 10.3. In this figure, C is called the camera centre and is placed in the coordinate origin, while *p* is the principal point of the camera. The points in space are projected onto a plane and the centre of projection (p) is located at distance z, being denoted as f. This plane is called image or focal plane. The point in space $X = (X, Y, Z)^T$ is mapped onto point x of the image plane.

By using the principles of similar triangles, it is calculated that the 3D space point X is converted into the 2D space point x while its coordinates on the image plane are: f*X/Z and f*Y/Z. Based on this principle, it is possible to map 3D points on 2D points using the following generic transform [3].

$$x = \mathrm{P} * \mathrm{X} \tag{10.1}$$

x is the point on the image plane and X is the point in the 3D space. The matrix P, is called the camera projection matrix. The content of P is as follows:

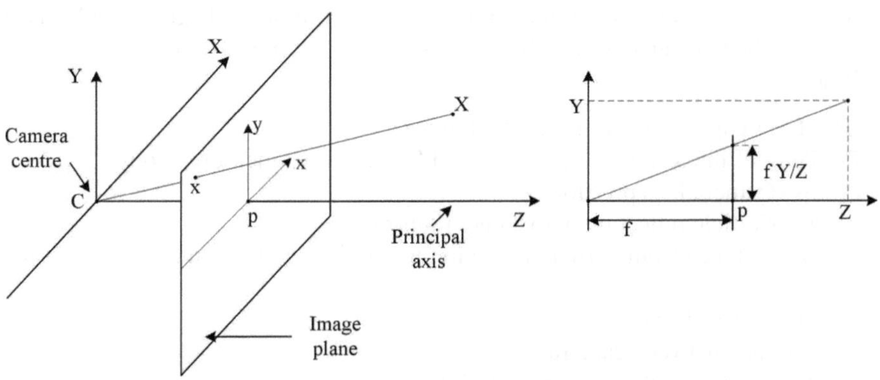

Fig. 10.3 Pinhole camera geometry

$$P = \begin{bmatrix} f_x & 0 & c_x \\ 0 & f_y & c_y \\ 0 & 0 & 1 \end{bmatrix} \tag{10.2}$$

f_x, f_y, is the focal length of the camera taking into consideration the structure of the camera pixels that may be non rectangle, while c_x, c_y, are the principal point coordinates with respect to the coordinate frame, which in a general case, may be in another location. In a generic case, the points in 3D space are expressed in terms of an individual Euclidean coordinate frame, which is called world coordinate frame. The camera's frame is related to the world coordinate frame in terms of a translation and a rotation. The translation is expressed by the translation matrix T, while the rotation by the rotation matrix R [2,3].

Due to distortions of the camera lens, which are caused by a mechanical misalignment in the process of building the lens and by the fact that the cameras use spherical lenses, a set of distortions is imposed on the image. Radial distortions arise as a result of the shape of a lens, whereas tangential distortions arise from the assembly process of the camera as a whole [2–5,11].

The process of calculating the camera matrix P, as well as the camera lens distortions, is called *camera calibration.* In the case that two cameras are used in a stereo rig, then the calibration process involves the calculation of the rotation and translation matrices that describe the relevant position of the two cameras [2,4,11]. Based on the outcome of this process, it is possible to correct the raw image taken by the camera and produce the undistorted image.

10.2.1 Stereo Triangulation Principle

Having performed the calibration process, it is possible for the position of a point to be calculated, having two images of the object in the 3D space. This is carried out with the use of the stereo triangulation method. The following diagram can be used for relating the two cameras and the object in their view (Fig. 10.4):

Where:

- IP—The image plane of the rectified images
- P—Point of interest which is also called Particle, on a measured body
- c_1, c_2—Camera lens position
- p_1, p_2—Particle reflection onto camera sensor
- v_1, v_2—distance from particle reflection to the center of the image plane on the sensor
- $v_2 - v_1 = \mathrm{d} =$ disparity
- b—Distance between cameras
- D—Distance from the camera lens ENP to the particle
- f—Distance from camera lens to camera sensor, focal length

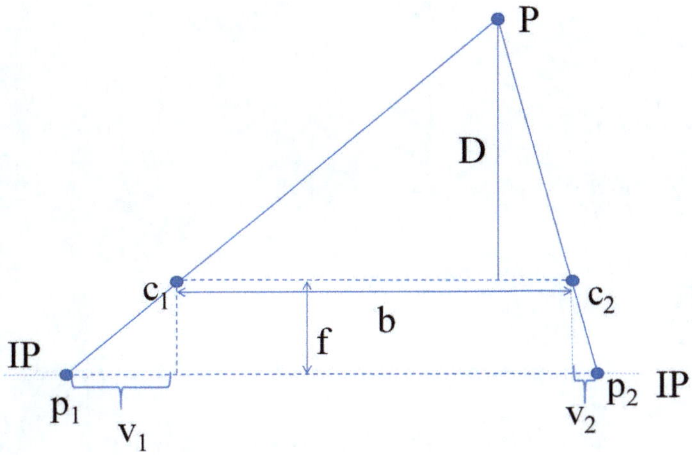

Fig. 10.4 Stereo vision triangulation

Using the relationship between similar triangles, it is possible to calculate the distance D.

$$D = \frac{b \times f}{d} \tag{10.3}$$

The camera calibration process calculates a correspondence between camera pixels and physical world dimensions. Then, the algorithm applies the triangulation calculation and the distance is calculated from the camera to the point of interest on the door. This is actually the distance from the origin of the camera coordinate system found in the projection centre with respect to the object's coordinate system.

The projection centre can also be referred to as the ENP, or entrance pupil position [9]. The position of the ENP is taken from the manufacturer's specifications and is used for transforming the distance, calculated by the triangulation method, into the distance from any other point that may be required.

10.2.2 The Correspondence Problem

In order to perform the triangulation process described in the previous paragraph, it is required that the coordinates namely x pixel location and y pixel location of the point of interest in both images be known. The identification of the point of interest in both images is known as the correspondence problem [7,8]. In this approach, the problem is addressed with the use of structured light with the help of laser projected lines, which is used to pinpointing the location of these points and making them distinguishable in the acquired images. The system is shown in Fig. 10.5.

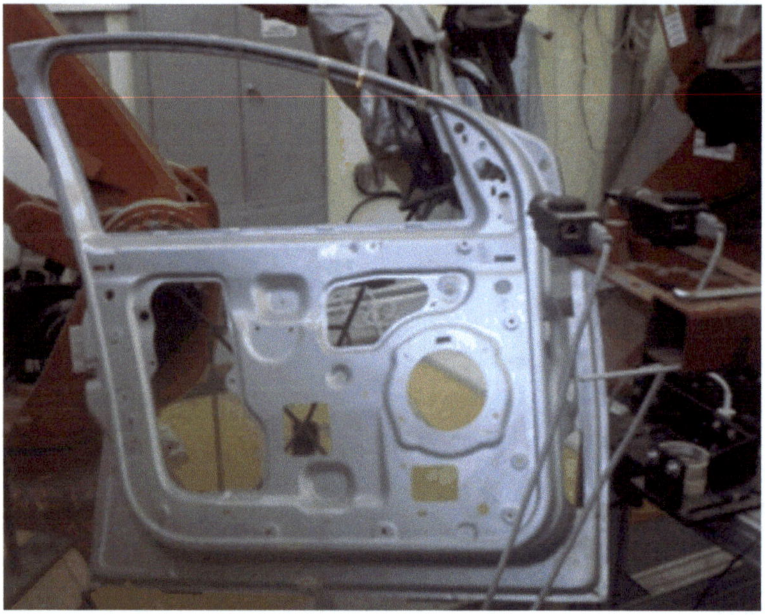

Fig. 10.5 Cameras and door experimental setup

10.2.3 Physical Setup for Stereo Triangulation

Structured light in the form of a laser diode line is projected on the area of the flange, where the welding is supposed to take place. Following a pair of cameras, whose absolute and relative position in space is already known based on the calibration process already discussed, is used to acquiring two images of the projected line. The laser line serves as a reference point so that the matching of the pixels, illuminated by the laser within the two images, can be possible. Additionally, the distortion/cutting of the laser line is used to determining the edge of the flange so that the welding offset can be calculated in case the robot is programmed to weld on a point being very close to this edge. The acquired images are then processed in the following steps [2]:

- Each image is undistorted according to the camera's intrinsic data that were calculated in the previous step of the camera's calibration.
- The images pair is rectified. This step actually corrects the errors of the camera's image planes that are not completely parallel. The result is a set of two computed images found in the same image plane.
- Finally, triangulation is used for calculating the three-dimensional coordinates of the new point. This calculated point should be used for providing the robot with a feedback in case that the pre-programmed position needs to be shifted in order for a good welding quality to be obtained.

Figure 10.5, shows an example for the physical setup of such a vision system. In particular, the following is shown:

- The four lines are projected on the door by laser diodes. Their projection position on the door is right where their coordinates in space have to be calculated. Four lines are projected on the door exactly as are the points that have to be calculated.
- The two cameras are mounted on a base and they are set at a relevant distance for experimentation. The distance between the cameras is referred to as the baseline.
- A laser-based measuring device is also included for measuring and comparing the calculated coordinates with the actual ones.
- One personal computer running the following software modules.

- NI Vision Acquisition Software. This is used for capturing the images from the cameras [10].
- Matlab calibration toolbox [5]. This is used for the calibration of each camera. This calibration step calculates cameras distortion coefficients as well as the relation between the two cameras of the stereo rig.
- OpenCV library and OpenCV implementation of the images un-distortion and rectification and finally, for implementing the triangulation method [2].

The specific setup is representative of similar methods employed in industry as well as research for building a vision system, which will allow the implementation of different algorithms and numerous technologies in the same setup. The implementation with the use of both the Matlab calibration toolbox [5] and the OpenCV framework [2] was enabled because of this PC based computing setup (Fig. 10.6).

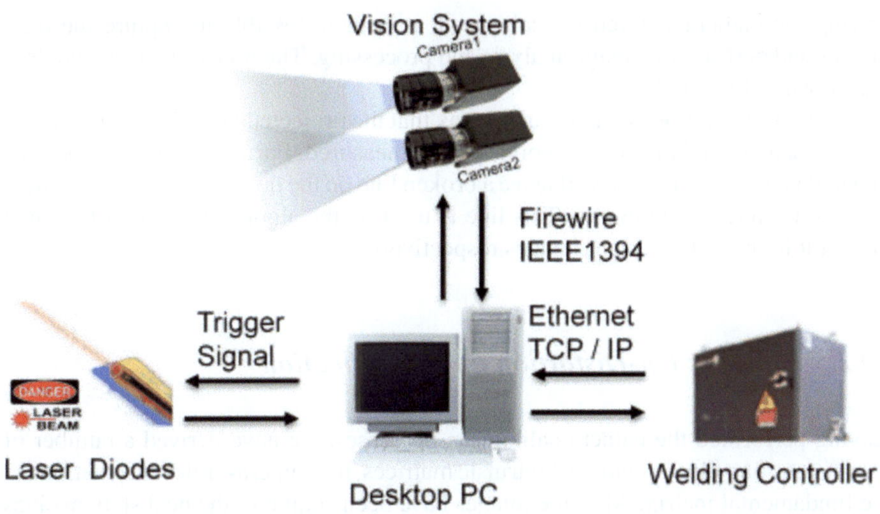

Fig. 10.6 Vision system deployment

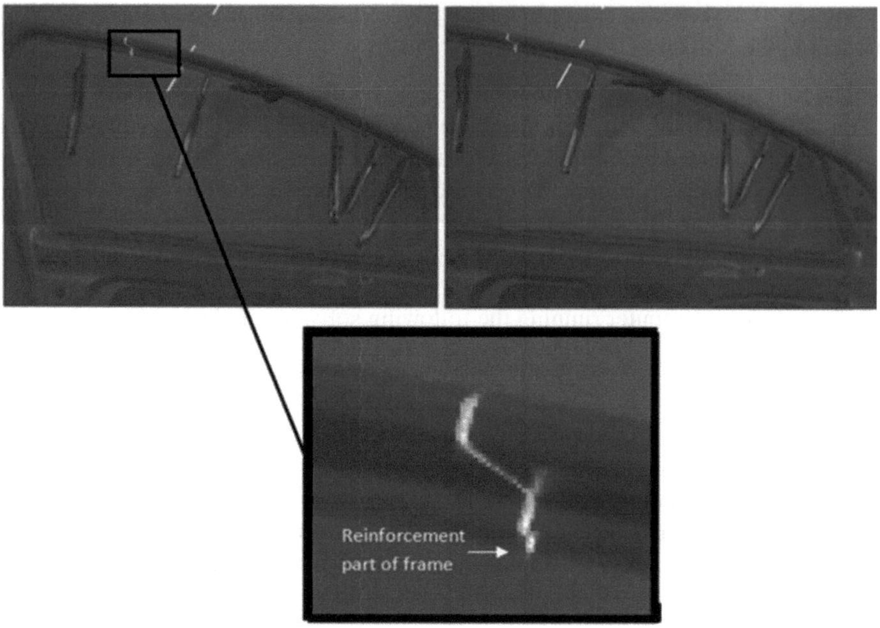

Fig. 10.7 Raw images captured by the vision system

10.2.4 Images Capturing

Having the camera's stereo rig calibrated, it is then possible to capture the door images and perform the image analysis and processing. The images that are acquired are shown in Fig. 10.7.

The two cameras are setup in such a way that the projected laser diodes are visible by each camera and the desired points can be measured. Furthermore, it can be seen in detail that the laser diode will leave a broken line on the door reinforcement, where there is a change of geometry. This line is used by the algorithm in order to match the point in the left and right images respectively.

10.2.5 Images Un-Distortion and Rectification

Having performed the camera calibration processes, we have derived a number of parameters, namely the cameras' intrinsic matrices, the cameras' relevant position via the fundamental matrix. After the images have been acquired, the next step involves the process of their correction so-called *rectification*.

The image set is first rectified, meaning that each pixel of the images is mapped to a new position in the rectified image [2,3]. As a result, the images from the two cameras

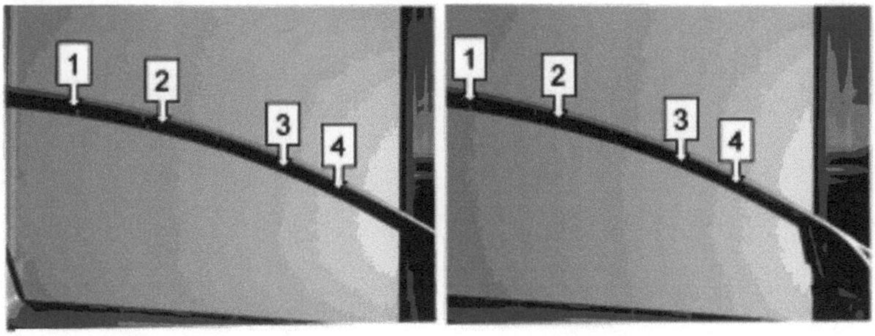

(a) Left camera image (b) Right camera image

Fig. 10.8 Rectified image set

are transformed to represent those taken by cameras with image sensors, perfectly parallel and aligned. The points become horizontally aligned and undistorted. The rectified images are shown in Fig. 10.8. At this point, each pixel that is visible in both images, is located in the same row as are both the left and the right-hand side images.

The projected lines from the laser diodes are still visible on both images and these are used for the identification of the common matches in both images. Towards this direction, the images are now converted into a simpler format, applying a threshold filter. The threshold filter actually eliminates all the pixels that have a lower value than the one set by the filter, while the remaining pixels are the ones that remain on the images. The principle is that each pixel has a value between 0 and 255, with 0 corresponding to the black colour and 255 to the white colour. Pixels values between 0 and 255 correspond to variations of grey. Due to the fact that lighting conditions may vary, it is possible that a pixel have a value i.e. 200 under given lighting conditions, it may have a value of 230 or 150 under different conditions. Therefore, in a general case, it is hard to pre-set the threshold value. In the case of the current research, an adaptive threshold technique in which the threshold level is a variable has been adopted. This technique is useful when there are strong illumination or reflectance gradients and a threshold relative to the general intensity gradient is required [2]. The results of the adaptive threshold method are shown in Fig. 10.9.

In the new images, the points of interest that occur from the broken laser diodes lines on the door reinforcement are still visible, as shown by the corresponding arrows for each of the points. Other edges are also shown on the images, which the developed algorithm has to neglect and identify only the points that are shown by the arrows. The process of identifying the proper pixels and their correspondence, in both images, is discussed next.

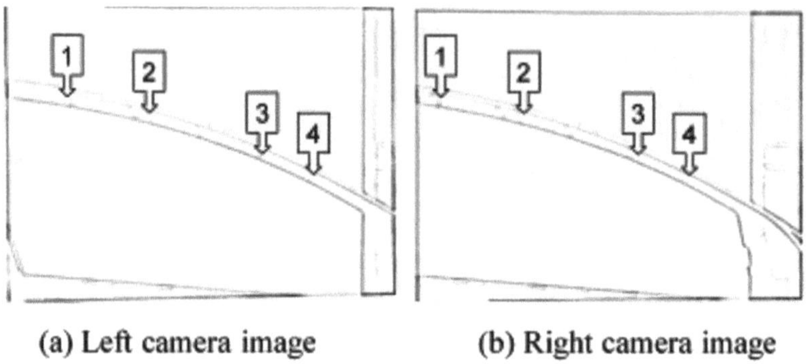

Fig. 10.9 Image set after rectification and adaptive threshold filter

10.2.6 Image Features Correspondence

Among a variety of methods employed for identifying corresponding points in two images [3], a rather simple approach is being discussed herein. The method helps in identifying the proper pixels and their correspondence in both images. It is based on the concept of the laser diode line that breaks on the changes of geometry and actually marks the area of interest that would help in tracing the correct pixel. The broken line was shown in Fig. 10.10.

After the process of rectifying and applying the threshold filter to the images, the broken line for point 1 is now shown in for the left and right camera view.

The images shown in Fig. 10.10, are rather unique for each point, because the laser lines will be projected on the door under a unique angle for each point. Therefore, the line break for each point and camera view will be unique. Based on that, the broken line for each point and camera view is used as a matching template by the algorithm that this research has developed, and based on this template, the pixel value, where the broken line is located on the rectified and threshold image, is calculated. The smaller the image used as a matching template, the better the accuracy that can be obtained in calculating the location of the pixels of interest. In this research, the templates were set to be in a size of 6×6 pixels. This step results in a list of pixels values for each one of the points of interest in each one of the images, thus a list of four points with the z coordinate is obtained for the left image and a similar list is

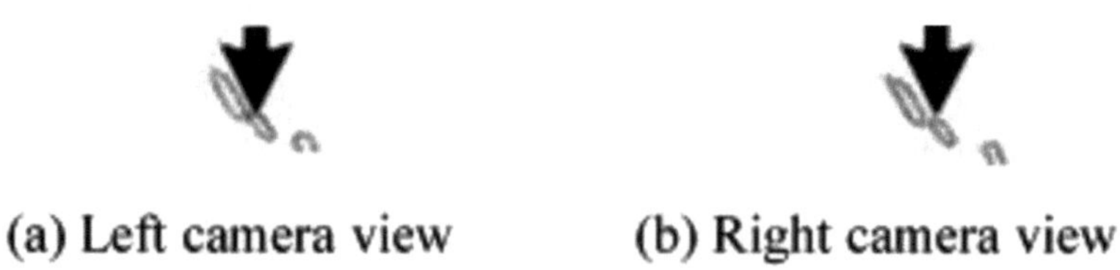

Fig. 10.10 Line breaks in the change of geometry-rectified and threshold image for left and right camera view for point 1

calculated for the right one, together with the correspondence of the points of the left image to the right image pixels.

Having the correspondence calculated, the next and final step is to simply perform the calculation of distance for each one of the four points from the cameras. This is achieved by performing the stereo triangulation-based calculation.

10.3 Calibration of Camera and Robot Base Frames

The concept for calculating the relation of the camera and robot base coordinate frames is based on the approach shown in Fig. 10.11. A global frame is used for the entire robot cell in order to reference any cartesian point or coordinate system in the 3D space, including the welding points. To achieve this, the identification of the transformation matrices between the different coordinate frames is needed.

By using a 3D vision system it is possible to calculate the coordinates of a point in relation to the camera reference system (CF). The Robot Base Frame (RB) is used for referencing all the movements of the robot. The first step is to transform the CF based coordinates with respect to the RB. Let's assume that coordinates of a point, P with respect to CF, are $P_c = [X_c\ Y_c\ Z_c]$ and $P_r = [X_r\ Y_r\ Z_r]$ with respect to RBF. These vectors are correlated with the function f, as follows:

$$P_r = f(P_c) \tag{10.4}$$

$$P_c = f - 1(P_r) \tag{10.5}$$

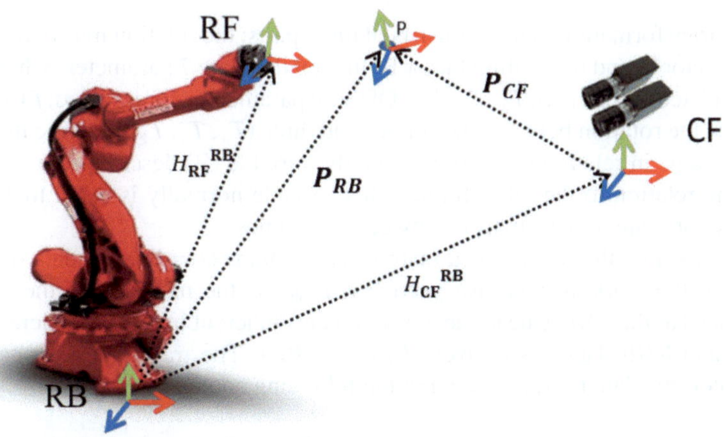

Fig. 10.11 Vision guided robot path correction

The definition of the transformation function f will be calculated with the calibration procedure that is described in the following sections and is the main research contribution of this work. In order to find a correlation matrix between the two corresponding frames, the basic idea is to measure and calculate the same data with respect to both frames, namely the frame of the robot base-RB and the frame of the camera-CF and then to calculate the transformation function.

Due to the fact that robot models are not accurate and well defined, it is not possible to define linear transformations in the entire working area of the robot, when vision systems are applied. This problem is addressed by dividing the robot workspace into narrow working areas, where the transformation from CF to RB can be considered linear with a high accuracy. In this case, it is possible to define a linear transformation between the camera and robot base coordinate frame.

The proposed method uses two cameras, in a stereo rig setup [3]. Assuming that there is a cartesian position in this working area, that is visible from both cameras, it is possible to calculate coordinates of this position in the CF using stereo triangulation. If the coordinates of the same object are available in relation to RB, the following equation describes the relation of two vectors [2,3].

$$\begin{bmatrix} X \\ Y \\ Z \\ w \end{bmatrix}_c = \begin{bmatrix} \cos(th)*\cos(ps) & \cos(f)*\sin(ps)+\sin(f)*\sin(th)*\cos(ps) & \sin(f)*\sin(ps)-\cos(f)*\sin(th)*\cos(ps) & T_X \\ -\cos(th)*\sin(ps) & \cos(f)*\cos(ps)-\sin(f)*\sin(th)*\sin(ps) & \sin(f)*\cos(ps)+\cos(f)*\sin(th)*\sin(ps) & T_Y \\ \sin(th) & -\sin(f)*\cos(th) & \cos(f)*\cos(th) & T_Z \\ 0 & 0 & 0 & B \end{bmatrix}$$

$$\begin{bmatrix} X \\ Y \\ Z \\ w \end{bmatrix}_c = \begin{bmatrix} R & T \\ 0 & B \end{bmatrix} * \begin{bmatrix} X \\ Y \\ Z \\ w \end{bmatrix}_r \tag{10.6}$$

The transformation matrix consists of three parts; the rotation matrix R, the translation vector T, and the scaling factor B. In total there are 7 parameters which need to be calculated $(th, ps, f, B, T_x, T_y, T_z)$. Of these parameters, three (th, ps, f) are used to describe the rotation between the frames and three (T_x, T_y, T_z) describe the position of one frame in relation to the other. Finally, variable (B) describes the scale of one frame in relation to the other frame. This variable normally is equal to 1, because there is not scale transformation between the frames.

Let's assume that P_{camera} is a matrix that contains coordinates' data of different points in the working area. Respectively, P_{robot} is the matrix with the same data referenced to the RBF. Due to the measurements inherent inaccuracy, there are errors to CF and RBF data, respectively $P_{E,C}$ and $P_{E,R}$. The transformation matrix that correlates this data is H, as shown in the following equation.

$$P_{\text{camera}} + P_{E,C} = H * \left[P_{\text{robot}} + P_{E,R} \right] \tag{10.7}$$

Using the measured matrices P_{camera} and P_{robot}, the optimization-calibration problem is to find the H matrix which minimizes the following error

$$E = P_{\text{camera}} - H * P_{\text{robot}} \tag{10.8}$$

10.3.1 Identification of Parameters

According to the previous paragraph, in order to find the matrix H, which describes the transformation between the two frames, it is needed to calculate the following matrices: P_{camera} and P_{robot}.

For these matrices, the basic idea is to attach a special marker on the robot flange and move the marker around the workspace, observing it through the vision system, and at the same time store the data. For this kind of problem, as a special marker a corner on a chessboard was used. The principle is shown in Fig. 10.12. The marker is attached on a known position on the robot, which is widely referred to as the Tool center point. The position and orientation of the marker frame is known and constant in relation to the robot flange H_{MF}^{RF}. Through the encoder of the robot axes combined with the robot kinematic model, it is possible to calculate the position of the marker frame in relation to the robot base frame. This is the one of the two measurements that is needed.

The other one is the position of the marker in relation to the camera coordinate frame. This is calculated using the stereo triangulation principle [6]. This algorithm has two basic steps. The first is to recognize the marker in both images from the left and the right camera. The second step is to use this information to calculate coordinates of marker in relation in the camera frame. This is the second vector that we need for each of the robot poses.

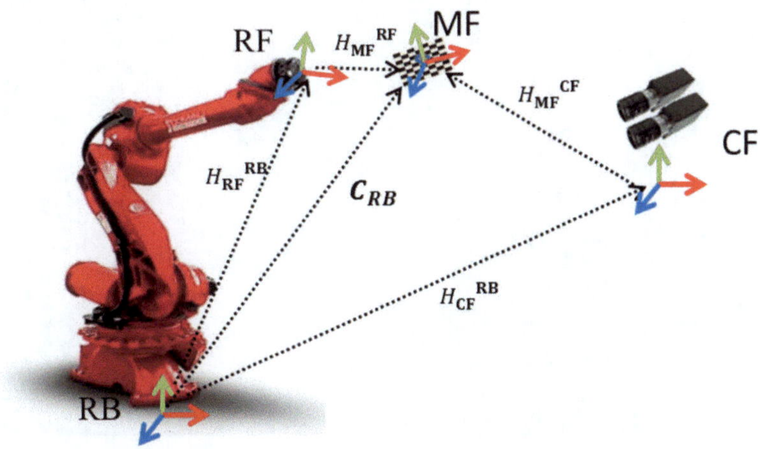

Fig. 10.12 Coordinate frames used for calibration of the system

In the following paragraph the basic steps for the procedure are presented. The setup of the vision system is the first step. In this step, a chessboard is placed and fixed at the robot flange as shown in Fig. 10.13. Using the robot controller built in functions it is possible to calculate the relative position of one of the chessboard corners in relation to the robot flange. Therefore, the robot controller can calculate coordinates of this corner in relation to the robot base frame. At the next step, the human operator defines the limits of a working space for the robot. This working space is a cubical space in which the robot can move without collisions. Following, the installation of the cameras in the robot cell is also carried out in this step. Extra attention is required to ensure that that the whole working area is visible through the cameras. Finally, 20 different positions of the robot are manually defined and stored in the controller of the robot. The concept of this setup of the system is presented in Fig. 10.12.

The calibration of the camera system is the next step of the procedure. An algorithm sends a message to the robot to move to the first calibration position and after the movement is complete a confirmation response from the robot controller is acquired to allow the process to further continue. Following, the camera system is triggered, and a pair of images is acquired and stored on the PC. This step is repeated for all the 20 different positions of the chessboard. A 3D calibration algorithm runs and calculates the matrix that describes the vision system, using the above data.

The final step of the procedure is to calculate the transformation between the measured and calculated data (CF and RB respectively). For this purpose, an algorithm starts robot motion in order to scan the pre-defined working area of the robot.

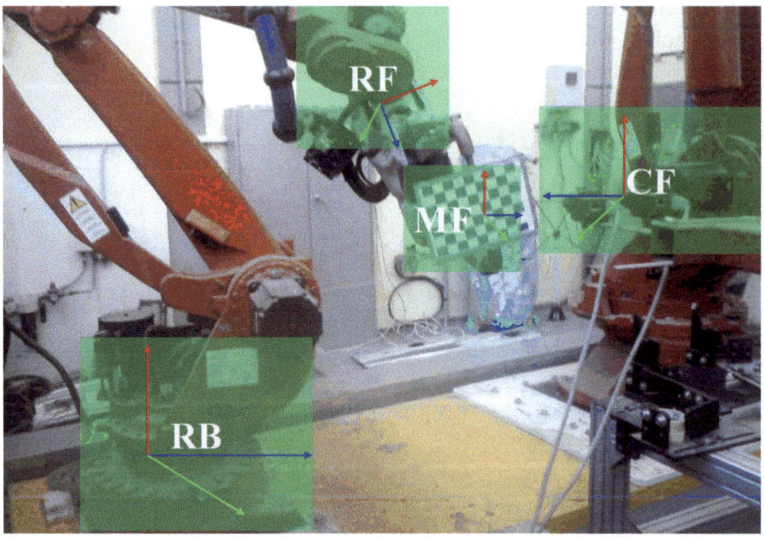

Fig. 10.13 Physical set-up for robot-camera calibration

The stereo vision system observes the chessboard corner and records the calculated positions. When the scan of the area is completed, the algorithm calculates the transformation between these matrixes.

10.3.2 Physical Setup for Calibrating Camera Frame and Robot Base Frame

For the identification of the method that was described in the previous paragraph, experiments were done in an industrial cell. In the following figure the setup of the cell is displayed including the main components of the such as the cameras, robot and the chessboard.

In the experiments, the working area was a cube of 150 mm size. The two cameras are placed so they both can observe the chessboard, the distance between them is about 100 mm. The distance of the workspace from the camera system is about 1000 mm. The software and hardware setup is similar to the one discussed in the section related to the triangulation based measurements and consists of two cameras in a stereo rig setup, a personal computer equipped with Matlab and OpenCV libraries and one industrial robot NJ 130 by COMAU which is the robot to calibrate.

10.3.3 Accuracy Aspects

For the evaluation of the method that has been described, a set of experiments was performed. For these experiments, a correlation space in the shape of a cube with an acme of 15 cm was used. As it has been discussed already, the robot joints are moving so that the marker position scans the working area of a 150 mm cube. With linear steps of one 1 mm per axis, the robot moved the marker in the working space.

This step affects the accuracy of calibration. Smaller step increments would lead to more poses and acquisition of more calibration data that would allow to define a more accurate transformation function. On the other hand more experiments are required on the expense of experimentation time.

In this experiment, the working space was divided in 1125 subspaces. Because the algorithms for 3d reconstruction of the space are relatively slow, about 3 s are needed for each step while, the whole calibration process requires about 1 h.

The data measured during the experiment are shown in the Figs. 10.14 and 10.15 contains the coordinates of the marker in relation to the robot base and the other includes the coordinates of the same points with respect to the camera.

The data in relation to the robot base and to the camera frame were analyzed with the method that had been described above and the transformation matrix H introduced in Eq. 10.7. If there was no difference between the two data sets, when applying matrix H to the camera measurements, the output should be equal with the

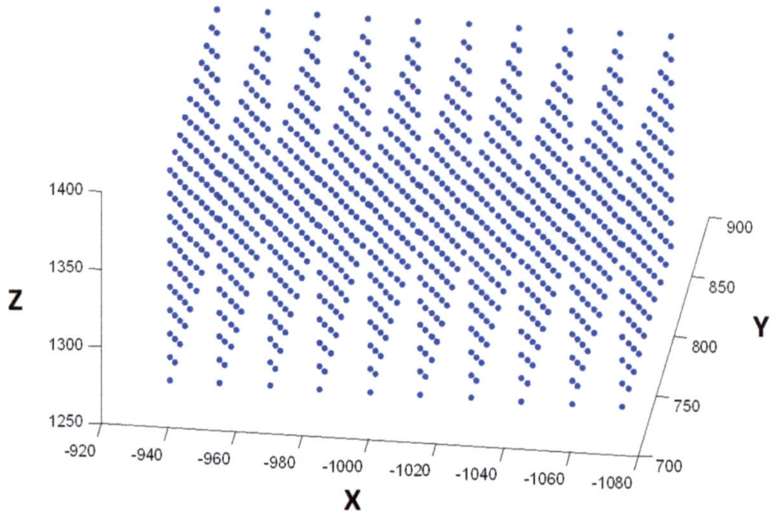

Fig. 10.14 Points with respect to robot base

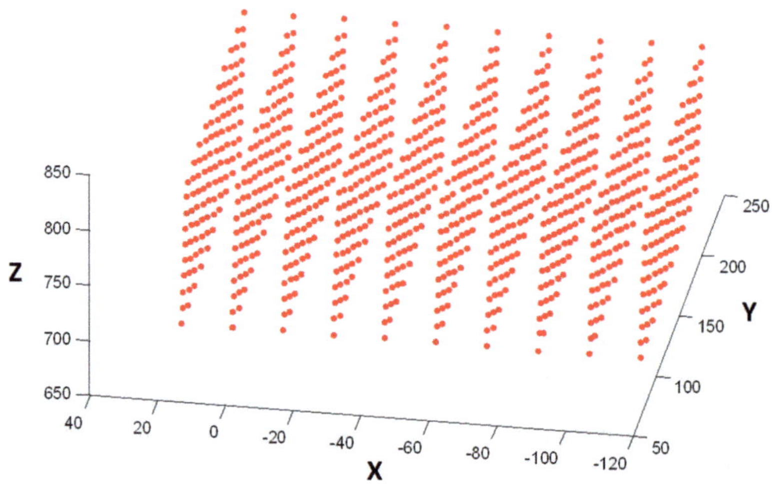

Fig. 10.15 Points measured by the stereo vision

measurements in relation to the robot base. According to Eq. 10.8, the matrix E is not a zero matrix and for the particular experiment, E has been calculated and shown in Fig. 10.16. It contains the error for all the iterations number and for each iteration the error for all axes x, y, z.

Moreover in Fig. 10.17, the mean square error for each measurement is shown as well as the overall mean square error. As the figure can show the mean square error of the transformation is less than 1 mm. The maximum error of the calibration is

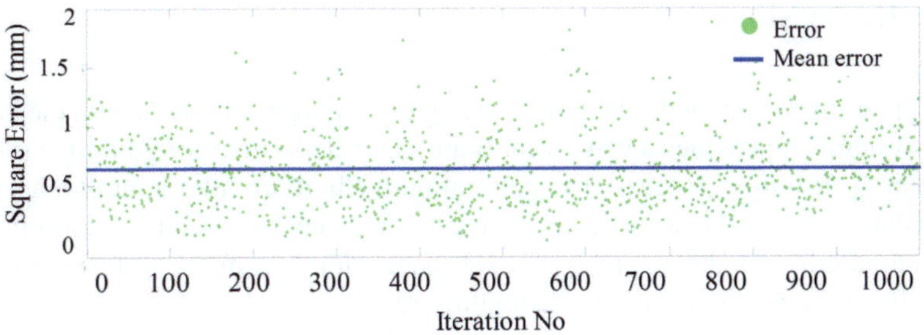

Fig. 10.16 Square error of transformation

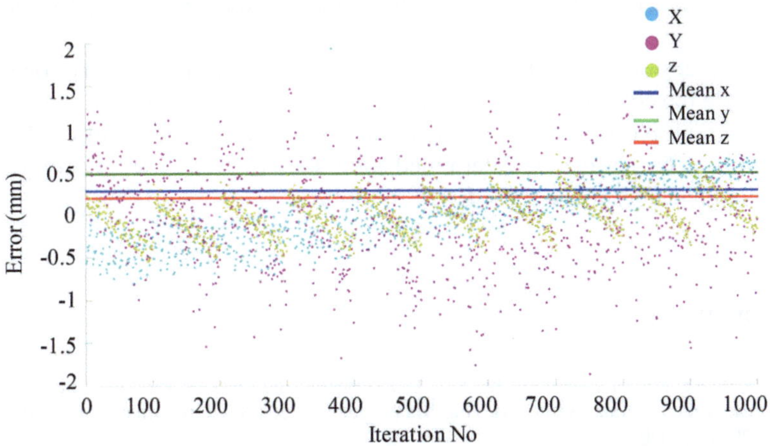

Fig. 10.17 Error of calibration method per iteration

less than 2 mm. This kind of accuracy is acceptable for most applications including robot handling and welding processes.

10.4 Robot Path Correction

Having the processes discussed throughout the chapter in place, it is now possible to measure the position of parts in the physical space and also to calibrate the measurements taken by a camera to match the coordinates measured from the robot base frame. When the robot is ready to receive the corrected coordinates for the welding points, the desktop PC client sends them, in a string form, to the specific port that the server listens to. The steps required to complete this process are well explained in detail in the literature [6].

10.5 Discussion

The chapter discussed methods for increasing the accuracy of robots with the help of vision systems. It provides a way to calibrate a camera system for guiding a robot in a welding or handling process. The calibration method provides a motion accuracy of less than 2 mm and can sufficiently be used with any robot/camera combination. The method relies on the mapping of a 3D space with the camera system coordinate frames based on multiple samples from this area.

The transformation function between the robot and the camera was determined for a very small area but the method can be generalized to cover all the working space of the robot. To achieve this the complete working space of the robot needs to be divided in much smaller spaces and the method needs to be applied for each one of these.

As an extension, the method can be used for a cell with more than one robot and increase the accuracy of frames exchange between the robots. By applying this method for multiple robots and one set of cameras, the transformation matrix between the robot base frames can be determined and therefore used to accurately translate point coordinates from one robot frame to another.

References

1. Abele E, Weigold M, Rothenbücher S (2007) Modeling and Identification of an industrial robot for machining applications. CIRP Ann Manuf Technol 56:387–390. https://doi.org/10.1016/j.cirp.2007.05.090
2. Bradski G (2008) Learning OpenCV: [computer vision with the OpenCV library], 1 edn. O'Reilly, Beijing; Köln[u.a.]
3. Hartley R, Zisserman A (2004) Multiple view geometry in computer vision, 2nd edn. Cambridge University Press, Cambridge, UK, New York
4. Heikkila J, Silven O A four-step camera calibration procedure with implicit image correction. IEEE Comput Soc 1106–1112
5. Jean-Yves B Camera calibration toolbox for matlab. https://www.vision.caltech.edu/bouguetj/calib_doc/. Accessed 14 Dec 2011
6. Michalos G, Makris S, Eytan A, Matthaiakis S, Chryssolouris G (2012) Robot path correction using stereo vision system. In: Procedia CIRP. Athens, pp 352–357
7. Scharstein D, Szeliski R (2002) A Taxonomy and evaluation of dense two-frame stereo correspondence algorithms. Int J Comput Vis 47:7–42. https://doi.org/10.1023/A:1014573219977
8. Trucco E (1998) Introductory techniques for 3-D computer vision. Prentice Hall, Upper Saddle River NJ
9. URL Computar lenses computar, factory automation lens. https://computarganz.com/misc/Computar_FA_2009.pdf. Accessed 14 Dec 2011
10. URL National instruments NI LabVIEW vision development module. https://www.ni.com/labview/vision/. Accessed 14 Dec 2011
11. Zhang Z (2000) A flexible new technique for camera calibration. IEEE Trans Pattern Anal Mach Intell 22:1330–1334. https://doi.org/10.1109/34.888718

Chapter 11
Cooperating Robots for Smart and Autonomous Intralogistics

11.1 Introduction

Nowadays, flexibility in manufacturing is considered as a key enabler for the production systems to be able to address the needs of the fluctuating market demand [1]. Intralogistics operations [2] refer to the material flow inside the factory that is needed for producing the required product volumes [3].

The majority of solutions proposed for the part supply in production systems is based in the Just—In—Time (JIT) method [4]. In the context if this method, the consumable material is transferred in the respective production stations just in the time when is needed. This practice aims to reduce the inventory size kept in factories facilities [5]. Building upon JIT, the Just-In-Sequence (JIS) method has been introduced [6]. Under JIS the suppliers are pre-sorting the consumable parts in bins based on the assembly sequence and then the workers retrieve these parts in the supplied order. This method helps on eliminating the scrap parts due to assembly errors.

In order to adopt in the later methods a new system has been proposed called Set Parts Supply (SPS). This system has been deployed and testing in Toyota Tsutsumi [7] and in Malaysian Automotive [8]. Based on SPS, two kinds of operators are considered in a production system, the human worker and the shop floor operator. The human worker executes the assembly process while the shop floor operator selects the required parts, sorts them in boxes and transfers them to the respective assembly workstation. However, due to the restricted communication between these two entities in the shop floor, it is common for failures to occur due to human errors (e.g. assembly worker picks the wrong parts) and lack of online monitoring the inventory levels in the workstation and the assembly process (e.g. shop floor man delivering wrong quantities of parts). Towards the direction of minimizing these failures, [3] proposed an integration of synchronized intralogistics and e—Kanban system with the Manufacturing Execution System (MES). These approaches are mainly focusing on the manual performance of the in-plant material supply tasks. However, in order to

S. Makris, *Cooperating Robots for Flexible Manufacturing*, Springer Series in Advanced Manufacturing,
https://doi.org/10.1007/978-3-030-51591-1_11

increase their sustainability, the manufacturing-based industries are currently facing the challenge of minimizing and balancing their human workforce [9].

In the past years, one of the research topics investigated was the to solve the part supply problem in a dynamic and autonomous way. To this direction, heuristics have been extensively investigated. The Tabu Search Algorithm has been proposed for planning part supply operations in a case study inspired by a hospital's supply department activities [10]. Moreover, [11] suggested the use of a combination of routine algorithms in order to control the execution of in-plant logistic operations implementing the "intelligent' warehouses concept. The replenishment of a single level assembly system problem has been addressed through a joint chance constrained model solved via an equivalent linear reformulation [12]. The target was to reduce the stoppages of production due to assembly components delay. The Ant Colony Optimization algorithm has also been used for optimal delivery paths generation for both indoors [13] and outdoors [14,15] part supply activities. However, the above-mentioned approaches have not yet been tested in industrial environments and thus lack in terms of validation in real world conditions.

The latest trends involve the integration of search algorithms for enabling mobile robot units to perform intra-factory logistics operations [16]. ICRA 2015 hosted the first Amazon Picking Challenge, where contestants' mobile robots were tested in several cognitive abilities such as object recognition, grasping, motion and task planning, execution monitoring and error recovery strategies. AUDI is currently testing the integration of two robots performing transportation tasks within its plan in Ingolstadt. STAMINA EU project, developed a mobile robot for executing kitting tasks driven by the requirements of PSA Peugeot Citroen factory in France. Multi-shuttle Move developed by Dematic, are mobile robots that can autonomously carry out transportation tasks communicating with each other through a decentralized based control system. KIVA systems developed a fleet of small robots navigating transferring shelves with consumables in different areas of the shop floor [17]. Autonomous Industrial Mobile Manipulators are also under extensive investigation [18]. Currently, the focus in on increasing the maturity of the individual technologies such as manipulation, tooling etc. and therefore there are no ready to use solutions.

In current industrial practice the Automated Guided Vehicles (AGVs) are of great use for intra-factory logistics operations. However, these devices face two limitations: a) they can follow fixed routes in the factory based on magnetic stripes places on the floor, and b) the loading and uploading of parts is performed by human operators. In addition, the scheduling of tasks for the AGVs is done manually since no automated decision-making system exists. The real time inventory levels are not considered during the manual planning and this creates constrains on the efficient material flow during production.

To address the existing limitations this chapter introduces the usage of Mobile Assistant Units (MAUs) for performing intra-factory material supply operations. These operations are dynamically and automatically planned by a dedicated Scheduler that uses intelligent search methods to generating efficient material supply plans. Integrating an online monitoring system allows the Scheduler to take decisions considering the real time shop floor needs in consumables [19]. Last but not least, the MAUs controller are also integrated in the system giving real time feedback on their availability as well as on the estimated time needed for performing different material supply plan alternatives. The integration of the involved components has been done based on a Service Oriented Architecture (SoA).

11.2 Approach

Current assembly systems consist of two parts, the assembly lines and the warehouses. The assembly lines comprise a set of assembly workstations where the products are assembled by using various consumable parts such as clips, nuts, screws, cables etc. These consumables are initially stored in dedicated boxes in the factory's warehouses located outside the line [20]. These boxes are used for suppling the assembly stations with new parts when the inventory in each workstation goes under depletion (Fig. 11.1).

Nowadays, in the era of mass customization, a typical assembly paradigm is the one of mixed model assembly. In the context of this paradigm, a set of different models of a product are assembled in the same line. One of the variations in their assembly process relies in the number and type of consumable used in each station.

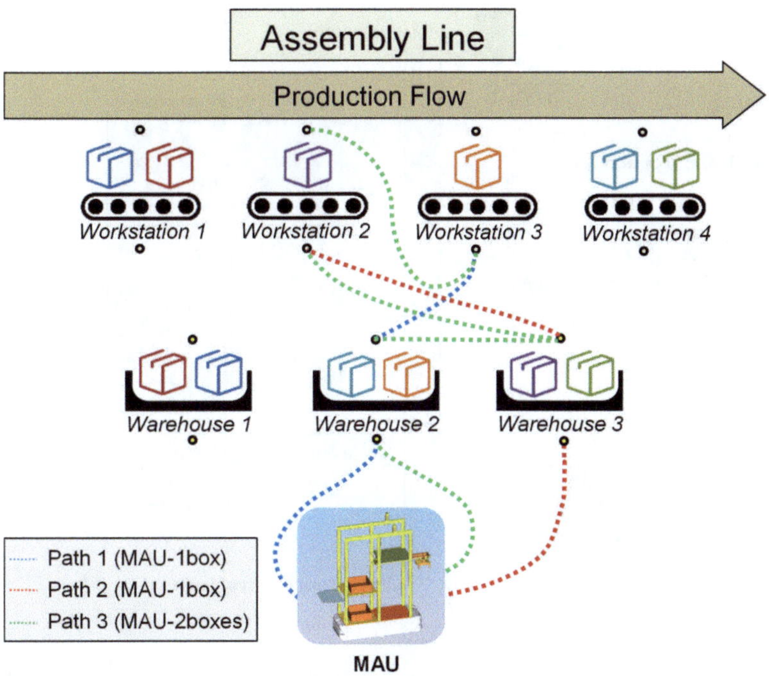

Fig. 11.1 Assembly system structure - MAUs alternative paths

This difference combined with the random production mix results in unbalanced inventory level in each workstation. Therefore, there is an evident need of supply the workstations with consumable parts in a dynamic way, depending on the production requirements. The online monitoring of the number of consumables in each workstation combined with the on-time response for replacement by flexible robot resources when these are under depletion are required.

Each time one or more consumables are close to depletion in any of the stations the system should be activated to replace the almost empty boxes with new ones. This chapter introduces the MAUs being able to autonomously navigate in the factory from the assembly stations to the warehouses and vice versa undertaking these intralogistics operations. The MAUs depending on the number of boxes that they can carry simultaneously as well as on the number of boxes being under depletion can follow different paths to realize the required tasks such as shown in Fig. 11.1.

The proposed architecture (Fig. 11.2) involves three levels: (1) the Decisional level, (2) the Execution Control level and the Physical Execution level. The Execution Control level integrates in a decentralized way the decision-making tools with the physical resources wrapped in the other two levels allowing the communication among the individual components of the system. A shared data repository has been deployed for enabling the seamless flow of information throughout the different levels.

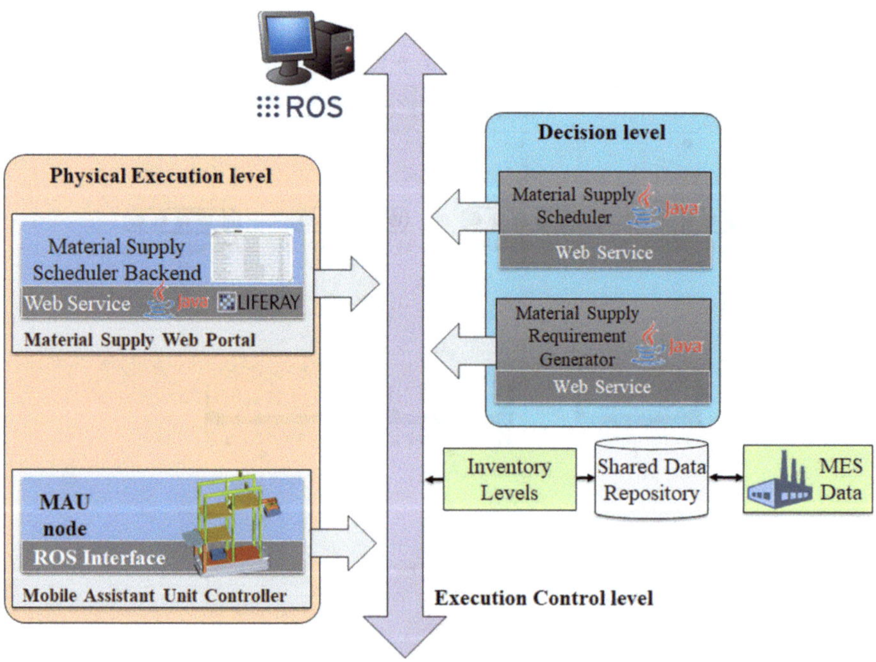

Fig. 11.2 SoA for autonomous intralogistics planning and control

11.2.1 Shared Data Repository

The information exchange between the different levels is achieved through a deployed share data repository. This repository, integrated in the Execution Control level, stores (a) the information provided by the Manufacturing Execution System, (b) the information regarding factories' resources online status, and (c) the real time inventory level of consumables in the different workstations.

The information stored in this repository has been modelled by an ontology, defining a set of specific objects and related properties. The relationship among these objects that represent the factory is visualized in Fig. 11.3. The Shop floor corresponds to the factory itself involving a set of Assembly Lines. These lines consist of several Stations and in its station a set of boxes with consumable with specific Box Positions are stored. In parallel, the factory has a set of Warehouses including various Shelves with specific Box Positions. Last but not least, the MAUs are mapped to the Shopfloor since they can move throughout the complete factory and are not dedicated to specific assembly lines.

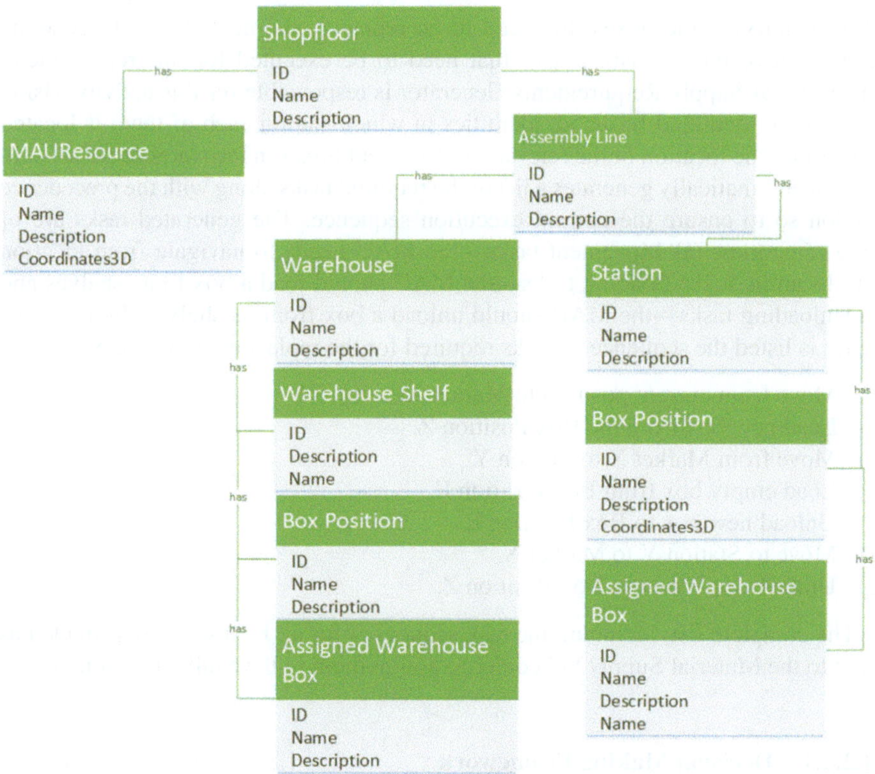

Fig. 11.3 UML shop floor data model

Each Box Position entity is characterized by the capacity and quantity property. While capacity represents the initial and maximum number of parts stored in a specific box, quantity represents the number of parts at the end of each cycle. The latter is updated in real time, knowing the parts consummation for each product variant in each station and monitoring the product variant assembled in each cycle. The modules included in the decisional level have access to this monitoring system so as to retrieve the information concerning the inventory levels. Under the Material Supply Scheduler module, specific quantity values have been defined as thresholds for each of the individual Box Positions. When the real time quantity of one or more Box Positions is lower than the defined thresholds, the need for replacement these boxes is identified by the Execution control system. Then, the latter triggers the Material Supply Scheduler so to generate the optimal schedule for replacing the boxes with new ones for the Warehouses.

11.2.2 Decisional Level

After identifying the boxes that need to be replaced, the next step is to generate the respective intra-logistics tasks that need to be executed for this replacement. The Material Supply Requirements Generator is responsible for this activity. Using as input the identified boxes id, identifies in which station each of them is located and defines the location of the relevant replacement boxes in the warehouses as well. Then, it automatically generates a list of the required tasks along with the precedence relation so to ensure the current execution sequence. The generated tasks are of three categories: (a) Movement tasks—the MAU needs to navigate from location X to location Y, (b) Loading tasks—the MAU should load a box to its shelves and (c) Unloading tasks—the MAU should unload a box from its shelves. Indicatively, please is listed the sequence of tasks required for the replacement of one box:

(a) Move from current position to Market X.
(b) Load new box from the Box Position Z.
(c) Move from Market X to Station Y.
(d) Load empty box from Box Position E.
(e) Unload new box to Box Position E.
(f) Move to Station Y to Market X.
(g) Unload empty box to Box Position Z.

The complete list, including the task sequences for all boxes is then provided as input to the Material Supply Scheduler to assign them to the available resources.

11.2.2.1 Decision Making Framework

The core decision making process is performed by the Material Supply Scheduler module with is responsible for generating efficient intra-logistics tasks schedules. The

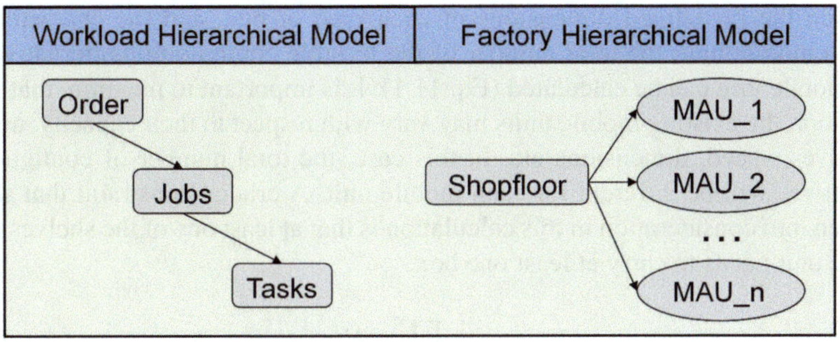

Fig. 11.4 Workload and facilities hierarchical modelling

internal model used in this module follows the principles of hierarchical modelling. As shown in Fig. 11.4 the workload activities have been modelled in three levels, Orders including Jobs that concern the replacement of a box. For instance, a Job may be "Box A replacement". Then this Job is broken down to tasks. These tasks are mapped with the level of tasks imported as input by the Material Supply Requirements Generator. The facilities related model follows a two-level breakdown. The MAUs are modelled as sub—components of the entire shop floor. The two hierarchies are connected in the Tasks—MAUs levels, indicating that the level of activities that is allocated to the MAUs is in task level.

Following this hierarchical modelling, the material supply problem has been formulated into a search problem using Artificial Intelligent techniques (AI). Whenever there is a requirement that one or more boxes have to be replaced, there are multiple alternative solutions that can address that requirement. There are different alternative pathways that a mobile unit may travel when considering possible material supply operations. These material supply operations are grouped into tasks for each mobile unit.

Additionally, the precedence relation between the different operations should also be one of the problem's parameters. For instance, the mobile unit cannot go to the station for the unloading of a box, unless this box has been previously loaded from the respective market. Such tasks are automatically generated by the system presented in this study for all combinations of stations and boxes that need to be served in a different sequence each time. In order for an efficient material supply plan to be generated for the tasks that need to be performed, the constraints described in this section, are taken into account.

A mobile unit may include a number of storage areas or shelves, where the boxes are located during their transportation. Since different types (dimensions) of boxes are used to storing the different parts, each storage area/shelf of the mobile unit can store only specific combinations, in terms of the number of boxes. For instance, it may accommodate only one big box, or 4 small boxes, or one medium sized box and two small ones. Each feasible combination of boxes that a mobile unit can carry simultaneously in all its storage areas/shelves is defined as a configuration.

Given the predefined dimensions of each type of box, and the characteristics of each mobile unit, the total number of the feasible configuration alternatives for each mobile unit can be calculated (Eq. 11.1). It is important to mention, that in the shop floor, the existing mobile units may vary with respect to their capacity, number of shelves, speed, dimensions etc. In this case, the total number of configuration alternatives will be different for each mobile unit. A crucial constraint that should be taken into consideration in this calculation is that at least one of the shelves of the mobile unit needs to carry at least one box.

$$n_c = \prod_{i=1}^{m} N_i \tag{11.1}$$

where, n_c represents the total number of feasible configurations, i (i.e. [1,2,3, …, $\gg m \ll$]) indexes MAU's shelf and denotes the feasible alternative combination of boxes that can be carried on the ith shelf.

Given the abovementioned representation of MAU's configurations, alternative configuration combinations may be formulated as a tree structure (Fig. 11.5) through the following steps [16]:

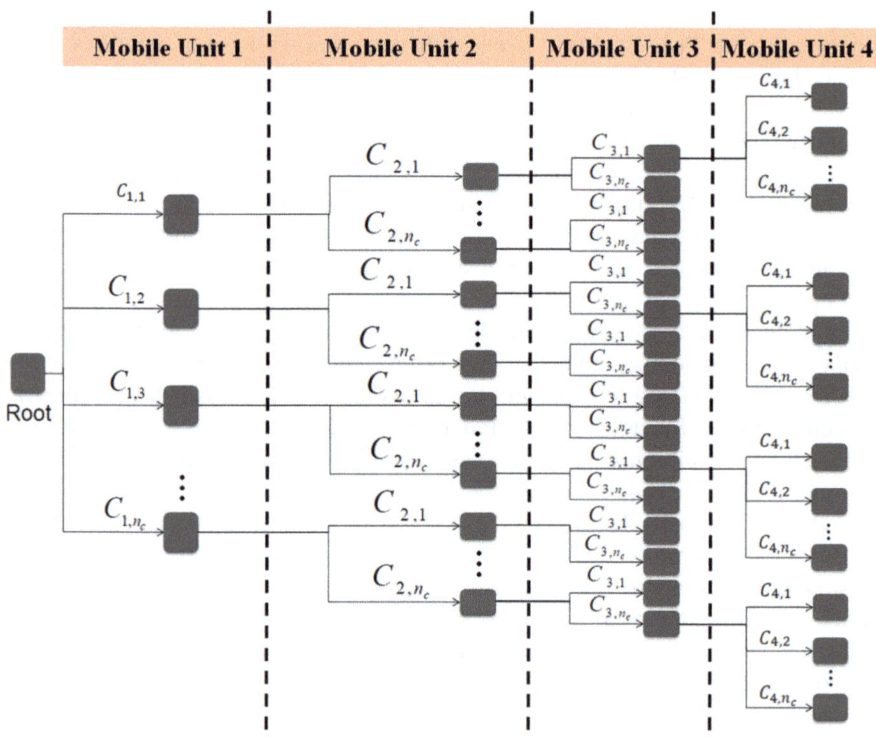

Fig. 11.5 Configuration alternative tree

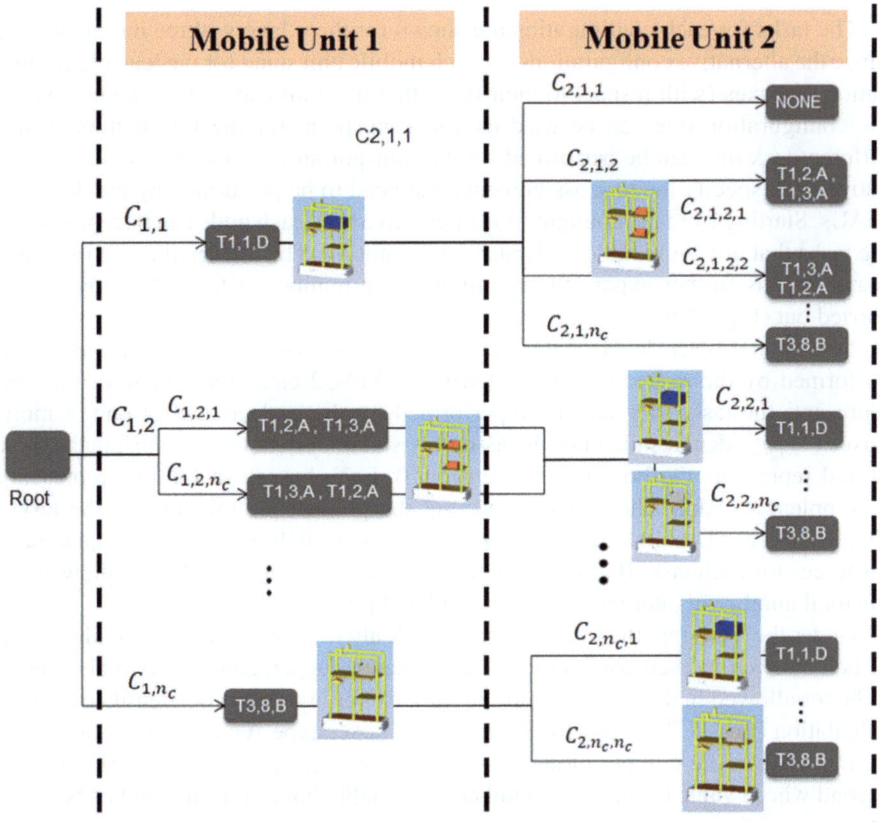

Fig. 11.6 Search formulation of the material supply problem

- Identification of the number and type of boxes that need to be considered in the planning process
- In layer 1, a mobile unit is selected (any of the available mobile units can be selected for each layer either sequential or randomized, having no impact in the formulation of the alternatives) and all possible configurations are listed as a tree's branches
- For each of the rest mobile units, all possible configurations are listed in separate layers considering the configuration/boxes of the previous layer to exclude any duplication of tasks in the same tree
- Each layer configuration is combined with the configurations of the next layers for the formation of a tree's branch, which is considered as a configuration alternative.

In Fig. 11.5, $C_{i,j}$, defines a configuration, where, i indexes the layer corresponding to the MAUs id [$i = 1$, …, number of mobile unitcs], j indexes the respective configuration for ith MAU [$j = 1$, …, n_c]. In the presented tree, the configuration for three MAUs (three levels) have been included.

The tasks for each configuration are shown for three layers (three mobile units). Since the alternative configurations for each mobile unit stand for the feasible combination of boxes (with respect to their type) that this unit can carry simultaneously, the configuration tree can be used as a starting point for the formulation of the different task that can be performed for the transportation of the boxes. Such alternatives will specify the task assignments that need to be performed by the different MAUs. Starting from the configuration alternative tree, each node can be replaced by the tasks that are compatible with each configuration. This means that the resulting search tree is further expanded according to the number of tasks that have to be carried out (Fig. 11.6).

In the latter, each level of the tree represents a group of tasks that should be performed by the respective MAUs (MAU 1, MAU2 etc.). Each node of the tree represents the tasks that should be performed for the replacement of one or more boxes ($T_{i,j,k}$: i denotes the id of the assembly station, j denotes the consumable box id and represents the box type (dimensions)). Each branch of the tree represents a complete schedule alternative respecting the precedence relations of the tasks. Finally, the MAU suitability constraints such as availability define the candidate resources for each task. Based on this formulation, two steps should be followed for the total number of alternatives to be calculated [19].

Under the first step, the total number of task alternatives for each individual type of box allowed in each configuration is calculated. Depending on the total number of boxes allowed in the specific configuration two cases have been identified for this calculation (Eq. 11.2—indicative example for boxes type A); the one whose value is smaller than the number of available boxes for this type in the scenario and the second whose value exceeds the number of available boxes for this kind of box.

$$N_{TAi} = \begin{cases} N_{TAi} = \frac{N_A!}{(N_A - na_i)!} & if\ N_A \prec na_i \\ N_{TAi} = N_A! & if\ N_A \geq na_i \end{cases} \tag{11.2}$$

where, i denotes configuration alternatives index [$i = 1\ldots, n_c$], NA denotes the total number of the available boxes type A, na_i denotes the total number of boxes type A allowed in the ith configuration and N_{TAi} denotes the number of task alternatives that can be performed for the replacement of boxes type A allowed in the ith configuration.

Next step is the calculation of the total number of task alternatives (for all the available types of boxes), through the aggregation of the number of tasks alternatives for each individual type of box, allowed in each configuration (Eq. 11.3):

$$N_T = \sum_{i=1}^{n_c} \left(\prod_{j=1}^{Z} N_{Tji} \right) \tag{11.3}$$

where, N_T denotes the total number of task alternatives, i indexes the configuration alternative while j indexes the type of box.

For the selection of an efficient task alternative among the set of the feasible ones, it is necessary to define criteria that can quantify the performance of each one of them. In the context of this work, these criteria concern the minimization of the time required for the performance of the tasks as well as the minimization of the distance that has to be covered by the MAU resources (cost criteria). These metric's minimization will ensure the fast serving of the line leading to reduction of stoppages due to lack of materials while maximizing utilization of resources [16]. For the calculation of both criteria for each alternative, real time estimations for each task duration and required distance to be travelled will be requested and received by the MAUs through the integration system. More specifically:

Time required for transportation (T): The minimization of the time required for each alternative realization by the involved MAUs is calculated by Eq. 11.4.

$$T = \sum_{i=1}^{N} \frac{d_i}{v_i} \tag{11.4}$$

where, N denotes the total number of task assignments, i indexes the task assignment $[i = 1 \ldots N]$, d represents the estimation of the required distance to be travelled by the assigned MAU and v for the velocity of the MAU.

Distance travelled (D): The total distance travelled by each involved MAU for each task alternative varies due to the different pathways followed and it may be calculated through Eq. 11.5.

$$D = \sum_{i=1}^{N} d_i \tag{11.5}$$

where, N denotes the total number of task assignments of the alternative, i indexes the task assignment $[i = 1 \ldots N]$, d represents the estimation of the travelled distance by the assigned MAU.

11.2.3 Execution Control Level

This level is responsible for integrating and allowing communication between the decisional and the physical execution level. On the side of the decisional level, the scheduling module integrated using services exposed by a client developed in Java. A web-based User Interface has been developed where the user can trigger the decision-making services for generating new material supply plans. On the physical execution side, the execution control level integrates the mobile robots employing

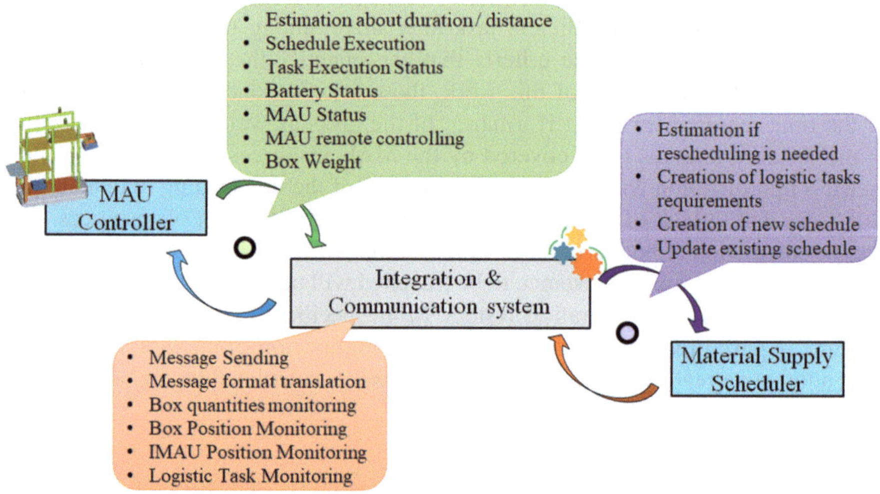

Fig. 11.7 Material supply main integration approach

the "publisher-subscriber" communication protocol. This integration interface from both sides has been deployed through the implementation of a set of Robot Operating System (ROS) services. The main functionalities and offered services are presented in Fig. 11.7.

11.2.4 Physical Level

11.2.4.1 Mobile Assistants Units (MAUs)

The discussed mobile robots comprise an autonomous mobile platform and an upper structure with a robot device. These MAUs can autonomously navigate in the factory from different workstations to different marker areas and vice versa. The upper structure involves three main parts: (a) the shelves that are used for loading the parts with consumable, (b) an additional shelf that is used for storing the first empty box loaded on the robot and (c) a gripping devices placed on a linear mechanism, used for loading and unloading the boxes. These robots can have different configurations considering the number of the included shelves. Figure 11.8 visualizes an indicative configuration. This device can grasp boxes up to 20 kg weight and 400 mm width.

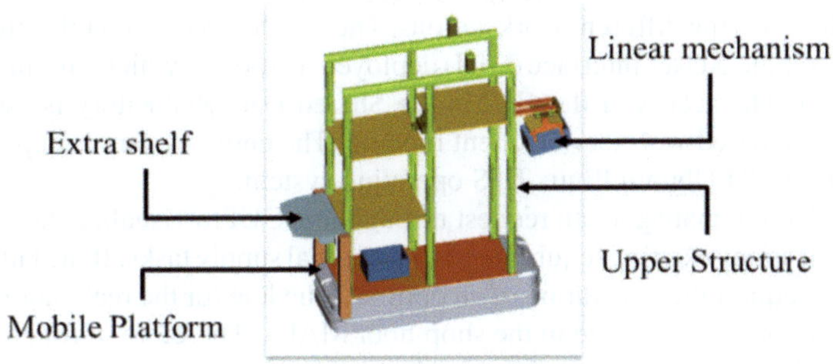

Fig. 11.8 Indicative MAU structure—3 shelves configuration

11.3 Implementation

Figure 11.9 presents the sequence of processes that take place for generating and executing a material supply schedule. The Shared Data Repository as described in the relevant section, stores a real time instance of the shopfloor status, including the

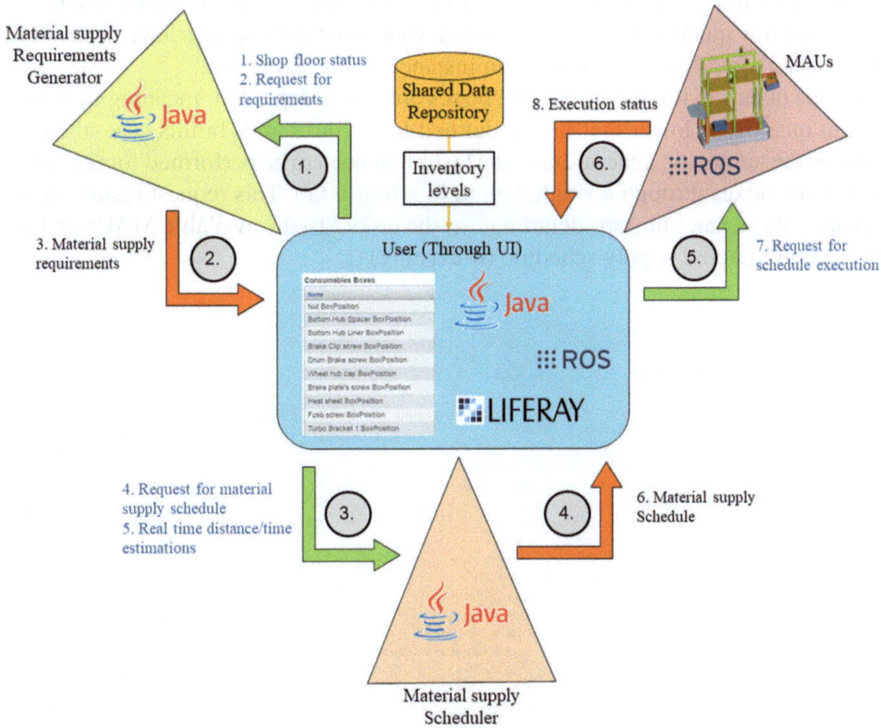

Fig. 11.9 SoA scheme implementation

inventory levels in the different workstations. The user has access on this information through a Graphical User Interface (GUI) deployed in a portal, which runs in a Tomcat 7.0.42 server. The access of the GUI in the Shared Data Repository is established through a Java based web service client module. The entire system is deployed in a PC that runs 14.04 Ubuntu Linux LTS operating system.

The production manager can request through this GUI to visualize the shop floor status as well as any existing requirements on material supply tasks. If the latter exists, the user can request the generation of an optimal schedule for the realization of these tasks considering the available in the shop floor MAUs. Therefore, through a simple button, the Material Supply Scheduler is triggered using as input the current instance of the shop floor. During the evaluation phase of the different alternatives internally in the Scheduler, online information is retrieved regarding the current position of the MAUs, their availability as well as task execution duration estimations by the different MAUs. The communication protocol is established through ROS and web services.

Figure 11.10 presents in presents the interfaces available in the GUI listing the assembly stations, the markets as well as the boxes with consumables mapped to the related assembly stations as well as market position. A dedicated tab is assigned for visualizing each of these three group of entities. With respect to the boxes, the user can monitor the current quantity of each one of them in comparison with the predefined low quantity threshold. The data visualized in these tabs have been saved in the shared data based to simulate an instance of the shop floor.

In this instance, three consumable boxes are under depletion, meaning that their current quantity is lower that the predefined lower threshold. In this case, the user requests the automatic generation of the tasks that need to be performed for replacing these three boxes through a single button in a fourth tab. This request results in the pop up of three new sub-tabs detailing: (a) the tasks, (b) the available MAUs and (c) any current material supply schedule (if existing).

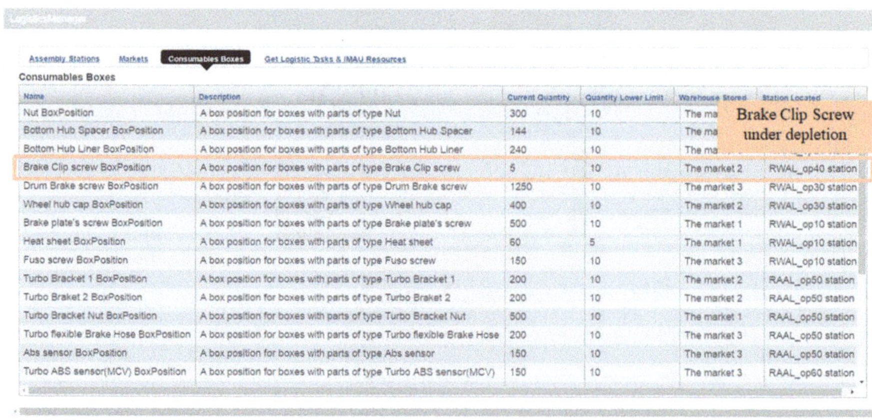

Name	Description	Current Quantity	Quantity Lower Limit	Warehouse Stored	Station Located
Nut BoxPosition	A box position for boxes with parts of type Nut	300	10	The ma	
Bottom Hub Spacer BoxPosition	A box position for boxes with parts of type Bottom Hub Spacer	144	10	The ma	
Bottom Hub Liner BoxPosition	A box position for boxes with parts of type Bottom Hub Liner	240	10	The ma	
Brake Clip screw BoxPosition	A box position for boxes with parts of type Brake Clip screw	5	10	The market 2	RWAL_op40 station
Drum Brake screw BoxPosition	A box position for boxes with parts of type Drum Brake screw	1250	10	The market 3	RWAL_op30 station
Wheel hub cap BoxPosition	A box position for boxes with parts of type Wheel hub cap	400	10	The market 2	RWAL_op30 station
Brake plate's screw BoxPosition	A box position for boxes with parts of type Brake plate's screw	500	10	The market 1	RWAL_op10 station
Heat sheet BoxPosition	A box position for boxes with parts of type Heat sheet	60	5	The market 1	RWAL_op10 station
Fuso screw BoxPosition	A box position for boxes with parts of type Fuso screw	150	10	The market 3	RWAL_op10 station
Turbo Bracket 1 BoxPosition	A box position for boxes with parts of type Turbo Bracket 1	200	10	The market 2	RAAL_op50 station
Turbo Braket 2 BoxPosition	A box position for boxes with parts of type Turbo Braket 2	200	10	The market 2	RAAL_op50 station
Turbo Bracket Nut BoxPosition	A box position for boxes with parts of type Turbo Bracket Nut	500	10	The market 1	RAAL_op50 station
Turbo flexible Brake Hose BoxPosition	A box position for boxes with parts of type Turbo flexible Brake Hose	200	10	The market 3	RAAL_op50 station
Abs sensor BoxPosition	A box position for boxes with parts of type Abs sensor	150	10	The market 3	RAAL_op50 station
Turbo ABS sensor(MCV) BoxPosition	A box position for boxes with parts of type Turbo ABS sensor(MCV)	150	10	The market 3	RAAL_op60 station

Fig. 11.10 Shop floor status viewer GUI

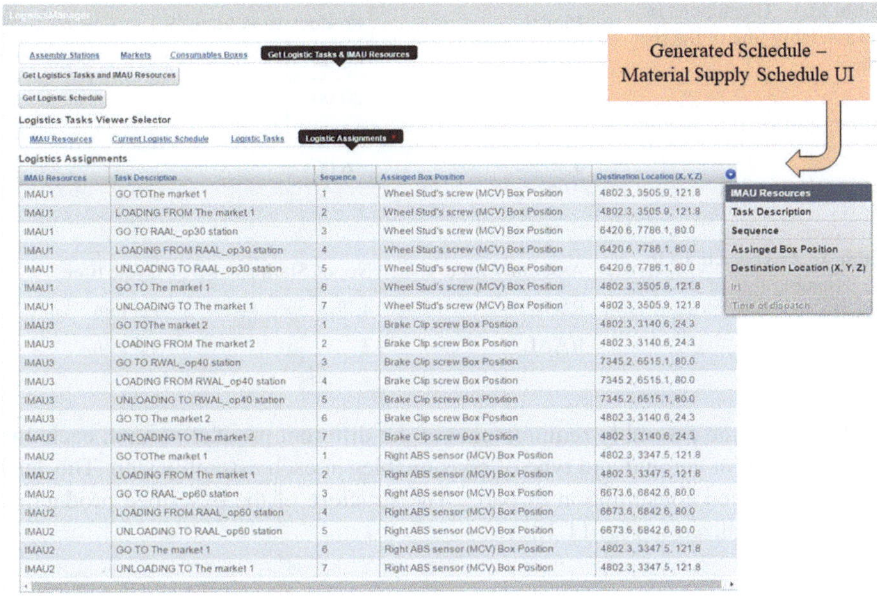

Fig. 11.11 Material supply scheduler GUI

Through the same tab, the user can request the calculation of an optimal schedule for executing the generated tasks. Upon this request, the Material Supply Assignments UI tab pops-up as visualized in Fig. 11.11, listing the tasks assigned to the MAUs. Finally, through a button in the same sub-tab, the user can request the dispatchment of the assigned tasks to the relevant MAUs so for the execution to start.

11.4 Industrial Example

The discussed approach has been validated in a case study inspired by the automotive sector. The validation focused on the final assembly of a passenger vehicle and more specifically in two assembly lines: (a) the vehicles wheel group assembly and (c) the vehicles rear axle assembly line. Four different vehicles models are produced in these lines as detailed in Table 11.1.

The number of stations as well as the cycle time for both assembly lines is presented in Table 11.2.

The total number of consumable boxes in both assembly lines is twenty-four. These are classified in five types of boxes varying their dimensions as listed in Table 11.3. The involved dimensions are similar to the typical dimensions of boxes used in the automotive for storing parts such as screws, clips and cables.

Table 11.1 The actions in the assembly line of the rear wheel

Model	Production volume (%)
1	46.00
2	20.00
3	28.00
4	6.00

Table 11.2 The actions in the assembly line of the rear wheel

Assembly Line	No. of Stations	Cycle time
RWAL	4	1.5 min
RAAL	7	1.5 min

Analyzing the assembly requirements of the different product variant, each one needs a different amount and type of consumables in each assembly step. This lead to an unbalanced consumption of parts. The structure of the described production system is visualized in Fig. 11.12.

11.4.1 Discrete Event Simulation (DES)

Towards analyzing the needs for material supply operations under the investigated scenario, a discrete event simulation model has been implemented using the simulation package Witness 2007. This kind of simulation analysis allows the representation and performance evaluation of the system with limited computation cost and effort. Each machine in this model (Fig. 11.13) represents an assembly station (seven machines for the RAA and four machines for the RWAL). Each of these machines receives input from multiple buffers (emulating boxes with consumables) and outputs the assembled part. The cycle time for each of the machines, for both lines is 1.5 min.

An additional machine element has been introduced in the model in order to emulate the MAU. This machine is responsible for replacing the boxes with consumables that are under depletion. In order to simulate this behavior, the machine is activated only when one (or more) of the existing buffers have less quantity of parts that the predefined thresholds. The cycle time for each operation of this machine

Table 11.3 The actions in the assembly line of the rear wheel

Type of box	Dimensions	Quantity
A	30 × 15 × 20 cm	14
B	40 × 30 × 30 cm	1
C	40 × 15 × 30 cm	4
D	60 × 30 × 40 cm	4
E	51 × 9 × 40 cm	1

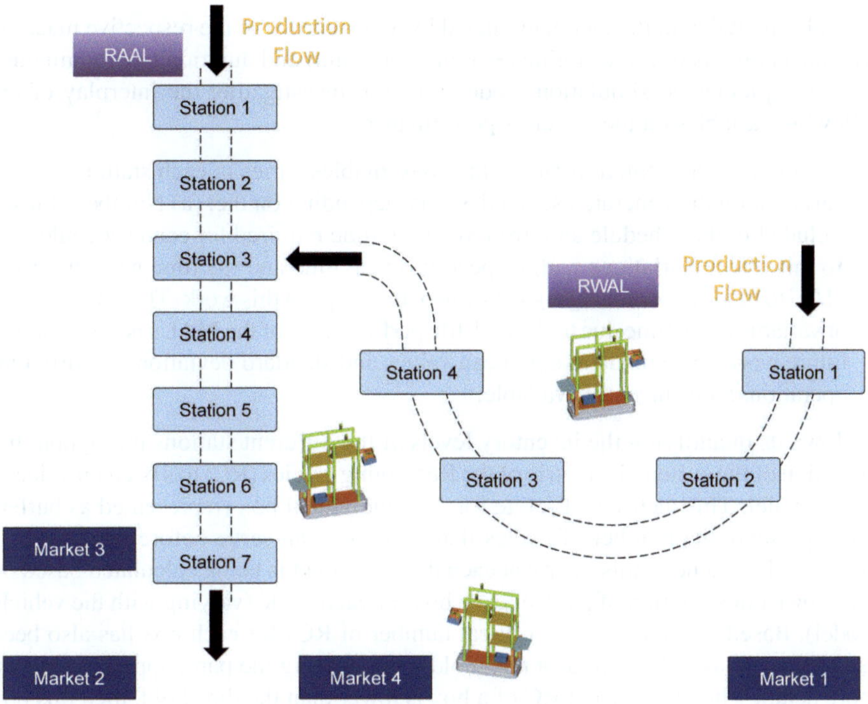

Fig. 11.12 Two assembly line based production system structure

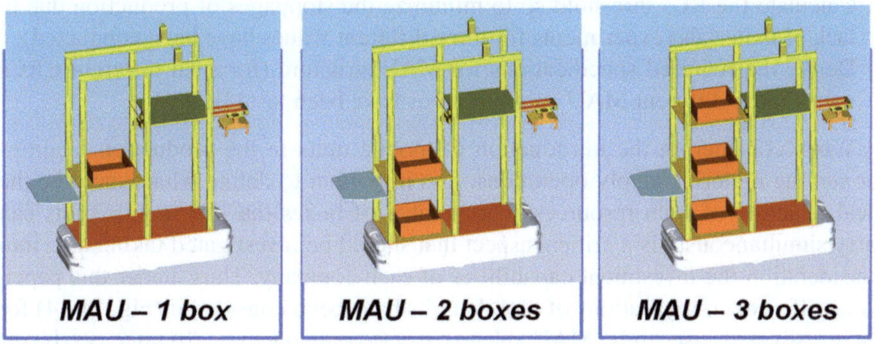

Fig. 11.13 DES model

varies depending on the distance that needs to be traversed by the MAU for each replacement. The distances between the different markets and stations are stored in the table (extracted by the shopfloor CAD model) and are retrieved by the model each time a replacement needs to be done. The time required for the loading/unloading operations has been fixed to 0.35 min. The requirements for MAU charging have been also included in the model emulated by a breakdown of the respective machine with mean time between charging ten hours (600 min) and duration twenty minutes.

The implemented simulation model aimed at investigating the interplay of the following variables on the system's performance:

- Variation on the depletion rate of the consumables varies in each station;
- Variation on the generated schedules vary depending on the: (a) Number of tasks included in the schedule and, (b) Execution time required for each schedule.
- To this effect, variations with respect the cycle time and the time needed for the MAU to load/unload/charging were out of the scope of this work. The investigation of variable cycle time due to the real-life performance of the MAUs is suggested as future work since no data (e.g. mean values and standard deviations for different operations) are currently available.

Towards quantifying the inventory levels of the different stations throughout the operation of the system the metric of the Remaining Cycles (RC) has been introduced in the model. This metric is discrete for each individual box (represented as buffer) and represents the number of cycles that each box can serve before its content is depleted. The value of this metric at each decision point is being calculated based on the known consumption of parts of each box for each cycle (varying with the vehicle model). Based on this metric, a critical number of RCs for each box has also been established so as to be used as a threshold for triggering the part supply process. In more detail, when the current RC of a box is lower than the threshold, then this box needs to be replaced. A total of nine experiments, with simulation time of one year (518.400 min) with 345.600 parts entering the system, were conducted aiming to:

- Calculate the RCs threshold so to minimize the stoppages of production due to lack of materials (experiments for three different values have been conducted);
- Derive the required specifications for MAU structure (for each different CRCs value, three different MAU configurations have been tested).

When considering the introduction of mobile units in the production facilities for serving material supply operations, it is important to define what would be the ideal structure of such resources. The number of boxes that the mobile units can carry simultaneously is a critical aspect that should be investigated taking also into consideration the investment capabilities of each company. Thus, under this paper, three different configurations of mobile units have been considered (Fig. 11.14) for the investigated case study: MAU able to carry (a) only one box, (b) up to two boxes and (b) up to three boxes.

Another important aspect to design an effective material supply system, is to define the critical RC values for the boxes that will used as threshold values for triggering the material supply scheduling system. Given the number of parts consumed by each box (4 parts average consumption) during one cycle, three different RC threshold values have been investigated: box quantity less than (a) 10 parts, (b) 15 parts and (c) 25 parts.

Using the designed simulation model nine set of experiments have been performed to derive the RC threshold value as well as the optimized configuration of the MAUs.

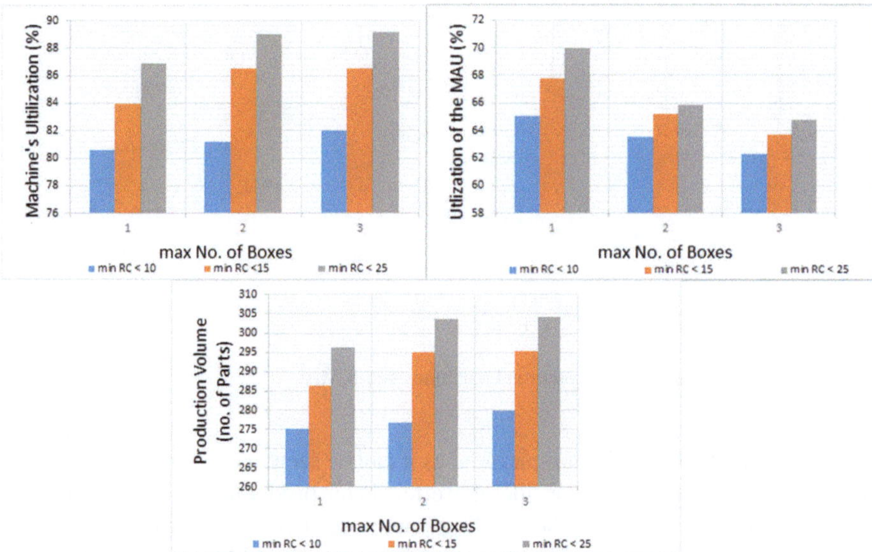

Fig. 11.14 MAU's components and alternative configurations

In Table 11.4, the criteria that were used for assessing each alternative and the acquired values are presented.

As it can be seen, the maximum production volume as well as the utilization of the operational machines can be achieved when introducing MAUs that can carry up to three boxes while the scheduling system is triggered when the quantity of one (or more boxes) is less than 25 parts.

Table 11.4 DES simulation experiments results

minRC < 10 parts	1 box	2 boxes	3 boxes
Utilization of operational machines (%)	80.6	81.2	82.01
Utilization of the MAU (%)	65.05	63.5	62.25
Production volume (vehicle's rear axles)	275.062	276.785	279.858
minRC < 15 parts	1 box	2 boxes	3 boxes
Utilization of operational machines (%)	83.96	86.49	86.5
Utilization of the MAU (%)	67.75	65.2	63.65
Production volume (vehicle's rear axles)	286.520	295.152	295.193
minRC < 25 parts	1 box	2 boxes	3 boxes
Utilization of operational machines (%)	86.87	89.02	89.17
Utilization of the MAU (%)	70	65.84	64.81
Production volume (vehicle's rear axles)	296.463	303.813	304.326

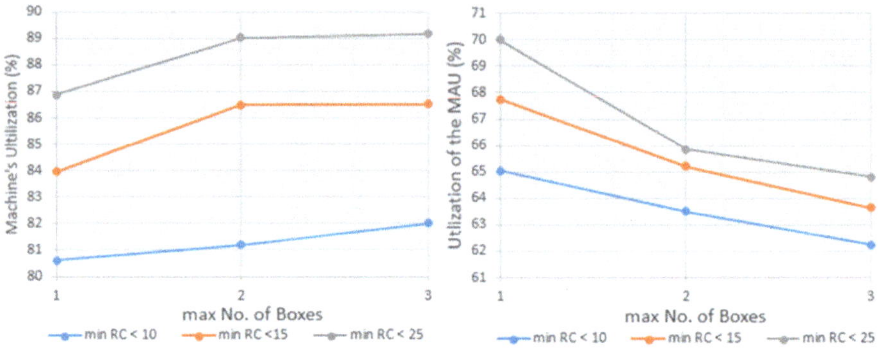

Fig. 11.15 Utilization of the **a** operational machines and **b** MAU

Concerning the MAU's configuration, as it can be seen, increasing the number of shelves from one to two has a considerable effect in the production volume and utilization of resources. However, when adding one more shelf the improvement is less. Thus, the investigation did not include a fourth self in the MAU since it is expected that this addition would be redundant, not providing any benefit in the system apart from increasing the investment cost for acquiring such MAUs. Regarding the utilization of the MAU, as it was expected, it keeps being reduced as the number of boxes that the MAU can carry simultaneously is increased (Fig. 11.15). This reduction allows the exploitation of MAUs capabilities in other workstations or assembly lines of the production system.

Thus, considering the above analysis, under this work, the RC value that will be used as threshold for the quantity of each box is 25 parts while the MAU considered is able to carry up to three boxes at the same time.

11.5 Discussion

This chapter discussed the implementation of a service-oriented architecture that would enable the dynamic scheduling of material supply operations in an assembly system, using MAUs. Driven by the requirements of an actual production line of the automotive industry the proposed system aims to eliminate the stoppages of the assembly lines due to lack of consumables. The deployment of the proposed system allows the efficient material flow in assembly stations while decreasing assembly error due to part depletion leading to a significant increase of the production volume of the line. Finally, the full automation of the material supply process will eliminate the need of human labour for planning or transportation actions allowing them to focus on more value adding activities. The proposed architecture is open and enables the integration of multiple and varying characteristics MAUs using service technology, providing the following advancements:

- Automatic generation of the material supply requirements, based on online feedback of shop floor's inventory levels;
- Dynamic creation of assignments of the required tasks to the available MAUs;
- Online retrieval of detailed estimations of the: (1) time required for the MAU to execute each task and (2) distance travelled by the MAU;
- Efficient handling of large instances of the material supply problem, involving multiple assembly lines, stations and boxes using AI techniques;
- Minimization of part depletion occurrences, leading to an increase in the production volume;
- Minimization of the distance covered by the MAUs, leading to the increase of these resources' utilization and to a reduction in the idle time.

Future study should focus on advancing and fine tuning the proposed system in order to render it applicable to real industrial environments. To this direction, integration with a factory's legacy system should be performed.

References

1. Chryssolouris G (2006) Manufacturing systems: theory and practice. Springer-Verlag, New York
2. Boysen N, Emde S (2014) Scheduling the part supply of mixed-model assembly lines in line-integrated supermarkets. Eur J Oper Res 239:820–829. https://doi.org/10.1016/j.ejor.2014.05.029
3. Jainury SM, Ramli R, Ab Rahman MN, Omar A (2014) Integrated set parts supply system in a mixed-model assembly line. Comput Ind Eng 75:266–273. https://doi.org/10.1016/j.cie.2014.07.008
4. Boysen N, Emde S, Hoeck M, Kauderer M (2015) Part logistics in the automotive industry: decision problems, literature review and research agenda. Eur J Oper Res 242:107–120. https://doi.org/10.1016/j.ejor.2014.09.065
5. Sugimori Y, Kusunoki K, Cho F, Uchikawa S (1977) Toyota production system and Kanban system Materialization of just-in-time and respect-for-human system. Int J Prod Res 15:553–564. https://doi.org/10.1080/00207547708943149
6. Werner S, Kellner M, Schenk E, Weigert G (2003) Just-in-sequence material supply—a simulation based solution in electronics production. Robot Comput Integr Manuf 19:107–111. https://doi.org/10.1016/S0736-5845(02)00067-4
7. Noguchi H (2005) A new mixed flow production line for multiple automotive models at Tsutsumi plant 51:16–33
8. Jainurya SM, Ramlia R, Rahman MNA, Omar A (2012) An implementation of set parts supply system in the malaysian automotive industry 59:19–24
9. Battaïa O, Delorme X, Dolgui A, Hagemann J, Horlemann A, Kovalev S, Malyutin S (2015) Workforce minimization for a mixed-model assembly line in the automotive industry. Int J Prod Econ 170:489–500. https://doi.org/10.1016/j.ijpe.2015.05.038
10. Lapierre SD, Ruiz AB (2007) Scheduling logistic activities to improve hospital supply systems. Comput Oper Res 34:624–641. https://doi.org/10.1016/j.cor.2005.03.017
11. Vivaldini KCT, Galdames JPM, Bueno TS, Araujo RC, Sobral RM, Becker M, Caurin GAP (2010) Robotic forklifts for intelligent warehouses: routing, path planning, and auto-localization. IEEE. pp 1463–1468

12. Borodin V, Dolgui A, Hnaien F, Labadie N (2016) Component replenishment planning for a single-level assembly system under random lead times: a chance constrained programming approach. Int J Prod Econ 181:79–86. https://doi.org/10.1016/j.ijpe.2016.02.017
13. Montemanni R, Gambardella LM, Rizzoli AE, Donati AV (2005) Ant colony system for a dynamic vehicle routing problem. J Comb Optim 10:327–343. https://doi.org/10.1007/s10878-005-4922-6
14. Arora V, Chan FTS, Tiwari MK (2010) An integrated approach for logistic and vendor managed inventory in supply chain. Expert Syst Appl 37:39–44. https://doi.org/10.1016/j.eswa.2009.05.016
15. Lanza G, Moser R (2014) Multi-objective optimization of global manufacturing networks taking into account multi-dimensional uncertainty. CIRP Ann Manuf Technol 63:397–400. https://doi.org/10.1016/j.cirp.2014.03.116
16. Kousi N, Koukas S, Michalos G, Makris S (2019) Scheduling of smart intra—factory material supply operations using mobile robots. Int J Prod Res 57:801–814. https://doi.org/10.1080/00207543.2018.1483587
17. Wurman PR, Andrea RD, Mountz M Coordinating hundreds of cooperative, autonomous vehicles in warehouses. p 29
18. Hvilshøj M, Bøgh S, Skov Nielsen O, Madsen O (2012) Autonomous industrial mobile manipulation (AIMM): past, present and future. Ind Robot Int J 39:120–135. https://doi.org/10.1108/01439911211201582
19. Kousi N, Koukas S, Michalos G, Makris S, Chryssolouris G (2016) Service oriented architecture for dynamic scheduling of mobile robots for material supply. Procedia CIRP 55:18–22. https://doi.org/10.1016/j.procir.2016.09.014
20. Heragu SS, Du L, Mantel RJ, Schuur PC (2005) Mathematical model for warehouse design and product allocation. Int J Prod Res 43:327–338. https://doi.org/10.1080/00207540412331285841

Chapter 12
Robots for Material Removal Processes

12.1 Introduction

Robot machining has been encountering difficulties, deriving from the fact that robots have been solely handling devices so far [1, 2]. Concurrently, the use of conventional material removal techniques, such as the CNC milling have the drawback of limited workspace and thus shape limitations, despite the high precision they offer [3]. In order to face these drawbacks, the industrial robots seem as a cost-effective solution with increased flexibility and versatility [4].

Robotic machining has been researched from different points of view, such as the compensation of cutting forces [5], vibrations minimization [6], compliance errors compensation [7], collisions prevention [8] and robots' configuration [9]. Example of commercially available tools and systems namely, the PowerMill robot and the KUKA tool are used for machining applications, when high accuracy is not required (e.g. drilling, brushing, deburring). These tools have focused on the programming of different robot platforms, enabling the shorter machining times and the higher throughput, compared to traditional solutions.

The main advantages of robotic machining methods, among others, are the lower position errors and the higher machining speed. Additionally, larger parts can be machined easier due to the flexible workspace and robot kinematics as well. Last but not least, the fact that a variety of robot configurations is available, many space and system layouts can be encountered for robot machining capable of replacing both manual and automated machining systems [1].

Despite these advantages, there are still problems related to the robot position accuracy required for machining. Additionally, one more issue related to the low robot body frequency is that of the vibrations in the produced surface quality. The lack of sufficient robot online and offline programming methods is another weakness and thus, the potential integration of robots into machining has yet to be realized [10].

This paper proposes a method for online robot program generation for robotic machining, by exploiting the traditional CAM tools. A tool is proposed for online

S. Makris, *Cooperating Robots for Flexible Manufacturing*,
Springer Series in Advanced Manufacturing,
https://doi.org/10.1007/978-3-030-51591-1_12

program generation by using as input code generated offline by a CAM tools. This is a general purpose tool and can be applied to multiple industrial robot platforms. Transformations among the different frames follow. This approach allows for the easy machining reconfiguration of three-dimension complex geometries. The method has been tested on medium-density fiberboard (MDF) for linear, circular and spiral geometries. A number of parametrical experiments have also been designed, for studying the relevance of different parameters to the quality of the final product.

12.2 Approach

The proposed method of robotic machining considers the use of the existing CAM tools that generate a G-code for the standard CNC machines considering different degrees of freedom (Fig. 12.1). Within these offline design tools, the final product geometry and the cutting paths are defined. Additional parameters, such as the depth of cut and the tool speed, which affect the machining result, are selected. The output of the offline code generation is exported in text files and can be eventually used from online robotic systems with the appropriate setup and cutting tool. Transformation between the physical setup and the offline generated code follows. These data are used as input to the online robot code generation tool. This tool is running on the robot controller side and allows the online generation of robot paths. The robot speed is adjusted to the robot side, as a parameter that affects the machining result. The final robot path is executed and tested on physical parts.

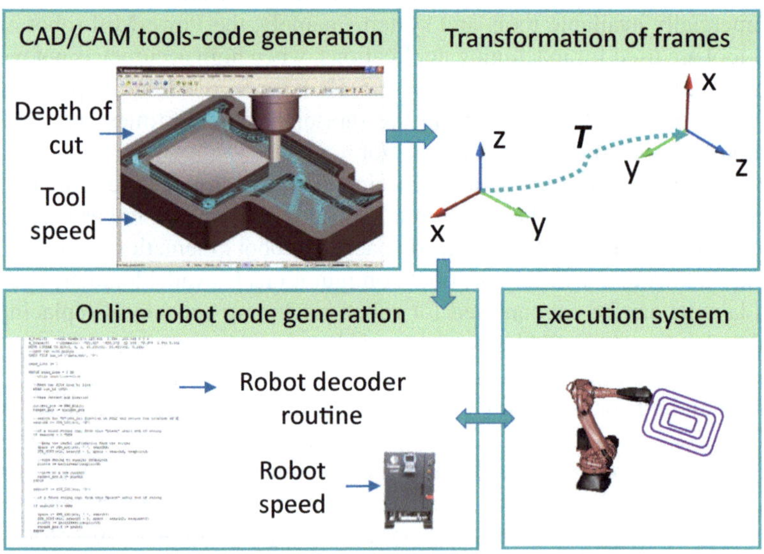

Fig. 12.1 From CAM data to online robot code generation

In order for the correct geometry to be cut by using the generated code of the CAM tools, transformation is required between the frames used in the offline tool and the robot cell. This transformation for the generation of the robot position in x, y, z from the tool position, in the offline CAM software tool, is calculated through the following mathematical expression:

$$\left[R_x R_y R_z 1\right]^T = T_w^f T_f^t T_b^f \left[T_x T_y T_z 1\right]^T \tag{12.1}$$

where

- $R_x R_y R_z$ are the calculated x, y, z coordinates of the robot;
- $T_x T_y T_z$ are the x, y, z coordinates of the tool, generated from the CAM tools;
- T_w^f is the transformation matrix from the workpiece to the flange frame;
- T_f^t is the transformation matrix from the robot flange to the robot tool frame;
- T_b^f is the transformation matrix from the robot base to the robot flange frame.

The last step of this method includes the online translation from the CAM code to the robot paths. This tool is developed on the robot's controller side and generally, it is to be applied to the different robot platforms (see Sect. 12.3).

The proposed method has been implemented as a robot decoder application into C4G and C5G versions of the COMAU robotic controllers. The CAD/CAD generated code, used as input, is generated with the help of CATIA (Computer Aided Three-Dimensional Interactive Application) from the Dassault Systèmes. Given the CAM code generation as input, in the form of a text file, the robot decoder application is developed (Fig. 12.2). This application reads the file line by line and decodes on each line the values of X, Y and Z variables. These variables are transformed from string to a real structure and are used for updating the current position of the robotic arm. Since the position is updated, the robotic motion is executed. This routine can be used in N number of lines, generated in the CAM code.

This method is quite general and can be extended for more degrees of freedom. It can be also applied as a generalized tool to different robotic platforms by establishing the robot decoder language format.

12.2.1 Case Study

The proposed approach has been applied to the machining of an MDF block of a complex geometry including linear, circular and spiral geometries. The experimental setup is presented in Fig. 12.3. The selected cutting tool has diameter 6 mm and can also support aluminum blocks. The maximum depth of cut is estimated to be 1.6 mm in order for overpressures and fragments to be avoided. This tool is integrated into a COMAU Smart Six robot of 6 kg payload and is placed vertically in the robot flange.

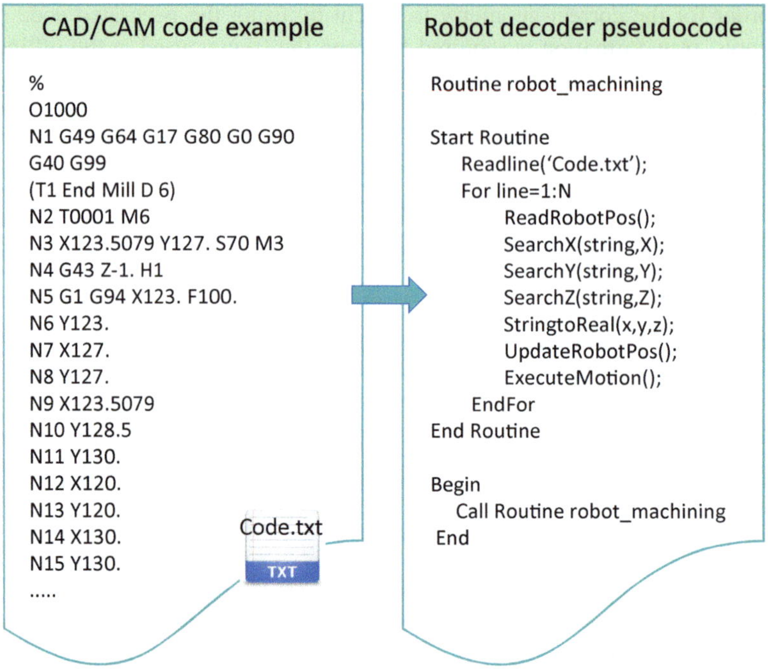

Fig. 12.2 Robot decoder-step by step

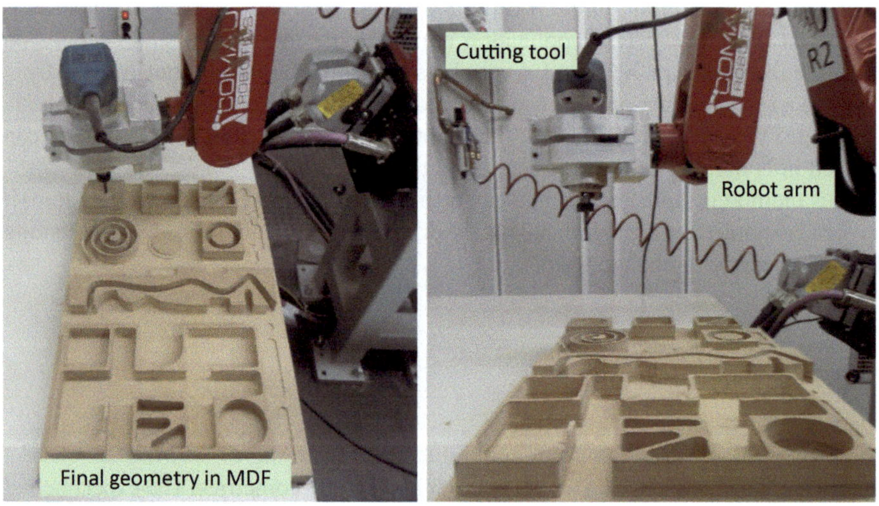

Fig. 12.3 Experimental setup

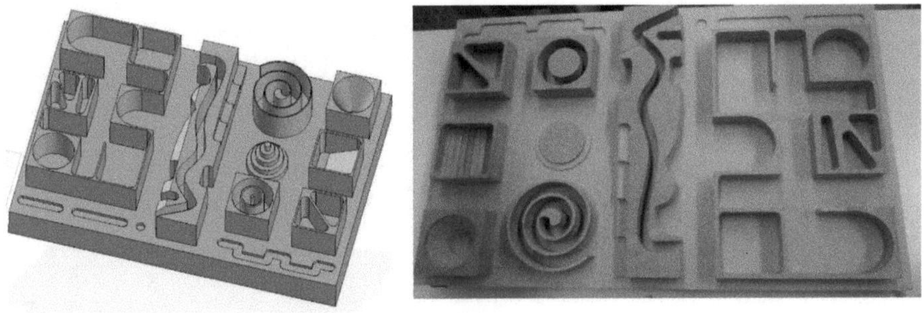

Fig. 12.4 Final part of the robot machining process

The block dimensions are 400 mm × 360 mm × 30 mm. The final part of the robot machining process is similar to a 3D CNC machine result (Fig. 12.4).

The generated code from the CAM/CAD is used as input to the robot's controller side. In the complex geometries namely, the spiral ones, the robot speed was reduced due to the increased robotic arm vibrations.

Besides the experiments, with reference to the feasibility of the robot's performance, in complex 3D shapes, experiments have been designed for investigating the influence of some parameters of the machining accuracy and efficiency. These parameters are the robotic speed, the cutting tool speed and the depth of cut. For this purpose, 16 experiments were performed with these different parameters. For each of these experiments, four linear geometries (L1-L4), three circular geometries (D1-D4) and five spiral geometries (S1-S5) were tested (Table 12.1).

The different geometries that were performed in each experiment are presented in Fig. 12.6. These geometries are part of the geometry, presented in Fig. 12.5. The distances L1 and L3 are equal and so are L2 and L4. The diameters D1-D3 and the densities S1-S5 are also the same according to the CAM files. In the 16 different experiments, during which the above geometries were cut by the robot, the measurements of the dimensions L1-L4, D1-D4 and S1-S5 showed some inaccuracies.

The variation between the minimum and maximum values, according to the expected result from the design files, is summarized in Table 12.2. The mean absolute error was estimated at 0.45 mm in the linear geometries L1, L3 and 0.225 and 0.275 in L2 and L4. In the circular geometries, this error ranged from 0.3 to 0.5 mm, while in the spiral geometries, the error ranged from 0.3 to 0.5 mm.

As an example, the measurements of the 16 experiments on the L1-L3 linear geometries and the S1-S5 spiral geometric dimensions are illustrated in Fig. 12.6. For each experiment, all the measured dimensions are visualized. Additionally, their minimum and maximum values are shown in red circles.

The circular geometries were the most difficult in comparison with the linear and spiral geometries, as regards the final quality. With the use of robot moving paths on the MDF block, the circular geometries had short linear lines around the expected circle, finally making it look as if it was regular a polygon geometry. This is more related to the increased roughness of the circular surfaces. The effect of the robot

Table 12.1 Set of experiments

Exp. No	Depth of cut (mm)	Cutting tool speed (rpm)	Robot speed (%)
1	1.6	10.000	5
2	1.6	15.000	10
3	1.6	20.000	15
4	1.6	25.000	20
5	1.8	10.000	10
6	1.8	15.000	5
7	1.8	20.000	20
8	1.8	25.000	15
9	2.0	10.000	15
10	2.0	15.000	20
11	2.0	20.000	5
12	2.0	25.000	10
13	2.2	10.000	20
14	2.2	15.000	15
15	2.2	20.000	10
16	2.2	25.000	5

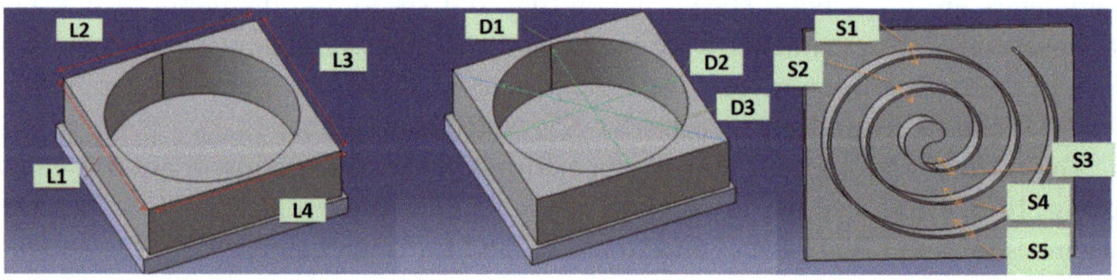

Fig. 12.5 Different geometries for parametrical experiments

Table 12.2 Errors during robotic machining in different geometries

(mm)	L1-L3	L2-L4	D1	D2	D3	S1	S2	S3	S4	S5
Expected dimension (mm)	81.5	79.7	68	68	68	3.1	3.1	3.1	3.1	3.1
Minimum measured dimension (mm)	81	79.35	67.9	68.2	68.05	2.9	2.55	2.55	2.55	2.45
Maximum measured dimension (mm)	81.9	79.9	68.5	68.8	68.75	3.5	3.15	3.2	3.15	3.45
Mean absolute error (mm)	0.45	0.275	0.3	0.5	0.4	0.3	0.3	0.325	0.3	0.5
Relative error (%)	0.55	0.35	0.44	0.74	0.59	9.68	9.68	10.48	9.68	16.13

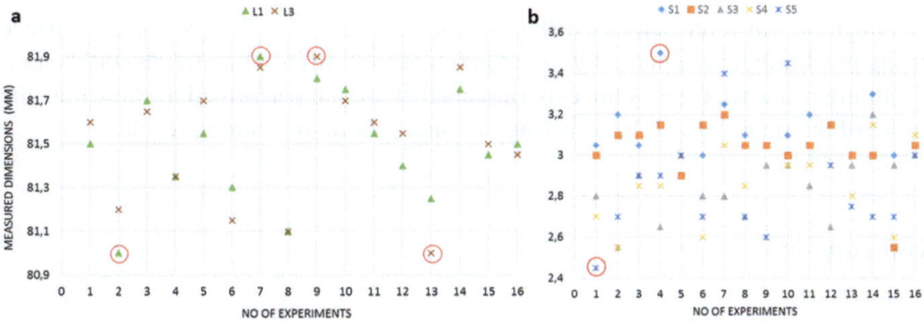

Fig. 12.6 **a** Linear geometries L1-L3; **b** spiral geometries—S1-S5

speed and the cutting tool's speed, in the final measured linear, circular and spiral geometries, was in order for the error to be increased since these parameters are increasing.

12.3 Discussion

This method was focused on developing a unified method for robotic machining in 3D geometries. Simple and complex geometries' cutting was tested with the proposed approach, while experiments were performed about the influence of the parameters. The main problems dealt with in robotic machining are oriented towards the final quality and the required accuracy. The main outcome of the proposed research study is summarized as follows:

- The generalized framework for robotic machining supported only the cutting on the X, Y and Z dimensions and was tested for complex geometries. The same approach with different transformations between the frames can be ex-tended for 6 or 9 degrees of freedom.
- The higher the robot speed, the cutting tool speed and the depth of cut, the more the errors in the expected dimensions
- The Robot Speed parameter is an important enough factor for machining accuracy, since the high feed rate speeds create vibrations onto the robotic arm as well as forces on the block surfaces. These facts can be absorbed but the stability and stiffness of the robotic arm will be reduced.
- Different cutting tool speeds generate different temperatures and different centrifugal forces within the tool, but the tool speed itself does not seem to affect significantly the robot's accuracy in machining. Despite that fact, this parameter is directly related to the quality of the final surfaces. As the tool speed in-creases, the surface gets smoother.

Future improvement of the proposed method is an extension to 6 and 9 degrees of freedom, by combing the multi-arm robotic system when it is required. Additionally, the integration of force sensors will allow the measurement of forces on the surface during machining. Last but not least, experiments on different materials will be part of a future research, besides the experiments that will be performed to show how the nature of the material will be affecting the final robotic machining result.

References

1. Karim A, Verl A (2013) Challenges and obstacles in robot-machining. IEEE ISR 2013. IEEE, Seoul, Korea (South), pp 1–4
2. Pan Z, Zhang H, Zhu Z, Wang J (2006) Chatter analysis of robotic machining process. J Mater Process Technol 173:301–309. https://doi.org/10.1016/j.jmatprotec.2005.11.033
3. Chryssolouris G (2006) Manufacturing systems: theory and practice, 2nd edn. Springer, New York
4. Pandremenos J, Doukas C, Stavropoulos P, Chryssolouris G Machining with robots: a critical review 9
5. Lehmann C, Halbauer M, Euhus D, Overbeck D (2012) Milling with industrial robots: strategies to reduce and compensate process force induced accuracy influences. In: Proceedings of 2012 IEEE 17th international conference on emerging technologies & factory automation (ETFA 2012). IEEE, Krakow, Poland, pp 1–4
6. Olabi A, Béarée R, Gibaru O, Damak M (2010) Feedrate planning for machining with industrial six-axis robots. Control Engineering Practice 18:471–482. https://doi.org/10.1016/j.conengprac.2010.01.004
7. Klimchik A, Bondarenko D, Pashkevich A, Briot S, Furet B (2014) Compliance error compensation in robotic-based milling. In: Ferrier J-L, Bernard A, Gusikhin O, Madani K (eds) Informatics in control, automation and robotics. Springer International Publishing, Cham, pp 197–216
8. Chen Y, Wei Y (2016) Simulation of a robot machining system based on heterogeneous-resolution representation. Computer-Aided Des Appl 13:77–85. https://doi.org/10.1080/16864360.2015.1059198
9. Huang H, Lin GCI (2003) Rapid and flexible prototyping through a dual-robot workcell. Robotics Comput Integ Manuf 19:263–272. https://doi.org/10.1016/S0736-5845(03)00022-X
10. Chen Y, Dong F (2013) Robot machining: recent development and future research issues. Int J Adv Manuf Technol 66:1489–1497. https://doi.org/10.1007/s00170-012-4433-4

Part III
Cooperating Robots: Human–Robot Collaboration

Chapter 13
Workplace Generation for Human–Robot Collaboration

13.1 Introduction

Easy reconfiguration and flexibility of assembly cells is required in case of new products or dynamic changes at the shop floor. On the one hand, the frequent release of new products to the market has to be realized in the shortest possible time in order for the competitiveness of a manufacturing company to be increased. On the other hand, Human Robot collaboration (HRC) is a promising concept for increasing the flexibility and reconfigurability of the production [1]. Human Robot (HR) involvement in assembly cells has been presented in [2–4].

This chapter addresses the need of evaluating multiple alternative layout designs and task allocation solutions for a hybrid HR design. The alternative solutions are automatically generated through a decision-making framework. The evaluation part of this framework includes various ways to measure how satisfying a human–robot collaboration workplace and task allocation is. A number of criteria are selected for the evaluation using both analytical models and simulation modules. The selection of the criteria is based upon the HR characteristics, as well as their unique capabilities. The final workplace layout is illustrated in 3D simulation environment.

The main benefits of the proposed research work involve the multiple criteria definition upon the user's requirements in order to enable the selection of a good result, regarding both the HRC layout and the task planning. Additionally, the consideration of humans and robots as a team is introduced in this work adopting a unified modelling of both active and passive resources. Last but not least, the integration of the decision-making framework with a 3D simulation tool allowed the calculation of criteria in a simulation mode as well as the visualization of the result. The main purpose of this research effort is decomposed in automatic design of hybrid HR layouts and task allocation in a short time frame.

S. Makris, *Cooperating Robots for Flexible Manufacturing*,
Springer Series in Advanced Manufacturing,
https://doi.org/10.1007/978-3-030-51591-1_13

13.2 State-of-the-Art Solutions

The facility layout problem of robots has been investigated in [5–9], but the task planning has not been considered in the problem formulation. The task planning algorithms for multi-agents teams, including both humans and robots have been investigated in [10, 11]. Decision making frameworks for task planning based on the evaluation of multiple criteria have been proposed in [12, 13], but have not been considered in the case of combining humans and robots. The problem of the robotic line design and the alternative configurations has been investigated in [14].

The evaluation of HRC has been investigated in [15], from an economical point of view, while research efforts in [16] considered metrics such as the robot and human idle time, the concurrent activity and the functional delay. Performance metrics for task-oriented human–robot interaction, involving humans and robots working as teams were presented in [17]. A human aware task planner was presented in [18], focusing on the "socially acceptable" plans for collaborative task achievement. The use of virtual simulation tools for the optimal ergonomic evaluation, considering the current safety regulations for HRC was presented in [19–21].

The problem of automatic generation of a workplace layout design has also been not new. It has been formulated as the 'facility layout' problem in [8, 9, 14], including only robots. The use of 3D simulation tools in those research works was restricted, especially in the case of considering the human tasks simulation. The lack of digital human models (DHMs) integration is the main reason, as well as the lack of methods for automatic path planning of HR tasks. An extended review on the simulation tools and their use in manufacturing has been presented in [22], while the role of DHMs has been investigated in [23].

13.3 Approach

The HR task planner for workplace and task allocation generation [24, 25] is implemented as a multi-criteria decision making framework. The overview of this framework is illustrated in (Fig. 13.1).

The basic steps that are implemented within this framework are explained as follows. The 3D modelling of the available resources and their characterization as active (humans, robots and their tools, grippers) and passive (working tables, equipment and the involved parts). This allows the framework to differentiate between the purpose and capabilities of each resource. In the same step a modelling of the workload is also carried out by considering a sequence of tasks where the active resources are assigned.

Every task can be executed by one or more active resources that are characterized as suitable resources. Suitability is a user defined constraint involving aspects that are not easy to be modelled in an automated way (e.g. ability of the gripper to grasp a part, maximum weight that can be lifted by the resource/human, ability to handle deformable objects etc.).

The decision-making framework uses as input the available active and passive resources, as well as the task list and their precedence constraints from the previous steps in order to generate alternative workplace and task planning solutions at the same time. As shown in Fig. 13.2 the decision-making takes places along several levels which involve:

- alternative position (x, y plane) and orientation (rotation around z) for each of the passive resources within the station;
- alternative allocation of tasks to suitable active resources;
- determination of active resource positions (x, y, z) inside the workplace.

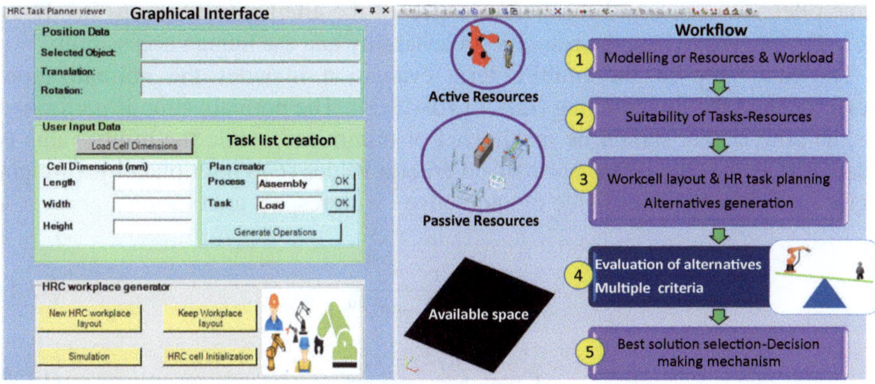

Fig. 13.1 Decision making framework overview

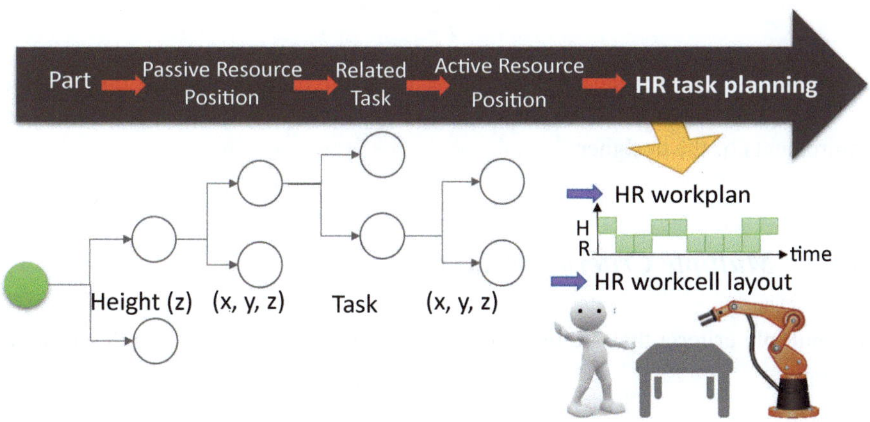

Fig. 13.2 Decision points of HR task planner

A number of criteria for the HRC evaluation have been selected and are estimated analytically or through the simulation result. The first one includes the optimal estimation of the passive resources' height based on the human height. This metric is based on the anthropometric analysis. Criteria for the human accessibility control as well as the HR reachability concerning the passive resources are also considered. The efficient floor space utilization, the task time to completion, the total lifted weight/resource and the investment cost are also calculated. Finally, ergonomics factors and human muscles strain are estimated in the simulation environment for each alternative workplace layout and planning.

The best alternative solution is selected following a multiple criteria evaluation process. Some of these criteria are characterized as benefit criteria and should be maximised in order for the alternative to be considered as good solution. The criteria whose values need to be minimized are considered as the cost criteria. Additionally, some rules are considered for the evaluation of the alternative solutions involving the anthropometric data, the resources reachability and the human accessibility.

The final step of the multi-criteria evaluation includes the normalization, weighting and the final ranking (Eqs. (13.1–13.3)). The normalization is performed for the criteria that should be maximized or minimized, while the utility value is calculated as the sum of the normalized criteria values (C_{ij}) multiplied by a weight factor (w_c). The C_{ij} is the value of alternative i with respect to criterion j.

$$C_{ij} = \frac{c_{ij} - c_j^{\min}}{c_j^{\max} - c_j^{\min}} \tag{13.1}$$

$$C_{ij} = \frac{c_j^{\max} - c_{ij}}{c_j^{\max} - c_j^{\min}} \tag{13.2}$$

$$U_i = \sum_{j=1}^{n} w_c C_{ij} \tag{13.3}$$

The final solution is selected on the basis of maximization of the utility value, U_i. Additional criteria can be implemented depending on other specifications or requirements of the designer.

13.3.1 *Multiple Criteria Evaluation*

The multiple criteria that are used for the alternatives' evaluation are described as follows:

Anthropometric data

The anthropometric data have been considered in the HR task planner for the design of the layout. The system is able to decide the height at which the tables and the

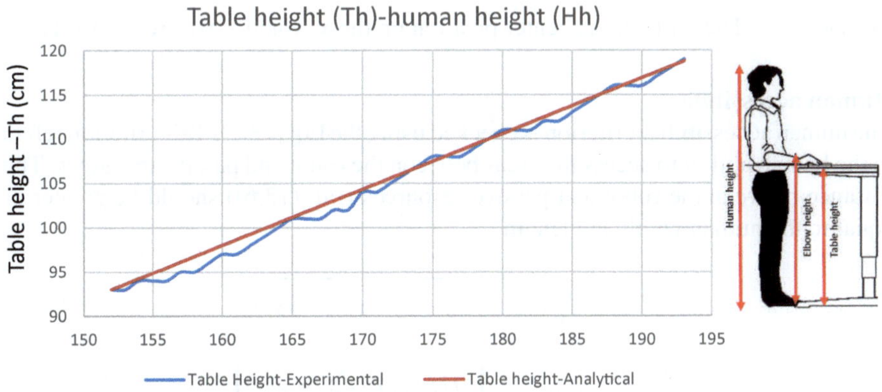

Fig. 13.3 Passive resource height with respect to human height (human is standing up)

parts must be located to ensure an ergonomic positioning in the case that human is standing up to work on them. The table height depends on the human elbows height, according to the anthropometric analysis [26]. According to the data represented in Fig. 13.3 the relation between the table height and the human height is almost linear, for human height range 150–195 cm. These data are analytically estimated using the Eq. (13.4).

$$T_h \cong 0.6286H_h - 2.543 \tag{13.4}$$

where

- T_h: Height of working table (cm)
- H_h: Height of human (cm).

Resources reachability

The reachability criterion of the robot resources is referred to their ability to move its joints and links in free space in order for a given target to be reached. The value of this criterion is given by the relation in Eq. (13.5).

$$Reachability = \begin{cases} true, if\sqrt{(x_R - x_P)^2 + (y_R - y_P)^2 + (z_R - z_P)^2} \geq WE \\ false, if\sqrt{(x_R - x_P)^2 + (y_R - y_P)^2 + (z_R - z_P)^2} \leq WE \end{cases} \tag{13.5}$$

where

WE	The robot's work envelope is the range of movement, measured in meters;
x_R, y_R, z_R	The robot's base frame position on the x, y and z axes respectively;

x_P, y_P, z_P : The part's mass center position on the x, y and z axes respectively.

Human accessibility
The human accessibility criterion is checked using the Eq. (13.7). This criterion refers to the human ability to access the areas between the robot and passive resources. The distance between the robot and passive resources (Eq. (13.6)) should be greater or equal to the human width and length.

$$D_{Ri}^{Rj} = \sqrt{(x_{Ri} - x_{Rj})^2 + (y_{Ri} - y_{Rj})^2} \tag{13.6}$$

$$\text{Accessibility} = \left\{ \begin{array}{l} true, if\, D_{Ri}^{Rj} \geq w_h \\ false, if\, D_{Ri}^{Rj} < w_h \end{array} \right\} \tag{13.7}$$

Floor space utilization
The following criteria that have been used in evaluation are characterised as cost criteria. The shop floor space (FS) utilization criterion for every active and passive resource is estimated with Eq. (13.8). This criterion should be minimized in order to have a good solution from the floor space utilization point of view.

$$FS = \left(x_i^{max} - x_i^{min}\right) * \left(y_i^{max} - y_i^{min}\right) \tag{13.8}$$

where

$x_i^{\max}, y_i^{\max}$ The maximum values of x, y positions for passive and active resources;
$x_i^{\min}, y_i^{\min}$ The minimum values of x, y positions for passive and active resources.

Task time to completion
The total time to complete human and robot tasks is one more criterion. It is estimated with Eq. (13.9) and should be minimized for the human resources during the evaluation process.

$$T_H = \sum_{i=1}^{n} T_c^i \tag{13.9}$$

where

T_c^i completion time of a task i that is assigned to a human or robot resource;
n the total number of tasks that have been assigned to a human or robot resource.

Total Weight of Lifted Parts/resource
The total weight of the lifted parts from human is the sum of the weight of all the parts that a human resource lifts (Eq. (13.10)).

$$W_H = \sum_{i=1}^{k} m * w_p \tag{13.10}$$

where

w_p the weight of a part that is lifted by a human resource;
m the iteration n of a part that is lifted by a human resource;
k the total number of parts that are lifted by a human resource.

Investment cost
The investment cost criterion (Eq. (13.11)) concerns the cost of the active and passive resources for a workplace layout. The target investment cost should be kept as low as possible.

$$C_{Assembly} = \sum\nolimits_{R=1}^{N_R} \prod C_R n_R \tag{13.11}$$

where

C_R Cost of active or passive resource;
n_R Number of active of passive resources;
N_R Maximum number of active and passive resources.

The simulation tool is also used for the estimation of the criteria values regarding the ergonomics analysis of human tasks. The following two criteria are considered in this framework.

Human muscles strain
The first criterion estimates the average human muscles strain (AMS) percentage (%) during a task execution using Eq. (13.12).

$$AMS = \frac{\sum_{i=1}^{n} MS_i}{n} \tag{13.12}$$

- whereMS_i refers to the maximum muscle strain percentage (%) for the human muscles that are involved in a specific task. This value is estimated during a task execution in the 3D simulation environment for each muscle.
- Where n refers to the number of muscles that are involved in a specific task.

Ergonomics factors
The second criterion includes the ergonomics factors (EF) level for the human tasks, taking into consideration the human poses during the simulation of each alternative planning. For the final score of the EF, 11 different human poses level are estimated in the 3D simulation environment. These levels are estimated as described in Table 13.1.

Table 13.1 Ergonomics factors

Human poses	Estimation of possible levels
Bending at the waist level (BW)	$\begin{Bmatrix} 1, BW \sim 0-15^\circ \\ 2, BW \sim 15-30^\circ \\ 3, BW > 30^\circ \end{Bmatrix}$
Waist rotation level (WR)	$\begin{Bmatrix} 1, WR \sim 0-15^\circ \\ 2, WR \sim 15-45^\circ \\ 3, WR > 45^\circ \end{Bmatrix}$
Arms height level (AH)	$\begin{Bmatrix} 1, if Armsarein Waistlevel \\ 2, if Armsareinshoulderlevel \\ 3, if armsareatuppershoulderlevel \end{Bmatrix}$
Knee bending level (KB)	$\begin{Bmatrix} 1, KB \sim 0-30^\circ \\ 2, KB \sim 30-60^\circ \\ 3, KB > 60^\circ \end{Bmatrix}$
Elbow restriction level (ER)	$\begin{Bmatrix} 1, ER \sim 0-90^\circ \\ 2, ER \sim 90-180^\circ \\ 3, ER > 180^\circ \end{Bmatrix}$
Parts removal level (PMR)	$\begin{Bmatrix} 1, for Easyhandling(withoutmoving) \\ 2, if \text{Arms can be withdrawn} \\ 3, for Difficultinsertion(needsattention) \end{Bmatrix}$
Working area level (WE)	$\begin{Bmatrix} 1, WE \sim 0-45^\circ \\ 2, WE \sim 45-90^\circ \\ 3, WE > 90^\circ \end{Bmatrix}$
Walking distance level (WD)	$\begin{Bmatrix} 1, WD \sim 0-4steps \\ 2, WD \sim 5-9steps \\ 3, WD > 9steps \end{Bmatrix}$
Handling level (HL)	$\begin{Bmatrix} 1, HL \sim 0-3kg \\ 2, HL \sim 3-5kg \\ 3, HL > 5kg \end{Bmatrix}$
Wrist rotation level (WRR)	$\begin{Bmatrix} 1, WRR \sim 0-15^\circ \\ 2, WRR \sim 15-45^\circ \\ 3, WRR > 45^\circ \end{Bmatrix}$

(continued)

Table 13.1 (continued)

Human poses	Estimation of possible levels
Neck rotation level (NR)	$\begin{Bmatrix} 1, BW \sim 0 - 15^\circ \\ 2, BW \sim 15 - 30^\circ \\ 3, BW > 30^\circ \end{Bmatrix}$

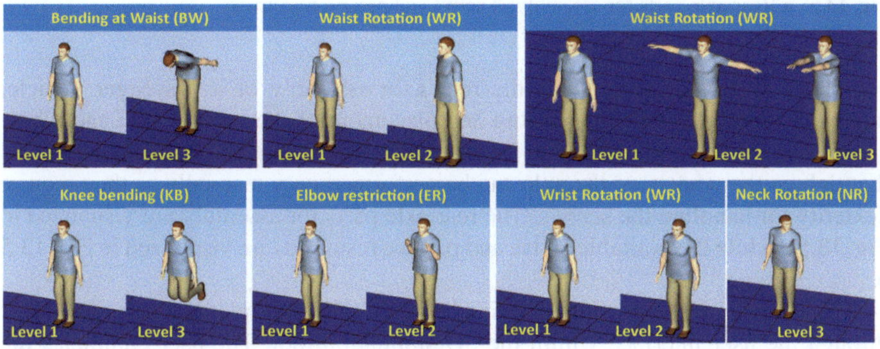

Fig. 13.4 Human poses and levels—an example

An example of some of the estimated levels based on the human poses in 3D simulation environment is illustrated in Fig. 13.4.

The total EF level is estimated for each alternative planning using Eq. (13.13).

$$\text{Ergonomics factors level} = \sum\nolimits_{j=1}^{k} EF_j \tag{13.9}$$

The research contribution of the proposed approach compared with existing works is mainly the concurrent generation and evaluation of both HR workplace layouts and workplans. This allows the design of the workplace to consider ergonomics implications coming from the positioning of resources as well as from the allocation of tasks to either humans or robots. New criteria can be easily integrated within the existing decision-making framework. The modelling of resources and workload is also important, enabling a new way of HR task planning. The integration of the proposed method within the 3D simulation tool can help researchers to overcome the use of spatial representation techniques for similar problems.

13.4 Industrial Example

The proposed framework has been applied to an automotive industry case study for the layout generation and the HR task planning. The assembly process to be

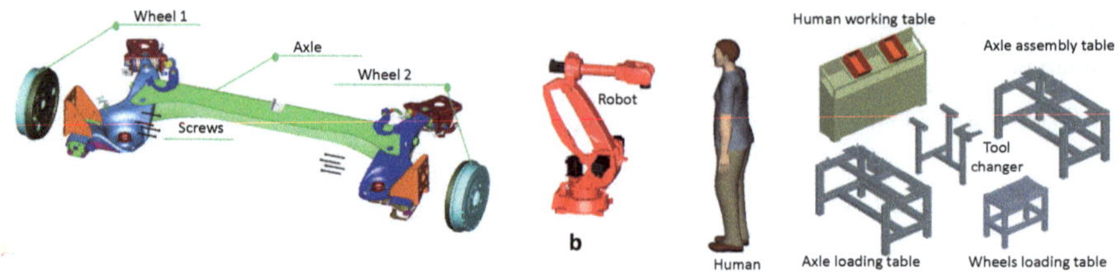

Fig. 13.5 **a** Rear axle parts for assembly; **b** active and passive resources

completed in this case includes the rear axle assembly of a passenger vehicle, involving thirteen tasks. The picking and placement of the vehicle axle and wheel groups can be performed by both human and robot resources. The screwing tasks that tighten the wheels on the axle can be performed only manually as they require dexterity for handling the screws. The rear axle parts for assembly are visualized in Fig. 13.5 a while the available active and passive resources are visualized in Fig. 13.5 b.

The alternative layout configurations and the HR task planning are generated through the user interface within the 3D simulation tool. The user completes the form with data such as the available floor space, the task list, as well as the available passive and active resources. The workplace layout is generated by pushing the button 'New HRC workplace layout' through the decision-making framework that has been implemented. The final result is visualized in the user within the simulation environment. The selected alternative has been evaluated against the criteria described in Sect. 13.2.

An example of four different good layouts that have been automatically generated by the proposed tool are visualized in Fig. 13.6. For these layouts, the human height is 170 cm and the human working table and the axle assembly table are estimated around 104 cm in height.

The reachability and human accessibility is satisfied in all these layouts as well. The total weight of lifted parts, the muscle strain and the investment cost is almost the same in the alternative solutions evaluated during the decision making process. The values of the floor space utilization, the total task time to completion and the ergonomics factors have significant changes during the evaluation process for a good solution selection. For the above 4 layouts, 150 alternatives are selected randomly as examples.

The floor space utilization is estimated in % value of the space that is occupied by all the active and passive resources. For the first layout, the 150 alternative layouts have average floor space utilization of 55.607%, while the average value is 66.163% for the second layout. The last two layouts alternatives have average values of 63,275 and 63,975% respectively. The decision-making framework evaluated a larger number of alternatives (around 30.000.000) and selected the alternative with the lower value of floor space utilization as the better one.

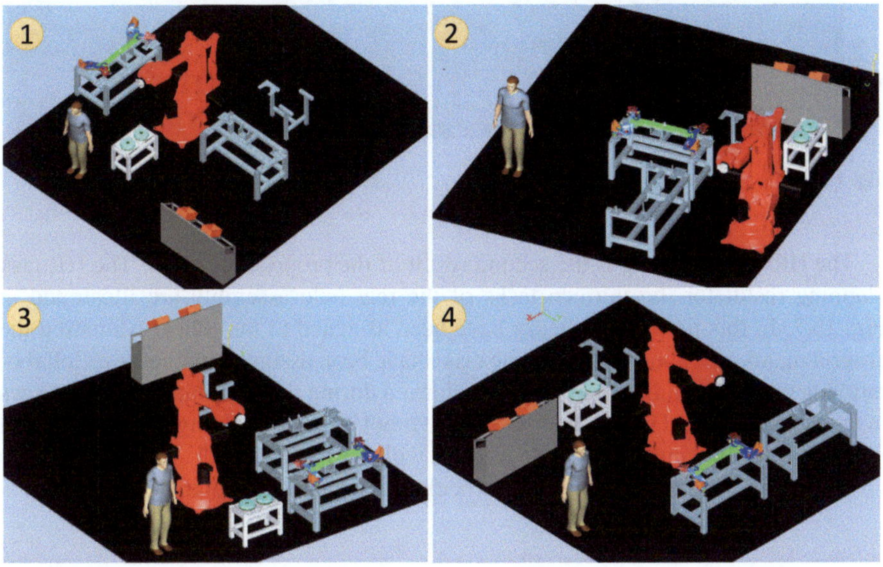

Fig. 13.6 HR workplace layout alternatives

The task time to completion for the 150 alternatives for each layout has values between 1.8 and 10.3 min. The solution with the lower task time to completion is selected as a good solution as well.

The last criterion that has significant changes in this demonstrator is the ergonomics factors estimation through the simulation environment. For the presented alternatives, there is a variance between 11 and 17 during the evaluation. The smaller that these values are, the better solution is selected from the ergonomics point of view.

The criteria values for the selected layouts presented in Fig. 13.6 are summarized in Table 13.2.

Table 13.2 Criteria values for 4 selected layouts

Criteria	Layout 1	Layout 2	Layout 3	Layout 4
Working tables height (cm)	104	104	104	104
Reachability (Boolean)	1	1	1	1
Human accessibility (Boolean)	1	1	1	1
Total weight of lifted parts (gr)	1500	1500	1500	1500
Floor space utilization (%)	54	53	59	66.5
Human muscle strain (%)	7	6.5	7.2	6.3
Task time to completion (min)	1.8	2.3	1.9	2.2
Ergonomics factors score	12	13	14	12

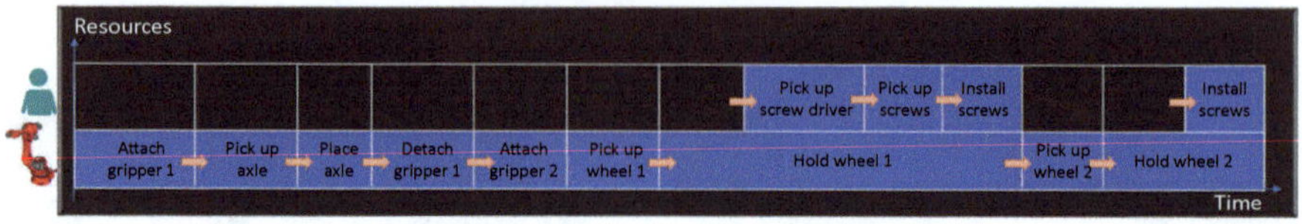

Fig. 13.7 HR task planning result

The HR task planning is the second result of the proposed method. The HR task planning result for the thirteen tasks of the rear axle assembly are illustrated in Fig. 13.7. In this figure some tasks have been assigned to the robot, evaluating the criteria mentioned above, while some tasks have been assigned to humans. Collaboration between the human and robot is achieved during the installation of the screws in the first and second wheel. In this task, the robot is active enabling the force control functionality allowing to human to interact with the robot as well. The human picks up the screws and install them in the wheels using a push-and-start screwdriver.

13.5 Discussion

The discussed method has been implemented within a 3D simulation tool, enabling the evaluation process with functionalities that are available in simulation mode. The selected 3D tool allows the simulation of human tasks as well, since digital human models (DHMs) are already integrated. The evaluation criteria for HR workplace includes anthropometric, space and cost efficiency, ergonomics issues etc. New criteria can be implemented involving other parameters according to the designer experience or new requirements and specifications. The advances of the proposed HR task planner are summarized as follows:

- The literature has not shown any method or tool for automatically generating a 3D Human robot workplace layout based on search algorithms and multiple criteria. The designers of workplaces are following empirical methods and steps to propose a layout, taking also into account the safety standards, but without evaluating the total number of feasible alternative configurations.
- In the past research some trials for 2D layout generation have been done using simple shapes. The proposed approach focused on the 3D layout generation using the dimension of the object and allowed 3D visualization.
- The design of a robotic cell as carried out by system integrators neglects the ergonomic positioning of components since the robots are not affected by such strains. Similarly, a workplace designed for humans fails to meet the reachability constraints of a robot since the operators can move inside the cell. The developed planner considers a more holistic perspective which allows to evaluate the complete production scenario, taking advantage of the capabilities offered by simulation.

- The existing simulation tools are not always open for integrating new applications for human robot tasks simulation. The task planner that is proposed is based on the development of a viewer that allows representation of the layout that has been generated and initial estimation of criteria values for the final evaluation of the assignments. The simulation of human and robot motions is possible but the lack of motion planners that automatically generate collision free paths for humans and robots are still open issues.
- During the selection of alternative configurations, the height of the passive resources and parts is selected according to the human height, in the case that human will work on this working table.

The HR workplace design and automatic task planning has been evaluated against nine different factors allowing easily to have initial results for the design and task assignment as follows:

- The reachability and human accessibility aspects in the layout of HR workplace allows in preliminary stage to automatically recognize if the position of passive resources allows the active resources to work efficiently in a task.
- The floor space utilization, the time to completion and the investment cost enables the selection of an efficient solution from the space, time and cost point of view considering a large number of possible configurations and task allocations.
- The anthropometric analysis regarding the selection of passive resources height based on human height, enables the efficient design of the layout from the human ergonomics point of view. Additionally, the total lifted weight, the muscles strain and the ergonomics factors are criteria enabling ergonomically efficient solutions for HRC.
- The feasibility of the evaluated HR task plans is also possible through the automatic rough simulation of human and robot tasks in simulation environment.

In the future, the integration of such methods in different 3D simulation platforms will help designers and engineers to propose easier a layout and a plan for HR tasks execution. Additionally, the integration of 3D simulation tools with automatic path planning applications for both humans and robots will allow automatic simulation of their tasks, improving the selection of alternative workplans.

References

1. Michalos G, Makris S, Spiliotopoulos J, Misios I, Tsarouchi P, Chryssolouris G (2014) ROBO-PARTNER: seamless human-robot cooperation for intelligent, flexible and safe operations in the assembly factories of the future. Procedia CIRP 23:71–76. https://doi.org/10.1016/j.procir.2014.10.079
2. Olsen R, Johansen K Assembly cell concept for human and robot in cooperation. 6
3. Krüger J, Lien TK, Verl A (2009) Cooperation of human and machines in assembly lines. CIRP Ann 58:628–646. https://doi.org/10.1016/j.cirp.2009.09.009

4. Tsarouchi P, Makris S, Michalos G, Matthaiakis A-S, Chatzigeorgiou X, Athanasatos A, Stefos M, Aivaliotis P, Chryssolouris G (2015) ROS based coordination of human robot cooperative assembly tasks-an industrial case study. Procedia CIRP 37:254–259. https://doi.org/10.1016/j.procir.2015.08.045
5. Kusiak A, Heragu SS (1987) The facility layout problem. Eur J Oper Res 29:229–251. https://doi.org/10.1016/0377-2217(87)90238-4
6. Meller RD, Gau K-Y (1996) The facility layout problem: recent and emerging trends and perspectives. J Manuf Syst 15:351–366. https://doi.org/10.1016/0278-6125(96)84198-7
7. Drira A, Pierreval H, Hajri-Gabouj S (2007) Facility layout problems: a survey. Annu Rev Control 31:255–267. https://doi.org/10.1016/j.arcontrol.2007.04.001
8. Aly MF, Abbas AT, Megahed SM (2010) Robot workspace estimation and base placement optimisation techniques for the conversion of conventional work cells into autonomous flexible manufacturing systems. Int J Comput Integr Manuf 23:1133–1148. https://doi.org/10.1080/0951192X.2010.528033
9. Tubaileh AS (2014) Layout of robot cells based on kinematic constraints. Int J Comput Integr Manuf 1–13. https://doi.org/10.1080/0951192X.2014.961552
10. Takata S, Hirano T (2011) Human and robot allocation method for hybrid assembly systems. CIRP Ann 60:9–12. https://doi.org/10.1016/j.cirp.2011.03.128
11. Galindo C, Fernández-Madrigal J-A, González J, Saffiotti A (2008) Robot task planning using semantic maps. Robot Auton Syst 56:955–966. https://doi.org/10.1016/j.robot.2008.08.007
12. Sabaei D, Erkoyuncu J, Roy R (2015) A review of multi-criteria decision making methods for enhanced maintenance delivery. Procedia CIRP 37:30–35. https://doi.org/10.1016/j.procir.2015.08.086
13. Buchert T, Neugebauer S, Schenker S, Lindow K, Stark R (2015) Multi-criteria decision making as a tool for sustainable product development—benefits and obstacles. Procedia CIRP 26:70–75. https://doi.org/10.1016/j.procir.2014.07.110
14. Michalos G, Makris S, Mourtzis D (2012) An intelligent search algorithm-based method to derive assembly line design alternatives. Int J Comput Integr Manuf 25:211–229. https://doi.org/10.1080/0951192X.2011.627949
15. Shen Y, Zastrow S, Graf J, Reinhart G (2016) An uncertainty-based evaluation approach for Human–Robot-cooperation within production systems. Procedia CIRP 41:376–381. https://doi.org/10.1016/j.procir.2015.12.023
16. Hoffman G (2019) Evaluating fluency in Human-Robot collaboration. IEEE Trans Hum-Mach Syst 49:209–218. https://doi.org/10.1109/THMS.2019.2904558
17. Saleh JA, Karray F, Morckos M (2012) A qualitative evaluation criterion for Human–Robot interaction system in achieving collective tasks. 2012 IEEE international conference on Fuzzy systems. IEEE, Brisbane, Australia, pp 1–8
18. Alili S, Warnier M, Ali M, Alami R Planning and plan-execution for Human–Robot cooperative task achievement 6
19. Ore F, Hanson L, Delfs N, Wiktorsson M Virtual evaluation of industrial Human–Robot cooperation: an automotive case study 9
20. Ore F, Hanson L, Delfs N, Wiktorsson M (2015) Human industrial robot collaboration—development and application of simulation software. Int J Hum Factors Model Simul 5:164. https://doi.org/10.1504/IJHFMS.2015.075362
21. Michalos G, Makris S, Tsarouchi P, Guasch T, Kontovrakis D, Chryssolouris G (2015) Design considerations for safe Human–Robot collaborative workplaces. Procedia CIRP 37:248–253. https://doi.org/10.1016/j.procir.2015.08.014
22. Mourtzis D, Doukas M, Bernidaki D (2014) Simulation in manufacturing: review and challenges. Procedia CIRP 25:213–229. https://doi.org/10.1016/j.procir.2014.10.032
23. Chaffin DB (2007) Human motion simulation for vehicle and workplace design. Hum Factors Ergon Manuf 17:475–484. https://doi.org/10.1002/hfm.20087
24. Tsarouchi P, Spiliotopoulos J, Michalos G, Koukas S, Athanasatos A, Makris S, Chryssolouris G (2016) A decision making framework for human robot collaborative workplace generation. Procedia CIRP 44:228–232. https://doi.org/10.1016/j.procir.2016.02.103

25. Tsarouchi P, Michalos G, Makris S, Athanasatos T, Dimoulas K, Chryssolouris G (2017) On a Human–Robot workplace design and task allocation system. Int J Comput Integr Manuf 30:1272–1279. https://doi.org/10.1080/0951192X.2017.1307524
26. Helander M, Helander M (2006) A guide to human factors and ergonomics, 2nd edn. CRC Taylor & Francis, Boca Raton, FL

Chapter 14
Dynamic Safety Zones in Human Robot Collaboration

14.1 Introduction

The recent automation technology advances have introduced the shared human robot (HR) workspaces concept for the future production lines. In this direction, the safety of human is a significant issue for research investigation. The need for real time communication in industrial safety systems enables the need for industrial certified safety solutions. Despite the fact that the HR safety aspects are really important for the industrial environments, there is no considerable research directed to such applications [1].

Safe HR interaction has been investigated from different points of view. Safety issues in physical HR interaction with lightweight robots [2–4], design of integrated sensors and systems [5, 6] and robot planning and control [7, 8] are some of these areas. The quantitative evaluation of safety limits for HR interaction and the identification of injury criteria for this purpose was investigated in [9, 10]. The above studies are mainly investigated for lightweight and humanoid robots.

Three industrial related safety strategies including (a) crash safety (e.g. robot power and force limiting), (b) active safety (e.g. vision systems) and (c) adaptive safety (e.g. online collision avoidance) have been identified [11]. In the same work, the integration of fenceless supervision systems in industrial applications was examined, following the related industrial standards [ISO 10218–1:2011, Robots for industrial environments—Safety requirements]. Despite the wide interest for HR collaboration in production lines, there are several drawbacks that prevent them from being widely adopted. Among them, the human safety issue is the most significant challenge for HR collaborative systems to be accepted.

Originally, ISO 10218 Part 1 ("Safety of Robots") and Part 2 ("Safety of Robot Integration") were intended to address workplace safety requirements for "assisting" robots working in a "collaborative workspace" with users. The two parts describe basic hazards associated with robots and provide requirements to eliminate,

S. Makris, *Cooperating Robots for Flexible Manufacturing*,
Springer Series in Advanced Manufacturing,
https://doi.org/10.1007/978-3-030-51591-1_14

or adequately reduce, the associated risks. In this standard, the following requirements are included: (a) Safety—related control system performance, (b) robot stopping functions, (c) speed control, (d) operational modes, (e) collaborative operation requirements and (f) axis limiting. The second part foresees further requirements on robotic integration such as: (a) general, (b) limiting robot motion, (c) integrated manufacturing system interface and (d) collaborative robot operation [12, 13].

The most recent ISO/TS 15066 (Robots and robotic Devices—Collaborative Robots) provides more concrete guidelines for collaborative robot operation in shared workspaces with humans: (a) establishing minimum separation distance, (b) establishing maximum safe speed, (c) tracking operator position and velocity, (d) determining and avoiding potential contact, (e) avoiding potential collision, (f) operator controls, (g) power and force limiting, (h) technological, medical/biomechanical requirements and ergonomic requirements and so forth [14].

Considering the HR safety related challenges and the indications by the EU standards, this chapter discusses the new generation of safety systems considered as enables for effective HRC. The analysis starts from the implementation of a 3D safety certified sensor combined with a non—certified depth sensor. This system allows the elimination of physical fences while constantly monitoring HR distance. As a next step, building upon this virtual fencing concept, this chapter also introduces the concept of dynamically changing virtual zones that allow the reduction of the allowed distance between humans and robots, boosting closer HR cooperation during assembly. Both systems have been deployed and validated through industrial examples from the automotive sector.

14.2 Approach

The discussed approach targets on facilitating fenceless environments where human operators can safely work close to robots. Specific focus is given on analyzing the safety related EU standards so to deploy HRC systems involving high payload robots. The safety concepts suggested have as a main enabler a 3D safety certified camera that can detect in real time any moving human or other object in the defined area. The discussion starts from a concept of replacing physical fences with virtual safety zones while monitoring human robot distance. As a next step, a more dynamic scheme is proposed by enhancing 3D safety zones with robot side safety functions while enabling the dynamic switching of zones to minimize the allowed HR distance.

14.2.1 Virtual Safety Zones' Configuration

The virtual safety zones configuration is the first safety system in the proposed method. A safe visual sensing device, the SICK SafetyEye [15], consisting of three cameras is used. This sensor is arranged in a fixed pattern, which jointly monitor a zone that can dynamically change. The performance of virtual zones is equal to the physical fences. The main difference is that with the physical fences, there is not option for easy reconfiguration of the HR cell with new safety setup and the human cannot work close to the robot. The virtual safety zones configuration through this safety device allows real time reconfiguration of the safety subsystem and fenceless HR cooperation (HRC). The overview of the safety zones configuration method is visualized in Fig. 14.1.

The monitoring area is the field of view of the sensing device. This sensing device allows to define different zone arrangements easily through a configurator tool. In each zone arrangement, a set of zones can be configured: (a) a warning (yellow) zone and detection (red) zone can be defined. A warning zone indicates that a human or an obstacle is present in the robot perimeter but not in a critical distance. In this case, the output signals of the sensor trigger an adjustment in robot behavior typically by speed reduction. A detection zone indicates that a human or any other moving object is near the robot. In this case, a safety rated monitored stop is triggered by the robot controller.

The distance between a detection zone and the danger source, in this case the robot, is called safety distance S. This distance is calculated by the following relationships [EN ISO 13855:2010].

$$\begin{aligned} S &= S_1 + S_2 \\ S &= (K * T) + C + Sa \end{aligned} \tag{14.1}$$

where

S Minimum distance in mm, measured from the start of the detection zone to the danger source.
S_1 Minimum width of the detection zone.
S_2 Distance between detection zone and danger source (robot).

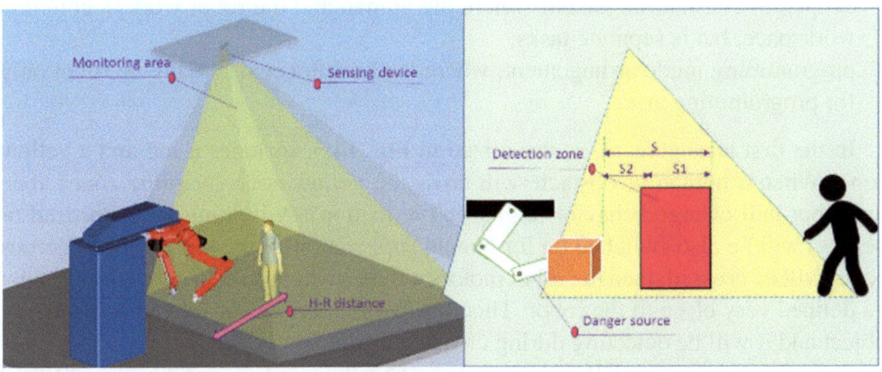

Fig. 14.1 HR monitoring area and detection zones

K The speed at which the object requiring detection (body) approaches the danger zone in mm/s. It is estimated as K = 1600 mm/s.
C Additional distance depending on the resolution. For the arm the resolution is estimated in 208 mm and for the body the resolution in 850 mm.
Sa Allowances for movement in x, y, z direction.
T Response time.

The response time T is estimated through the Eq. (14.2):

$$T = t1 + t2 \tag{14.2}$$

where

t1 sensing device response time. This time is depending on the high of the detection zones and the distance of the danger source. It has a range between 165 and 1715 ms;
t2 robot response time, t2 = 450 ms (in 75% speed for industrial robots).

14.2.2 *Real Time Human Robot Distance Monitoring—Static Virtual Zones*

The environment where the safety zones were applied involves a dual arm robot and a human working in cooperation for the assembly tasks. The view from the top of the sensing device, about 5.5 m from the floor, is visualized in Fig. 14.2. At least four markers for calibration and recognition are used in the cell as shown in the same figure.

Three different zones arrangements have been defined, depending on the type of the task that is executed. These arrangements are:

- robot task arrangement, where the human and robot work in separate workspaces;
- cooperative task arrangement, where the human and the robot work in common workspace, but is separate tasks;
- programming mode arrangement, where the human coexists with the robot only for programming tasks.

In the first arrangement, as visualized in Fig. 14.3, includes a red and a yellow zone. When a human or obstacle will cross the yellow zone (warning zone), then the robot will change behavior (e.g. speed reduction). A lighting indicator (red or yellow) will be also switched on for human supervision. If the red zone (dangerous zone) will be crossed, then the robot motors will be switched off. These zones cannot be defined very close to the robot. This is why the robot is considered as a moving object and it will be detecting during crossing a zone.

During a cooperative HR task execution, a walkway of warning zone is defined in order for the operator to approach the robot (Fig. 14.4). In this case the motors will not be switched off, allowing both human and robot being active. The dangerous

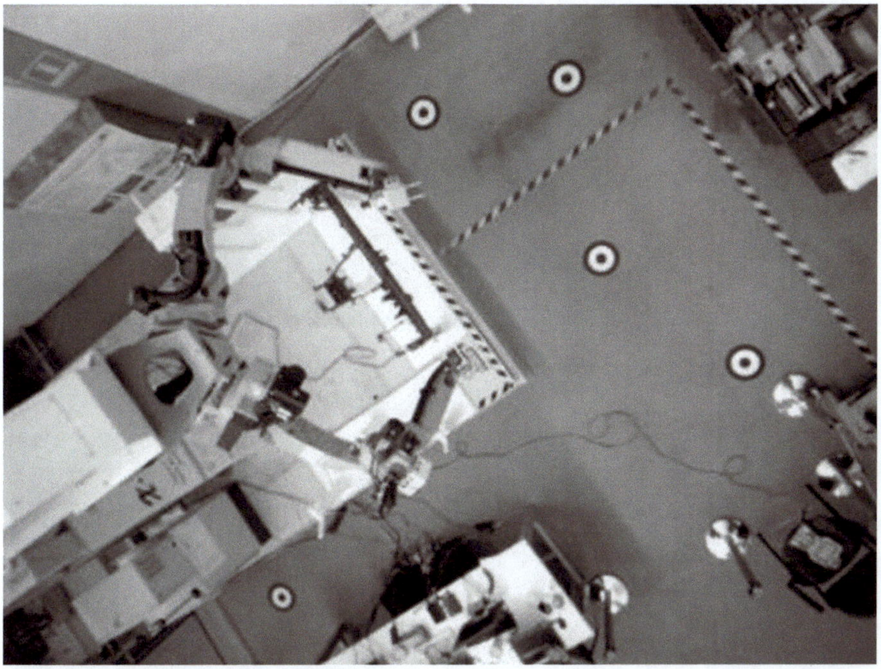

Fig. 14.2 SafetyEYE system—top view

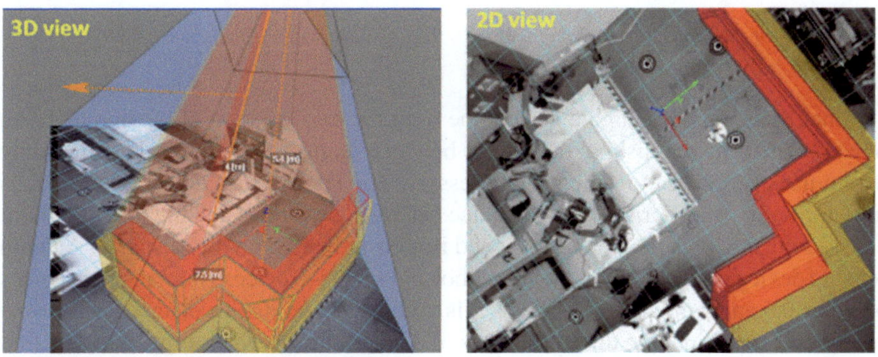

Fig. 14.3 Robot task arrangement

zones are defined for detecting a second human or object that will try to approach the robot.

The programming mode arrangement allows the human to work at the same area close to the robot during the programming of a new task. This zone arrangement includes only the warning zone (Fig. 14.5), where the supervisor is aware of the HR coexistence during programming.

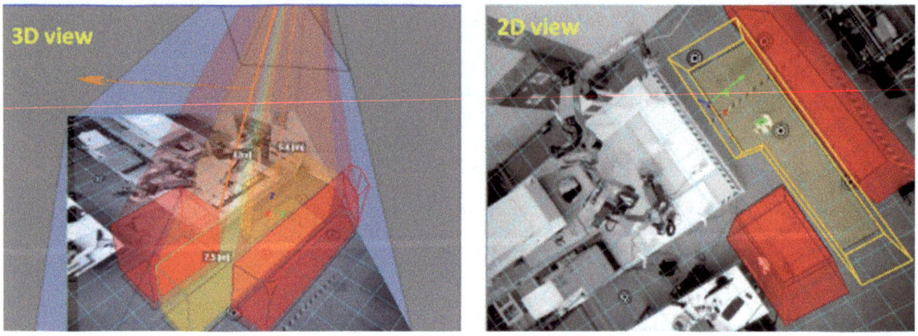

Fig. 14.4 Cooperative task arrangement

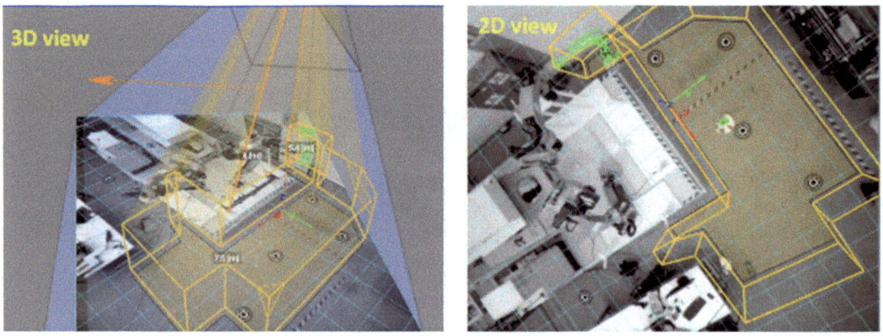

Fig. 14.5 Programming zone arrangement

Since the safety sensing device can be used for the detection of the human or an object without distinguishing difference between them, a module for detecting only the humans and their movements is necessary. The HR distance detection module is used for online monitoring of the distance between the human and the robot, without considering any objects. The robot speed is adjusted based on this distance. A depth sensing device is placed in front of the robot and is able to detect a human on distances between 0.5 and 4 m. A skeleton tracker is used for detecting this distance. The robot speed is adjusted in three different levels of distance, as shown in Fig. 14.6. The first level is when the HR distance, D, is equal of less than 1 m. In this case, the robot speed override value is adjusted from 0 to 10%. For distances between 1 and 1.5 m the robots peed changes between 10–20%, while for bigger distances the robot speed changes between 20 and 100%.

The relationship expressing the robot speed adjustment based on the distance D is summarized in Eq. (14.1).

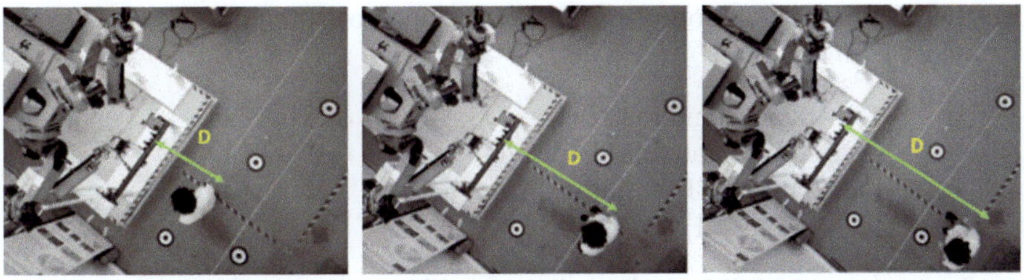

Fig. 14.6 HR distance monitoring through depth sensor

$$Robot\ speed\ adjustment = \begin{cases} 0 - 10\%\ SPD, & when\ D \leq 1\text{m} \\ 10 - 20\%\ SPD, & when\ 1\text{m} < D < 1.5\text{m} \\ 20 - 100\%\ SPD, & when\ D \geq 1.5\text{m} \end{cases} \tag{14.3}$$

The proposed multi-layer safety systems can be used either separately or concurrently, depending on the application area and the user preferences.

14.2.3 Dynamically Switching Safety Zones

The concept presented in the previous may increase the dynamic capability of the system and minimize the setup costs, since the industry can easily update the virtual fences based on its needs. However, the operator is still constrained out of the cell while the robot is working, without being able to actually collaborate with the robot by sharing a common task. The latter case, creates more challenges, indicating the need for a more flexible safety system being able to be adapted in the different needs of an assembly process.

The concept discussed in this chapter, is an effort of going towards to a more dynamic capability of the safety zones. The main idea behind this approach is to create monitoring zones around the robot instead of creating them around the cell. This minimizes a lot the covered area that is monitored by the safety camera, allowing the operators to work in areas inside the cell, closer to the robot.

The procedure followed for the design of dynamically changing safety zones system involves the following steps:

(1) Identification of the collaboration levels involved in the assembly process at defined in Fig. 14.1,
(2) Regulation of the layout of the cell based on the findings of a Risk Assessment procedure. This Risk Assessment is being performed focusing on the potential risks of human safety concerning the involved hardware in the cell such as the robot, fixtures, involved parts etc.

(3) Selection of the appropriate safety methods among the following as defined under ISO/TS 15066:

   (a) Safety-rated Monitored Stop—SOS
   (b) Speed and Separation Monitoring—SSM
   (c) Power and Force Limiting—PFL
   (d) Hand Guiding—HG.

(4) In the cases of SSM, analyze the assembly process and the required robot motions so to effectively design the different safety zone arrangements.
(5) Selection of the safety certified hardware devices that will be used for implementing the discussed functions such as cameras, enabling devices, safety PLCs, emergency buttons etc. are selected.
(6) Deploy the overall safety control logic that integrates the different components. It responsible for regulating the activation of each safety function at each point of the execution depending on the current phase of the scenario (Fig. 14.7).

According to the SSM method, the minimum separation distance between the human and the robot should be ensured at all time when this function is active. For the efficient monitoring of the area, the workspace has been divided into three zones (a) the Warning zone as defined in the previous section, similarly (b) the Danger zone and (c) the Robot safe zone. The latter is a safety function provided by the robot

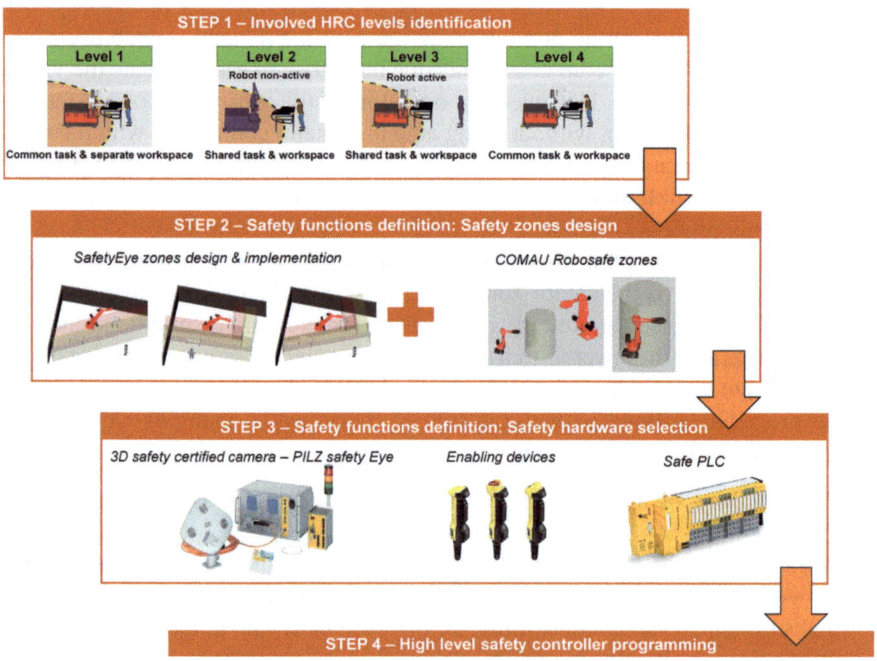

Fig. 14.7 Dynamically switching safety zones—Concept

controller providing cartesian position monitoring and constraint. During the entire execution of the assembly scenario a dedicated function of the robot, is responsible for monitoring and constraining the robot motion inside predefined Cartesian volumes. These volumes are user defined as 3D cubes for monitoring and/or constraining in real time the Cartesian position of the robot. The type of these volumes varies depending on the required functionality (monitoring, constraining, Safety-related Monitored Stop on violation etc.). For this work, the defined zones were set to issue a category 1 stop (cut power on the robot motors) in the case any joint of the robot approaches and tries to overcome the boundaries of the volumes.

Their geometrical shape may vary depending on the layout of the cell as long as their design respects the following two factors based on ISO/TS 15066: (a) the width of the warning and the danger zone should be calculated through an equation provided in the standard ISO 13855 (Eq. 14.1) based on the maximum speed of the robot in each case and b) when the zones are activated, their dimensions should ensure that there is no free space in the perimeter of the cell from which the operator could enter the workspace without being tracked by the monitoring system. Based on the manual of the SafetyEYE the minimum values for the zone length and height should be 200 mm and 1200 mm respectively, in order for the camera to be able to detect the human.

During each point of the assembly process one zone arrangement including the three described zones is active. The process followed for defining when each arrangement will be active includes (a) the definition of the range of rotation of robot axis 1 during the process and the division of this range (measured in degrees) into the desired of sectors. The number of sectors is mapped with the number of the desired zone arrangements. Then, during execution, while the robot is moving, passing from one sector to the other sequentially, the respective zone arrangements are activated. Indicatively, Fig. 14.8 shows three different zone arrangements for three different robot Axis 1 rotation points. 3D simulation is used for designing the zones respecting the minimum dimensions indicated by the EU legislation.

Based on human presence implementing the risk reduction functions, three different cases have been identified as follows:

- Case 1—Operator working outside the Warning zone (ISO 10218–2 Clause 5.11.5.3): In this case a *Safety Rated Monitoring Speed function* is applied that monitors the correct speed of the robot while the latter is moving in automatic mode. The robotic arm that is used in the current application can achieve the maximum speed of 2 m/s.
- Case 2—Operation working inside the Warning zone (ISO 10218–2 Clause 5.11.5.4): In this case two functions are applied. The *Safety Rated Reduced speed function* is applied to ensure the reduction and maintenance of the correct speed of the robot. According to ISO 10218—1 Clause 5.6, the maximum speed of the robot's TCP allowed when the operator is inside the collaborative workspace is 250 mm/s. In addition, a *Space Limiting function* is used to limiting the robot axe's motion inside pre-defined areas based on the needs of the execution. If the robot exceeds these areas a Category 1 Stop is generated.

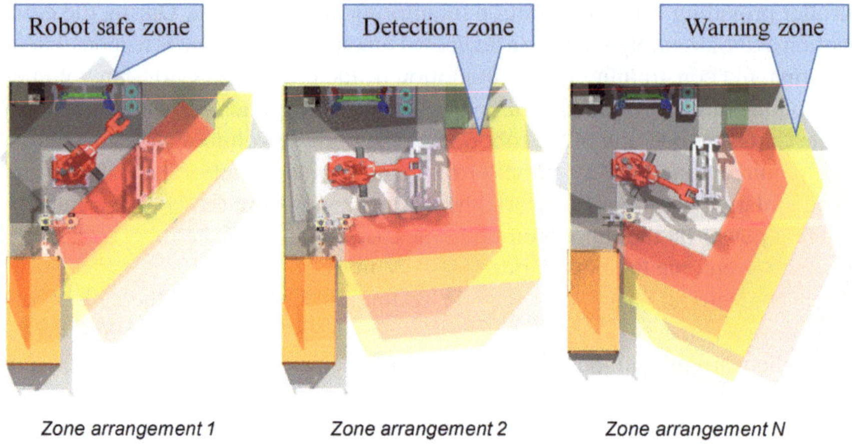

Fig. 14.8 Dynamic safety zone arrangements examples

- Case 3—Operator entering the Danger zone: In this case, the minimum separation distance is violated from the human. Thus, a *Safety Rated Monitored Stop function* is directly generating a Category 2—Protective Stop.

14.3 Implementation

14.3.1 Real Time Human Robot Distance Monitoring—Static Virtual Zones

The architecture of the proposed system (Fig. 14.9) includes the safety sensing that is the SafetyEYE provided by the company PILZ, connected through fiber-optic cables to the analysis unit and the second through Ethernet in the PSS 3000 control unit [16]. There is also a connection between the PSS and the robot controller through the SDM board, an Ethernet connection to a user interface for the configuration of safety zones and a connection on a lighting indicator. Within the robot controller, a program runs for enabling the change between the dynamic zones' configuration. Interface with the standard teach pendant buttons has been developed allowing to change zones by using three different buttons on it. The safety sensing device has been setup around 5.5 m from the floor, having an adequate field of view in front of the HR workcell.

For the robot speed adjustment module, there is a depth sensing device (Kinect Xbox) connected through USB cable in an external Linux PC. Within this PC there is the software module with the skeleton tracker and the depth monitoring functionalities. A connection between this software module and the robot controller is established through Ethernet. On the robot side, a server program runs, decoding the

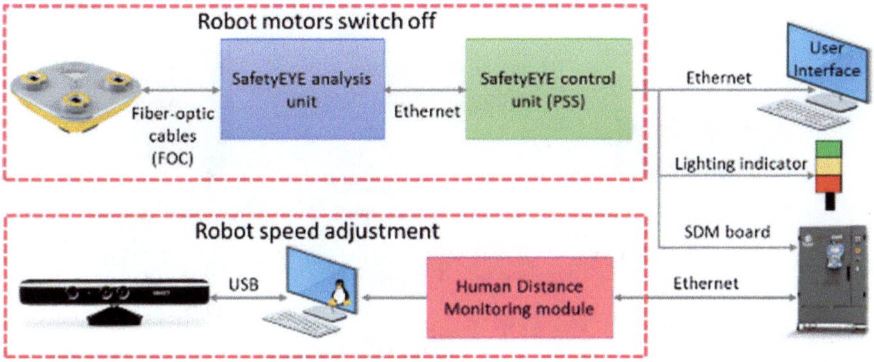

Fig. 14.9 Multi-layer safety system- System architecture

information received from the Linux PC and adjusting the robot speed accordingly to this information. Regarding the communication time for the depth sensor, there is a response time of 42 $\pm$ 4 ms, while the communication time with the robot controller is measured in 113 $\pm$ 6 ms.

14.3.2 Dynamically Switching Safety Zones

To enable the discussed concept, the safety architecture (Fig. 14.10) implemented involves the following three main modules:

(1) B&R Safe Logic PLC [17]—RoboSafe [18]: A PLC is integrated in the robot controller, providing embedded safety functions such as the robot safe zones. The output of this PLC is directly sent to the safety controller. In the current implementation two functions have been deployed. The first one focuses on the robot position monitoring. In more detail, a set of zones is designed and deployed in the system. The zones include the entire robot and are used for constraining robot movement. These zones are dynamically switched during the execution of the scenario with respect to the robot's Axis 1 position. The second safety function focuses on the robot speed monitoring. The speed limit for the execution of the scenario is 250 mm/s. When the manual guidance functionality is enabled, the robot speed is reduced to 20 mm/s.
(2) PILZ SafetyEye—Safe camera system: A certified safe camera system is used for real time monitoring of the collaborative workspace. This system includes the same component as described in the previous section. A set of specific zone arrangements has been deployed in the SafetyEye camera system. The zone arrangements are dynamically switched during the scenario's execution according to robot Axis 1 position, mapped to the respective RoboSafe zone.
(3) Safety Controller: This controller runs an external PLC that is used for handling all the signals coming from the safety modules of the scenario (SafetyEye,

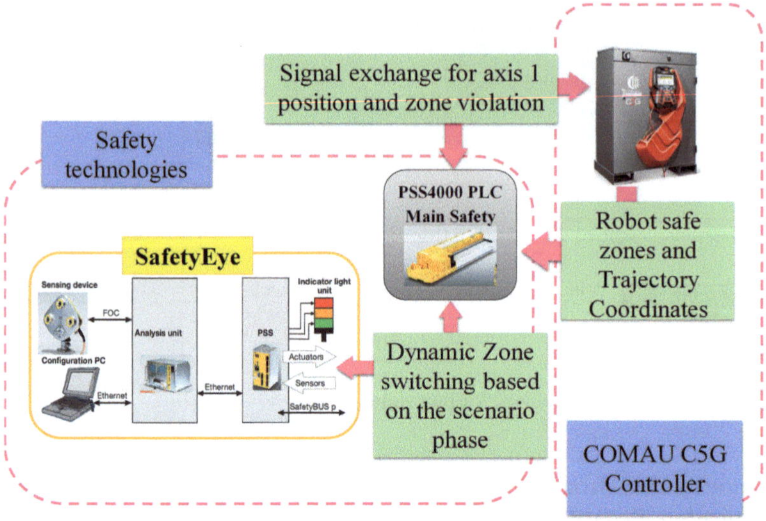

Fig. 14.10 Dynamically switching safety zones—System Architecture

B&R PLC, external safety devices). This PLC retrieves in real time the robot Axis 1 position and triggers the activation of the respective Robosafe and SafetyEye zones. Moreover, it receives all the inputs from the modules such as zone violation, external emergency button pushed, enabling device pushed, etc. After receiving such input, it translates it to requirement and sends the required input to the C5G controller for requesting Category 1 or Category 2, or speed reduction.

14.4 Industrial Example

14.4.1 Real Time Human Robot Distance Monitoring—Static Virtual Zones

The proposed multi-layer safety system has been tested for the HR collaborative assembly in an automotive industry case. The first task of this assembly was a robot task for the pick and place of a vehicle traverse on an assembly working table. The robot task arrangement has been used in this case and the robot can work on the 100% of its speed. As presented in Fig. 14.11, since a human is monitored in the red zone, the robot motors are switched off.

The second task was executed by a human for the placement of a fuse box on the traverse slot. In this case the cooperative configuration is enabled by the teach pendant, as well as the robot speed adjustment module. The human can cross the

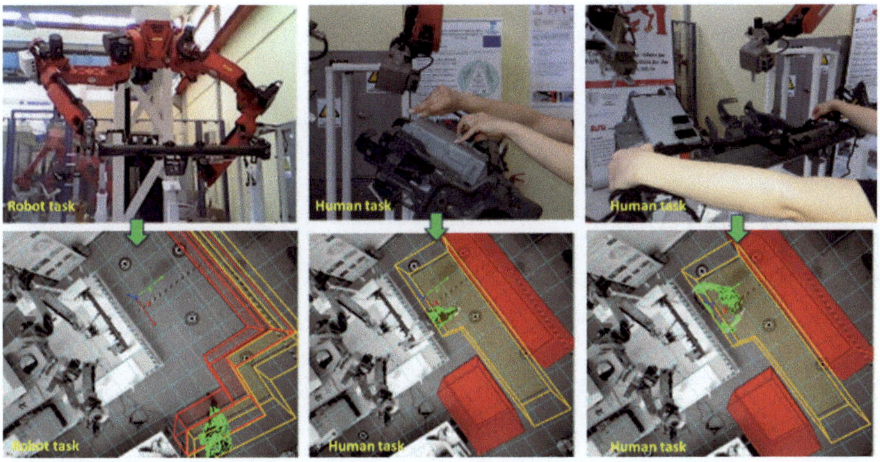

Fig. 14.11 Multi-layer safety system on HR collaborative case study

yellow walkway (Fig. 14.11) in order to approach the robot, while the robot is still moving. Since the human approaches, the robot in less than 0.5 m, then the robot speed is set to zero. When this task is completed and the human moves away, the robot speed is increased according to the detected distance. Similar to the second task, the third task execution follows by the human. A wire harness is assembled manually, while both safety and depth sensors monitor the human during execution. The robot speed is set to zero with the human works in distance less than 0.5 m.

The minimum safety distance S is calculated in 1.43 m and the longest reaction time is 265 ms for the dashboard assembly scenario. The experimental measurement of the robot speed adjustment based on the HR distance is presented in Fig. 14.12. When the human approaches the robot the distance is measured as being negative as visualized in the figure. The robot speed is reduced until the 0.5 m, while after this distance the robot speed is increased as the human moves away from the robot.

14.4.2 Dynamically Switching Safety Zones

The control architecture design and system implementation were applied on a case study derived from the automotive industry and more specifically the assembly line of a passenger's vehicle rear axle. The scenario was divided in two phases (Fig. 14.13).

During the first phase the robot is working by itself, using the two grippers that were designed specifically for this use case, an axle gripper that brings the axle from the kitting area (axle table) to the assembly area and the drum gripper that brings the drums from the kitting area (drum table) to the assembly area next to the axle. The robot is able to switch between these two grippers by using a pneumatic tool

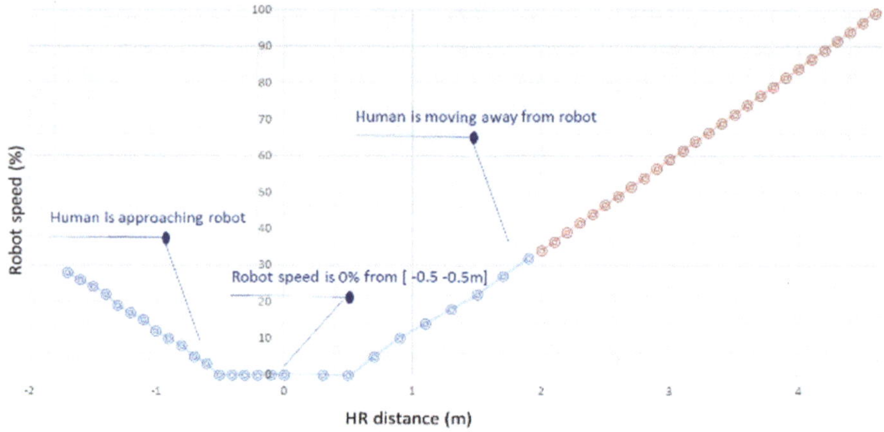

Fig. 14.12 Robot speed adjustment vs HR distance

changer mechanism. During this phase, the robot is moving freely inside the robot safe zones and the zone arrangement switching is taking place.

In the second phase, the human enters the red zone causing a deliberate violation in order to approach the assembly table and manually guide the robot that holds the drum and align its holes with the respective of the axle in order to screw them together. This action is performed for both drums, left and right. Since the human and the robot are very close, the SafetyEYE cannot be used to ensure safety. For this reason, COMAU C5G robot safety zones are used, that monitor the tool-center-point (TCP) position of the robot and constraining its motion inside these zones. The activation of the monitoring zone is performed by pressing two enabling devices that are located in this area, ensuring at the same time that both hands of the operator are occupied and that they are not between the drum and the axle by accident, having the danger of getting injured. Once the two parts, drum and axle, are perfectly aligned the screwing operation is taking place.

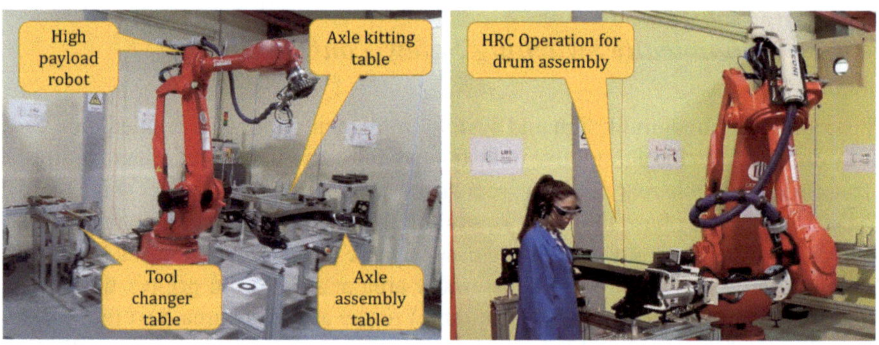

Fig. 14.13 HRC Rear axle assembly cell

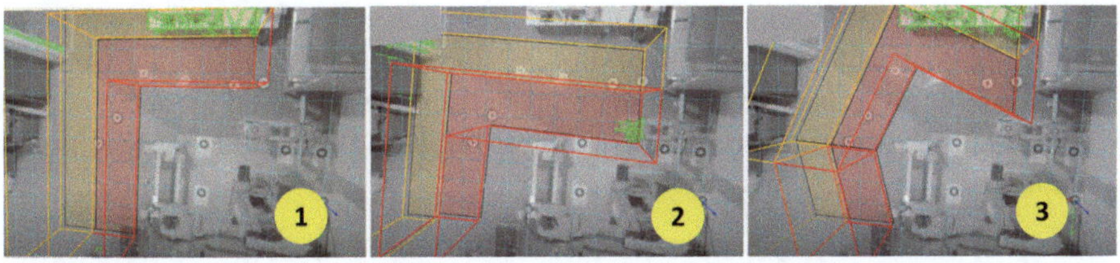

Fig. 14.14 Implemented dynamically changing zone arrangements (three out of the total nine)

During the first phase, nine zone arrangements have been deployed in the SafetyEye camera system. Each zone arrangement includes a detection zone (red zone) and a warning zone (yellow zone) and they are dynamically switched during the scenario's execution according to robot Axis 1 position, mapped to the respective RoboSafe zone. Each red zone is the extension of the respective Robosafe zone. When the human operator violates the warning zone a request for robot speed reduction to 250 mm/s is sent to COMAU C5G controller [19] through the PILZ PSS4000 PLC [20]. If the operator enters the detection zone, then a request for CAT 2 robot stop is sent to C5G controller (Fig. 14.14).

To evaluate the ability of the safety-related parts of control systems to perform the deployed safety functions under foreseeable conditions the metric Performance Level (PL) has been defined. The implemented system has been validated against this metric resulting in "d" PL. This value indicates that the Probability of Dangerous Failures for Hour (PFHd) is higher than 10^{-7} and lower than 10^{-6}.

14.5 Discussion

This chapter discussed the deployment of enhanced safety systems able to support fenceless human robot collaborative assembly. The combination of 3D certified sensors with non-certified sensors has been investigated for achieving seamless common workspace monitoring. Building on top of this static virtual zones' setup, a advancements has been proposed, integrating certified sensors with the robot's safety PLCs targeting in more flexible safety control schemes. Under this concept, the dynamically switching safety zones approach has been introduced for optimizing the human robot common workspace allowing their close co-existence and collaboration.

Except from the advantages of the proposed system, there are some issues that could be improved such as:

- There are disturbances related to the different lighting conditions in the environment. In this case, there is a possibility that an object or a human shadow is recognized as a real object or human. In this case, the PSS needs a reset, while there is not any problem with the human safety.

- The issue of zone violations due to disturbances created by reflections or extra light on equipment, objects on the floor and the floor itself needs also improvement. This is strongly depended on the material and surfaces involved in the work cell.
- The reduction of the camera sensitivity towards lighting conditions changes would improve significantly this problem.

The proposed safety system may be further enhanced by involving more stable and industrial sensors for the human monitoring. One direction could be the use of laser scanners, for the detection of both humans and moving objects in combination with 3D vision systems. Safety skins, light curtains, safety mat systems are some potential solutions that could be investigated as well. Since the response time is very important in safety systems, investigation on more reliable solutions apart from the SafetyEye that is already certified is one more issue for the future.

References

1. Tsarouchi P, Makris S, Chryssolouris G (2016) Human–robot interaction review and challenges on task planning and programming. Int J Comput Integr Manuf 29:916–931. https://doi.org/10.1080/0951192X.2015.1130251
2. De Santis A, Siciliano B, De Luca A, Bicchi A (2008) An atlas of physical human–robot interaction. Mech Mach Theory 43:253–270. https://doi.org/10.1016/j.mechmachtheory.2007.03.003
3. Vogel J, Haddadin S, Jarosiewicz B, Simeral JD, Bacher D, Hochberg LR, Donoghue JP, van der Smagt P (2015) An assistive decision-and-control architecture for force-sensitive hand-arm systems driven by human–machine interfaces. Int J Robot Res 34:763–780. https://doi.org/10.1177/0278364914561535
4. Cherubini A, Passama R, Crosnier A, Lasnier A, Fraisse P (2016) Collaborative manufacturing with physical human–robot interaction. Robot Comput-Integr Manuf 40:1–13. https://doi.org/10.1016/j.rcim.2015.12.007
5. Bdiwi M (2014) Integrated sensors system for human safety during cooperating with industrial robots for handing-over and assembling tasks. Procedia CIRP 23:65–70. https://doi.org/10.1016/j.procir.2014.10.099
6. De Luca A, Flacco F (2012) Integrated control for pHRI: Collision avoidance, detection, reaction and collaboration. 2012 4th IEEE RAS & EMBS International Conference on Biomedical Robotics and Biomechatronics (BioRob). IEEE, Rome, Italy, pp 288–295
7. Kulić D, Croft E (2007) Pre-collision safety strategies for human-robot interaction. Auton Robots 22:149–164. https://doi.org/10.1007/s10514-006-9009-4
8. Kulić D, Croft EA (2006) Real-time safety for human–robot interaction. Robot Auton Syst 54:1–12. https://doi.org/10.1016/j.robot.2005.10.005
9. Heinzmann J, Zelinsky A (2003) Quantitative safety guarantees for physical human-robot interaction. Int J Robot Res 22:479–504. https://doi.org/10.1177/02783649030227004
10. Cordero CA, Carbone G, Ceccarelli M, Echávarri J, Muñoz JL (2014) Experimental tests in human–robot collision evaluation and characterization of a new safety index for robot operation. Mech Mach Theory 80:184–199. https://doi.org/10.1016/j.mechmachtheory.2014.06.004
11. Michalos G, Makris S, Tsarouchi P, Guasch T, Kontovrakis D, Chryssolouris G (2015) Design considerations for safe human-robot collaborative workplaces. Procedia CIRP 37:248–253. https://doi.org/10.1016/j.procir.2015.08.014
12. URL ISO 10218–1:2011. https://www.iso.org/standard/51330.html
13. URL ISO 10218–2:2011. https://www.iso.org/standard/41571.html

14. URL ISO/TS 15066:2016. https://www.iso.org/standard/62996.html
15. URL PILZ Safety Eye. https://www.pilz.com/en-INT/eshop/00106002207042/SafetyEYE-Safe-camera-system. Accessed 25 Dec 2019
16. URL PILZ PSS3000 Function Block. https://www.pilz.com/en-INT/eshop/00102002117039/PSS-3000-function-blocks. Accessed 18 Dec 2019
17. URL B&R Industrial Automation. https://www.br-automation.com/en/. Accessed 15 Dec 2019
18. URL COMAU. https://www.comau.com/en. Accessed 17 Dec 2019
19. URL COMAU C5G Controller. https://www.comau.com/en/our-competences/robotics/controls/c5g
20. URL PILZ PSS4000 PLC. https://www.pilz.com/en-INT/products/automation-system-pss-4000. Accessed 27 Dec 2019

Chapter 15
Seamless Human–Robot Interaction

15.1 Introduction

Assembly processes that are executed in current production lines, require flexible systems that could handle parts with different characteristics and materials. Since these parts can behave differently during the handling operations and show unpredictable behavior (upholstery, rubber, fabric etc.), human operators, either single or group, have been selected to process them, using their cognitive and handling capabilities [1, 2]. Additionally, the use of human operators enables the companies to update their products more often, introducing new models and more variants per model, compared to the use of automation technologies and mass production equipment, which don't share the flexibility of humans [1, 3].

A production system can be defined as flexible and adaptable based on its sensitivity to changes, either internal or external, while several paradigms have been used to manage these dynamics such as holonic, flexible, lean, reconfigurable, evolvable, self-organizing and autonomous assembly systems that have been partially realized in the last decades [4–7]. Apart from the above, industries aim to high quality products for their customers and for this reason they target on the quality of the production processes they are using, in terms of precision and repeatability. This would help them, reduce throughput time, increase task traceability and minimize ergonomic stress from the operators.

Automation systems seem the key solution that is been researched and has been introduced to the assembly lines. Nevertheless, poor robot acceptance has been detected in 90 European companies (61% use 1 to 10 robots, 32% use 11 to 50 robots), according to the European Business Test Panel (EBTP), due to the challenges that exist for adopting the automation solution, such as lack of sophisticated safety systems [8].

For this reason, both the research and the industrial communities focus a lot on hybrid production systems, where robots can work alongside humans, exploiting the full potential of both entities [9]. More specifically, intelligence and cognitive skills

S. Makris, *Cooperating Robots for Flexible Manufacturing*,
Springer Series in Advanced Manufacturing,
https://doi.org/10.1007/978-3-030-51591-1_15

offered by the human operators are combined with the repeatability, dexterity and strength of the robots creating a promising co-existence, sharing both workplace and tasks [10]. New projects [11–13] and products [14, 15] have been introduced for the exploitation of the flexibility and productivity potential [16] of these hybrid systems. Additionally, the latest trend is the use of mobile dual arm robots that can be used as human operators, supporting the latter ones and increasing the reconfigurability of the system [16]. In order to achieve true HRC a number of different technologies should be combined together to support humans as a valuable asset in the production lines [17], as explained above due to their intelligence and dynamically in small-scale production [18]. Nevertheless, this chapter focuses on a key technology that enables the seamless bilateral communication among the human and robot resources.

Augmented Reality (AR) is an impressive technology that gains public interest increasingly during the last few years. This is based on a general concept of Mixed Reality (MR) where real and digital information are merged together in the user's field of view, seaming as a unique environment [19, 20]. Until few years ago, AR application focused mainly on mobility either for commercially available products or for custom applications [21–23]. Nevertheless, in the last years more and more manufacturing applications appear, supporting engineers and operators on different tasks. Indicative examples of such application are in the collaborative product and assembly design [24] where easier product personalization can be achieved through specifically developed AR tools [25, 26]. Another example involves the delivery of remote assistance and instructions to the operators during the execution of an assembly process [27] or during the execution of remote maintenance operations [28]. Additionally to the above, AR has been also used to increase human safety feeling in HRC environments [29] or increase his awareness about the next production steps through the provision of text based [30] and 3D CAD model [31] based assembly instructions. Finally, additional information that has been sent to the human include, but are not limited to, visual/audio alerts, upcoming robot trajectories and robot working zones [32, 33].

All the aforementioned applications are still on research level, despite the progress done a HRI a two-way communication channel should be established between human and robot.

15.2 HRI Functionalities for the Programming Phase

15.2.1 User Calibration

The proposed tool has been designed and implemented in such way to adapt easily to the specificities of the workstation and the operator that is using it. Since each operator usually needs to work in several workstations, the application should be initiated for the specific workstation and provide the related information. Additionally, the operators have different heights having different field of views when looking at the

Fig. 15.1 Field of view of AR glasses—User calibration

same object. For this reason, the digital objects should be adapted to this deviation, appearing in the correct scale and position related to the other objects. In order to address the aforementioned, a calibration process takes place when the application is launched as displayed in Fig. 15.1. A marker is used for this purpose by the human, which is positioned on a specific and already known by the program location in the assembly area. Based on the distance between the marker and the camera, operator height is calculated. Following this, a QR code is used to read information about the specific assembly area to help the system understand the exact station where the operator is located. Following these two recognition processes, the user is ready to start using the AR application.

15.2.2 Programming Phase

In current approach, robot programming is performed offline by robot experts. They are programming each task with high accuracy, based on the station characteristics at the time such as the layout or the parts involved. The limitation that exists on this approach is the lack of dynamic changes. If the assembly process or the product changes then the production needs to get offline again and the experts perform again the programming or fine-tune manually the previous one. This leads the industry to great cost and time losses, impacting their productivity. In order to avoid this, below two different robot control mechanisms are provided to enable human operators to easily and quickly instruct the robot without needing any expertise in robotics. In order to achieve this, two functionalities have been provided to him/her, comprising the programming phase of the application.

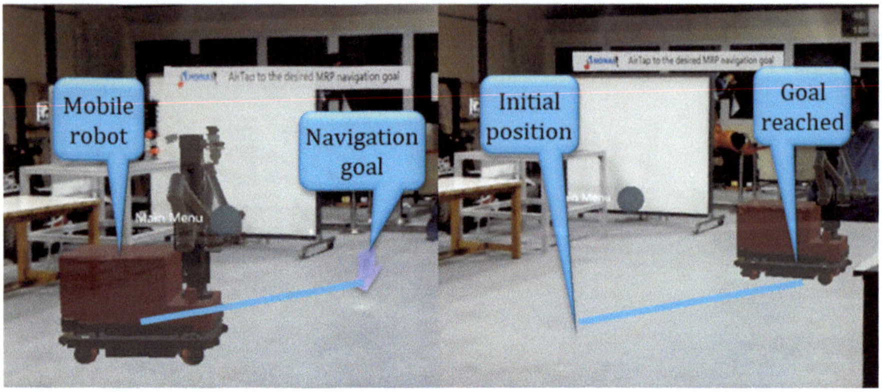

Fig. 15.2 Initial and final position of mobile platform

15.2.2.1 *Direct* Robot Navigation Instructions

In the first functionality, the operator is able to give new navigation goals to the mobile platform, different from the initially programmed. As a result, the robot could go to different workstations online, even if it has not been programmed to do so, to cover other production requirements. In the following Fig. 15.2, is visualized this procedure. The operator is able to make a simple AirTap in the desired location he/she wants to move the mobile platform to. This destination is translated to mobile platform coordinates and is fed to the path planner as input, which generates the optimized path the platform should follow. Then this path is sent to the platform for execution.

15.2.2.2 Robot Teleoperation

The second functionality is similar to the first one, with the only difference now is that the operator is able to move the mobile robot based on a predefined offset, enabling him/her to make small adjustments and real time corrections on its position. The operator sees in his/her field of view a cross pad composed of four virtual buttons. Using AirTap gesture, the user is able to press one them and the platform moves towards this direction. It is worth mentioning that the pad follows platform's translational and rotational movements in order to have aligned orientation with it. The four buttons refer to the following movements (a) forward movement, (b) backward movement, (c) left rotation, d) right rotation. The AR environment as it is visualized in operator's field of view, is displayed in Fig. 15.3.

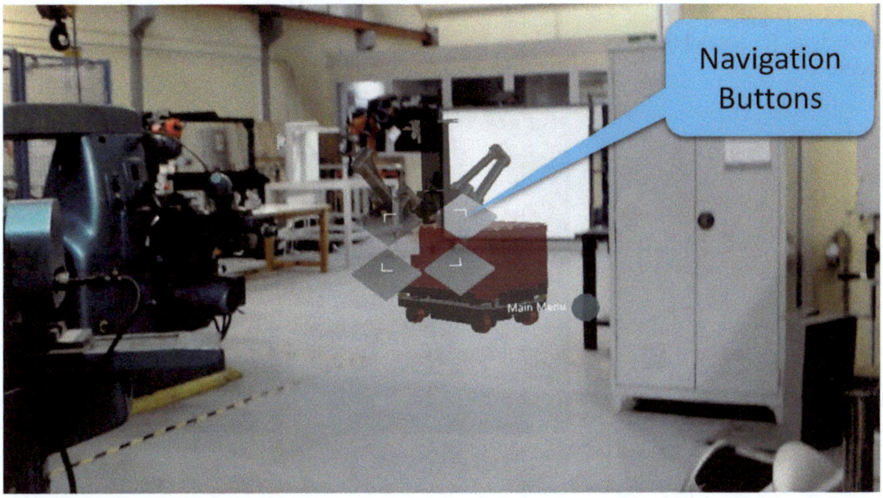

Fig. 15.3 Mobile platform direct position corrections

15.3 HRI Functionalities for the Execution Phase

Apart from the programming phase, several functionalities are provided to the operators for the execution phase. The aim of those functionalities is not focusing on changing the robot configuration, program or task sequence but to support the operator in his/her tasks, increase his/her awareness for hazardous cases in the station and enable the collaboration with the other resources. These functionalities are explained in detail in the following sections.

15.3.1 Assembly Process Information

This functionality focusses on providing information in the form of 3D models and textual instruction to the operators about the task they need to perform. More specifically, the real components the operator needs to use, such as screws, tools, other parts etc., are superimposed on the real ones on the position they need to be installed. For example, in Fig. 15.4, are visualized a drum's 3D model, 4 bolts and the way they should be placed onto the real object (axle).

15.3.2 Robot Motion and Workspace Visualization

Apart from the assembly process, the operator should be informed about the status of the other resources in the cell and more specifically about the robot. Providing

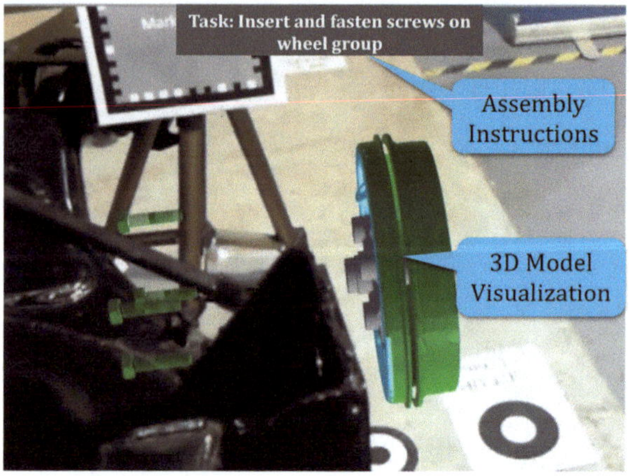

Fig. 15.4 Digital instructions for assembly operation

this information to the operator, he/she is more aware about what is going on in the cell, anticipates pending movements by the robot and is more willing to accept working next to big industrial robots. This is achieved by showing in operator's field of view information that derive from robot's controller, namely its trajectory and the workspace zones. This information is presented in detail in the following figures. More specifically, in Fig. 15.5 the operator can see a red line in front of robots end effector, which represents its trajectory that would follow to position the drum next to the axle, before the robot starts moving. Additionally, in Fig. 15.6, are visualized two semi-transparent boxes, one green and one red. The red one represents robot's restricted working area while the green box represents an area where the robot is not allowed to enter, enabling the human to work in there safely. It must be pointed out that these boxes have been programmed inside robot controller and the AR system is used for visualization purposes only. There is no sensing mechanism to check if the human operator is inside the red area to stop the robot from moving.

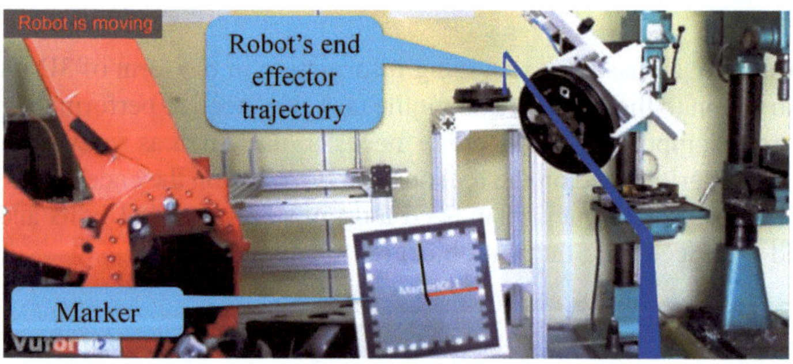

Fig. 15.5 Robot's end effector trajectory

Fig. 15.6 Safety volume (green cube) and robot's working area (red cube)

Fig. 15.7 Visual alerts

15.3.3 Visual Alerts

In addition to the previous functionalities, this one focuses on alerting the operator, when he/she is not looking at the robot or its working area, that a potentially hazardous activity takes place. More specifically, a warning message appears in operator's field of view, informing him/her when a robot moves, a machine is activated, a process takes place etc. Those messages derive from the main controller that controls the execution of the tasks in the cell. In Fig. 15.7 an example message is shown.

15.3.4 Assembly Status Reporting

In current assembly lines that are fully automated and robot based, the process execution is monitored and coordinated by dedicated devices, namely Programmable Logic Controller (PLC) or other service-based approaches [34]. However, a hybrid production system, where humans are also included, has the following requirements:

- Create a bilateral human–machine interface that, on the one side, would provide information to the human operators regarding the task they should perform and on the other side enable them report back its status, such as completed or paused etc.
- Provide information about the production status in real time, such as which resources are active, what operation is executed, how many have been completed etc. In this way the operator would be alerted about potential dangers or details in the workstation.

15.3.5 *Running Task Information*

Once the main controller instructs the robot resources to start the execution of their tasks, the operator is able to request to see information about the tasks that are in progress in each workstation. Additionally, when the main controller wants to instruct the human resource to initiate a task, a notification is sent to him/her about which task he/she should execute, alongside with a virtual button that should be pressed once the operator completes his/her task. This "Task Completed" button is superimposed in his field of view and can be pressed with the AirTap functionality as explained above. In this way the main controller can have the control of the whole assembly process, following the task sequence. In the following figures are presented the aforementioned functionalities. In Fig. 15.8 is presented the field of view of the operator where information regarding the current task, either robot or human task, is visualized.

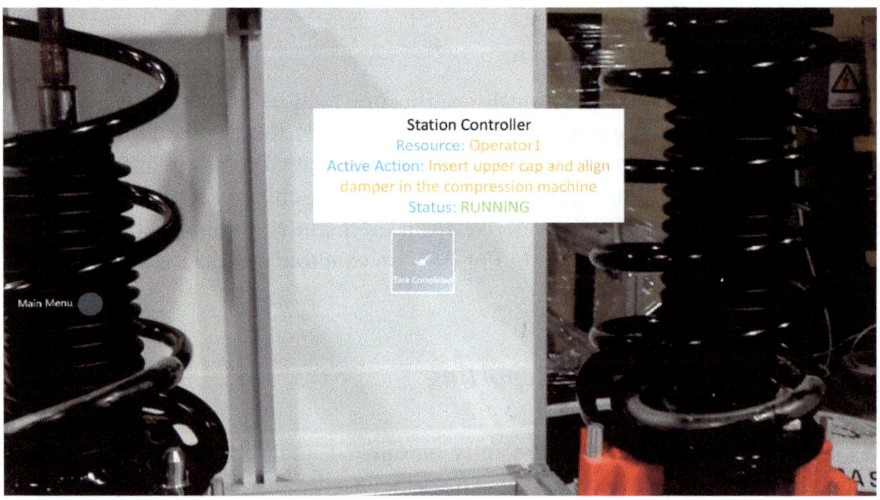

Fig. 15.8 Human operators' field of view—human task information

15.3.6 Production Line Data

Apart from the above functionality, which dispatches information to the operator based on the resource type, there is another functionality that could provide information to the operator regarding the production line upon request. The message contains information about the current and the next model to be assembled, the average remaining cycle time to complete current operation and the status of the successfully completed operations versus the targeted ones. The information is visualized in Fig. 15.9, it is updated automatically from the main controller and can be appear/disappear in operator's field of view upon his/her request.

15.4 System Architecture—Control System

In this section the control system will be described in detail. This is a library that is responsible to track the status of each task, dispatch instruction to each resource and follow the production schedule. In the following sections two cases of station controllers are described, one without and one with a digital twin implementation.

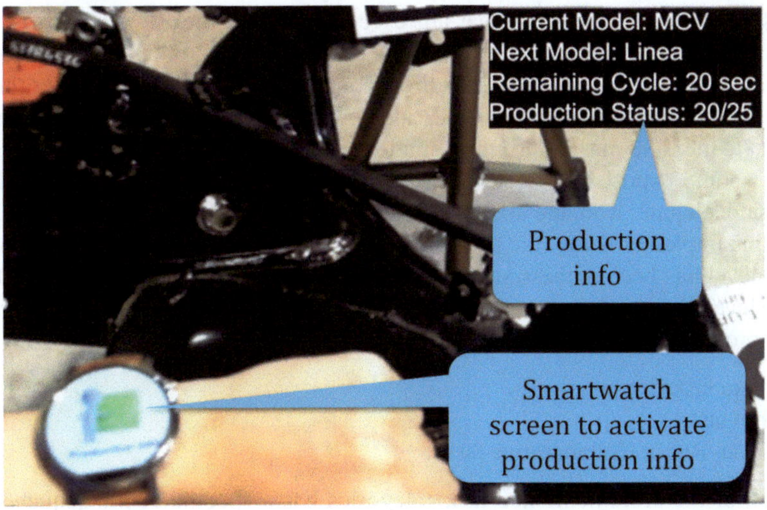

Fig. 15.9 Production information messages

Fig. 15.10 Frontend of the control system

TASK	RESOURCE	STATUS
Pick up part 1	Human	Completed
Guide robot	Human	
Screw	Human	In progress
Reset position	Robot	Completed
Pick up part 2	Robot	Completed
Release part 1	Human	
Release part 2	Robot	

15.4.1 Control System Without Digital Twin

The control system that doesn't involve a digital twin simulation has been implemented using Java, both in the backend and the frontend. In Fig. 15.10 is visualized part of the frontend interface that gives the capability to the user to access the manufacturing schedule and execute it either step by step or all at once. In the backend, the control system is using ROS topics and services to communicate with the respective resources and dispatch the necessary messages.

More specifically, the ROS-Java implementation has been used for registering ROS services and publishing topics. Additionally, a JSON API has been used, namely ROSBridge, for the programs that are executed outside the ROS master. In other words, ROSBridge is a server which exposes its API to the robot controller, the AR application and others, to enable the direct connection with the ROS master through a web socket. The benefit of the ROSBridge protocol is that allows for many devices connect to the ROS master through a single WebSocket and port. Otherwise the ROS master should expose a different port per device, which makes the current solution more safe, elegant, lean and able to be integrated to an industrial environment.

In this implementation, the production has been split in multiple operations and in this level has been orchestrated by the control system, as seen in Fig. 15.10. In a higher level, the operations are grouped in tasks containing one or more operations. Therefore, there are three kind of manufacturing tasks:

- Human tasks
- Robot tasks
- Human–Robot Collaborative tasks.

Each of the aforementioned task categories has a specific action sequence and message exchange among the control system and resources. For the human tasks, the control system sends to the AR application the necessary instructions for the imminent task, which from its side automatically informs the control system of the initiation of the operation. Upon its completion, the human operator notifies the control system. For the robot tasks, the control system automatically communicates with robot's controller, confirming the initiation, the execution and the completion of the operation. In parallel, the control system sends to the operator information related

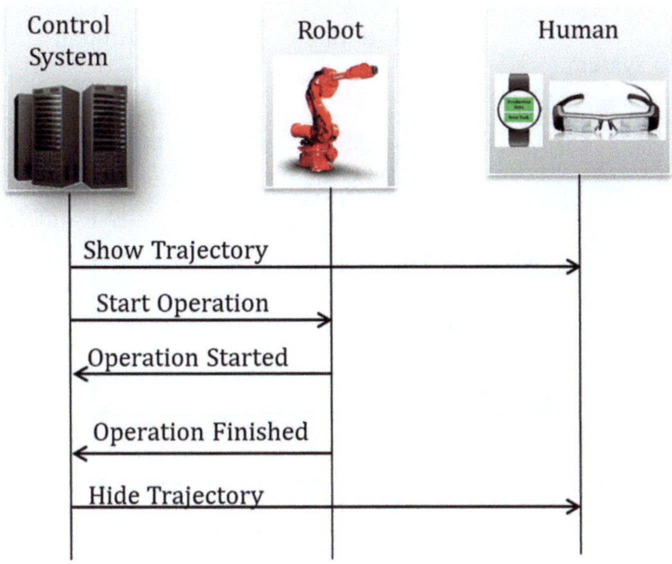

Fig. 15.11 Execution of robot operation sequence

to robot's movement, such as its trajectory and alerts as described above. This information is dismissed automatically upon robot's operation completion (Fig. 15.11). Last but not least, for the HRC tasks, the robot is set to manual guidance mode to allow the human operator to manipulate bare-handed the robot. Additionally, in the AR application, robot's workspace and instructions to execute a specific task are visualized. Finally, when the task is completed, the visualized information disappears and the robot enters normal operation (Fig. 15.12).

A system design architecture is presented in the following figure (Fig. 15.13).

15.4.1.1 Station Controller with Digital Twin

Another example of a station controller that has been implemented involves a digital twin simulation of the production station. This helps to integrate and test the AR tools in a realistic environment [35], while the station controller algorithm is still responsible to monitor the task execution and dispatch the necessary operations to the respective resources [32].

Similar to the above implementation, the station controller is based on ROS [36] and for the communication of non-ROS application the ROSBridge API has been used. Regarding the simulation environment, it was created in Gazebo which is compatible to ROS. All the information exchanged among the AR application, digital twin, robot's planner and station controller is transferred using ROS messages. In this way, the AR application is able to visualize a digital model of the robot having access to robot's Universal Robot Description File (URDF), base position and execution

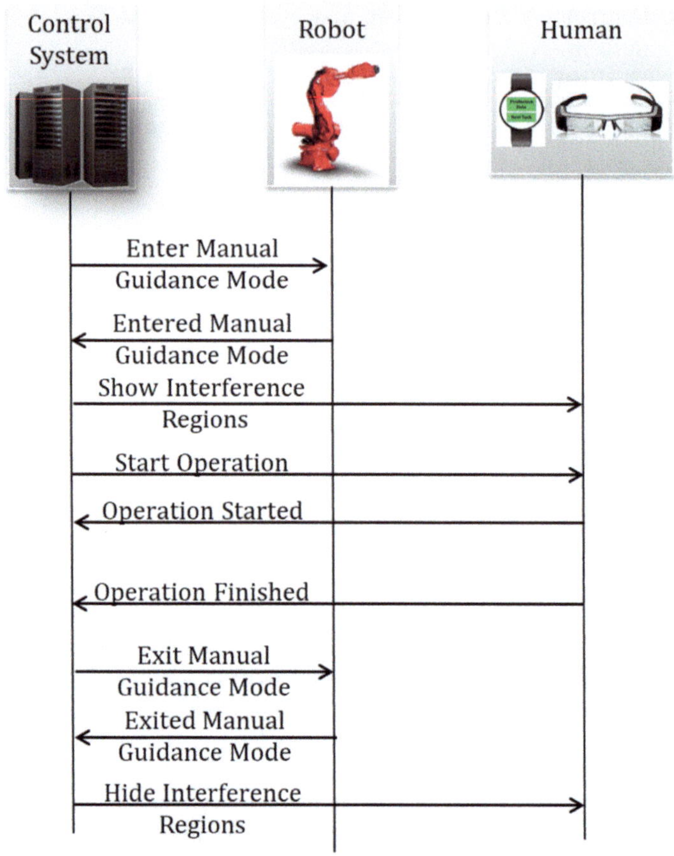

Fig. 15.12 Execution of collaborative operation sequence

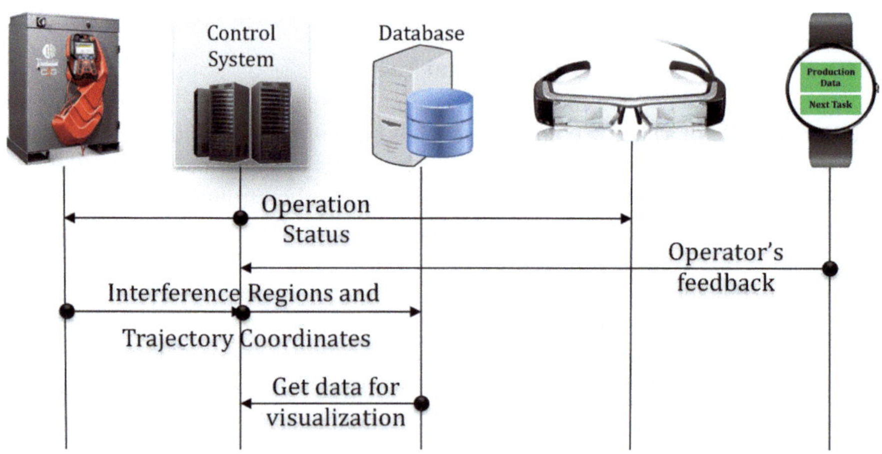

Fig. 15.13 System implementation

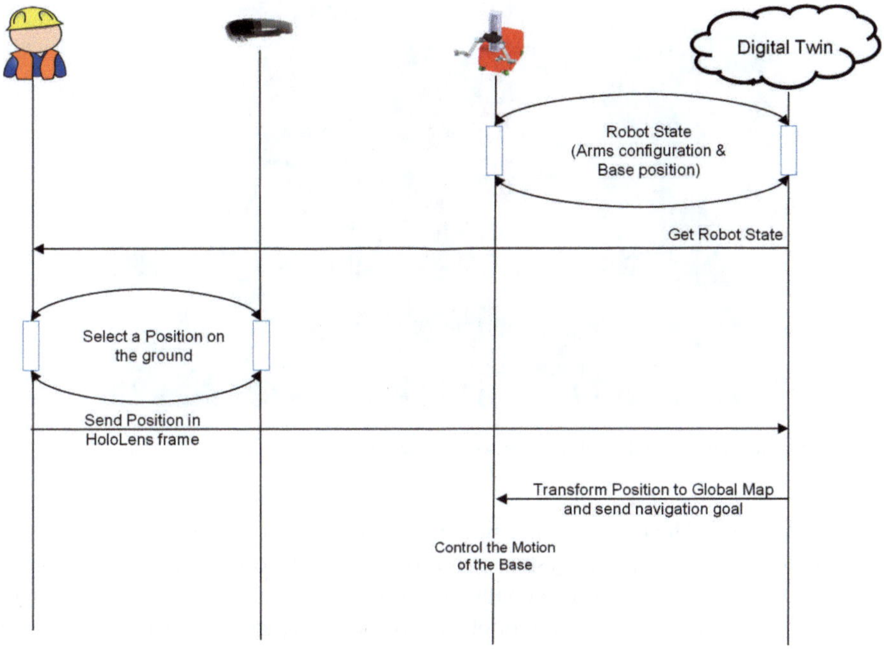

Fig. 15.14 Sequence diagram for robot direct navigation instructing

status. Data exchange for the direct robot navigation, as explained in the previous section, is visualized in Fig. 15.14.

15.5 Industrial Example

The aforementioned technologies have been tested and validated in industrial use cases that stem from the automotive industry. More specifically, both case studies derive from the assembly lines of passenger vehicles, and more specifically from the rear-axle assembly station and front suspension preparation station.

In Fig. 15.15, is demonstrated the first use case, namely the rear axle assembly station. It contains a high payload robot (COMAU NJ 130), part supporting bases, a rear axle from an actual car model and a wheel group that should be assembled on the aforementioned axle. The robot is used to load the different parts on the assembly fixture, since the axle weighs 25 kg and each wheel group 11–12 kg. One the other side the operator is making the more delicate operations by adjusting the relative position of the parts to perform the assembly, performs the screwing operation to tighten the parts together and inserts the cables. In order to ensure operators safety, a safety certified 3D camera is used, namely PILZ SafetyEYE, following the safety strategy

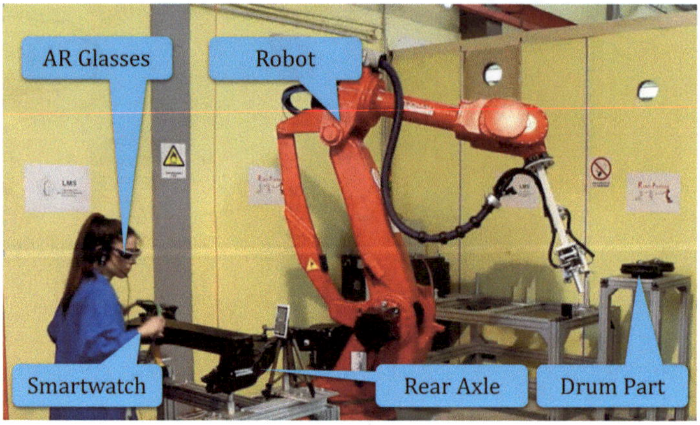

Fig. 15.15 The cell where the automotive scenario was tested

described in [37]. The demonstrator proved that the time needed by the operators to retrieve information from the system and provide their feedback directly from their work posts is minimized to milliseconds. Additionally, operators are more reluctant to work with high-payload industrial robots without safety fences, due to risk awareness tools they are using. Furthermore, stoppages have been reduced while the training process has been enhanced by using more intuitive instructions, sent directly to the production lines. Lastly, the monitoring of the production process is facilitated by tracking the information exchanged among the control system and the resources.

Regarding the front suspension preparation station, the current approach is based on manual work using at least one operator per workstation. As described in [17], the use of mobile dual arm robots could be beneficial by increasing production flexibility and decreasing ergonomy issues. As a result such robots can be considered as co-workers to the current operators in the three workstations that focus on the preparation of the damper and its final assembly to the disk break. Similar to the previous case, the robot performs the heavy lifting and all the non-ergonomic activities, while the human performs more delicate tasks, like small part insertion, cable fixing etc.

In order to test in a 3D realistic environment, in terms of layout, Gazebo simulation [38] has been created to reproduce the assembly environment, as shown on Fig. 15.16. Apart from the environment, the 3D models of the human and the dual-arm mobile robot have also been inserted, as well as the sensors that are used, namely two laser scanners that are attached on the mobile platform and a Kinect that is positioned on robot's torso. Finally, the simulation environment is constantly updated based on the data received from the aforementioned sensors, showing obstacles, moving objects etc.

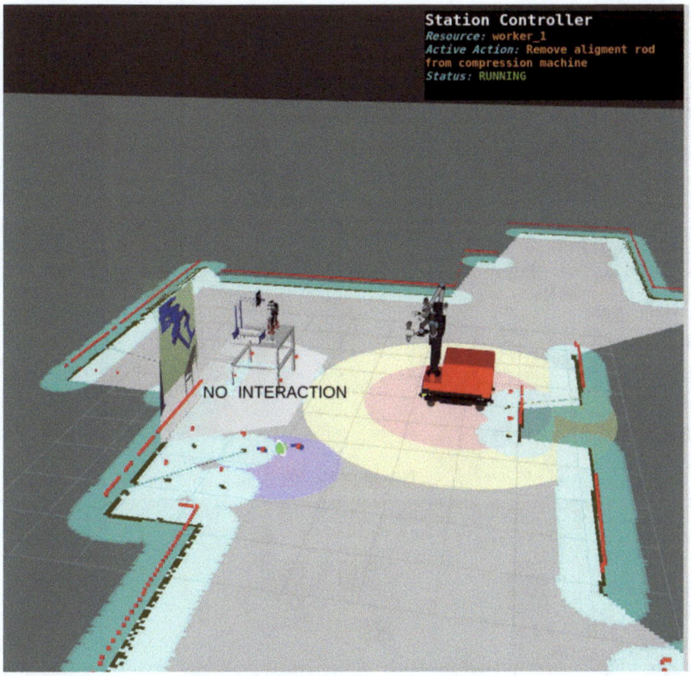

Fig. 15.16 Digital Twin scene re-construction

Regarding the AR application, which is running on HoloLens, it implements the functionalities that have been described in the above sections. In other words, the operator is able to teach the robot to move in space and get information about active or assigned tasks. Additionally, using the Digital Twin environment, all the instructions sent to the mobile robot are tested there to ensure their safe and collision-free execution. The steps that are followed to do that are shown in the following figure where from the initial environment, the operator instructs a new position through the AR application, the path is calculated in the simulation environment and finally in the real world the mobile robot executes the command (Fig. 15.17).

15.6 Discussion

Nowadays, flexibility plays an important role in EU manufacturing, requiring more advanced systems, to adapt the fluctuating demand on the market. Under this notion, flexible resources, mobile dual arm robots for example, in collaboration with human cognitive capabilities, will be the central resources in the factories of the future. Therefore, effective and efficient human robot interaction is the key objective of EU research activities. Driven by that, this section aims to present the latest trends in AR

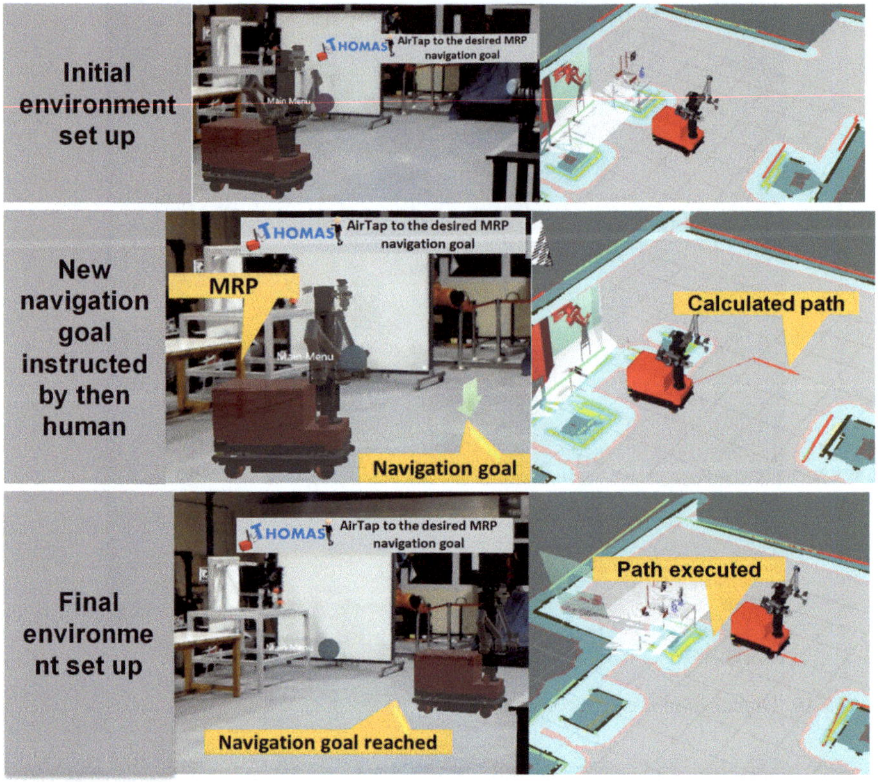

Fig. 15.17 Direct navigation instructions functionality workflow

technology, used for HRI purposes. The main goal of such applications is to enable human workers to understand, feel comfortable and interact with the robot resources that share their working area with, without prior technical knowledge. This would increase their safety feeling and acceptance levels to work with industrial robots.

The functionalities described in this section, that have been included in the AR application, involve the robot behavior visualization, the possibility to instruct to move to unknown positions, the capability to inform the human resources about the active tasks in real time and report back to the main control system about the execution status of the manual tasks. Additionally, on the functionality list is included the visualization to human operators of instructions about the current task, the notification through alerts about potential hazards on the workstation and the presentation of robot's working areas.

Furthermore, the execution control system has also been presented, including a digital twin simulation. It is based on ROS, using topics and services for message exchange with the resources and the sensors. The task orchestration for three different operation types, namely human operations, robot operations and hybrid operations, is achieved through a web interface that has been implemented in Java and enables the individual or mass operation execution.

Last but not least, it can be verified that the markerless approach, using the HoloLens, facilitates the programmer in terms of deployment and setup time while at the same time it provides a better experience to the operator due to higher stability and more intuitive experience. Both case studies demonstrated an enhanced HRI capability, using AR glasses and sensor data from the shopfloor.

References

1. Chryssolouris G (2006) Manufacturing systems: theory and practice. Springer, New York
2. Michalos G, Makris S, Papakostas N, Mourtzis D, Chryssolouris G (2010) Automotive assembly technologies review: challenges and outlook for a flexible and adaptive approach. CIRP Journal of Manufacturing Science and Technology. 2:81–91. https://doi.org/10.1016/j.cirpj.2009.12.001
3. Giordani S, Lujak M, Martinelli F (2013) A distributed multi-agent production planning and scheduling framework for mobile robots. Comput Ind Eng 64:19–30. https://doi.org/10.1016/j.cie.2012.09.004
4. Zhao F, Hong Y, Yu D, Yang Y, Zhang Q (2010) A hybrid particle swarm optimisation algorithm and fuzzy logic for process planning and production scheduling integration in holonic manufacturing systems. Int J Comput Integr Manuf 23:20–39. https://doi.org/10.1080/09511920903207472
5. Houshmand M, Jamshidnezhad B (2006) An extended model of design process of lean production systems by means of process variables. Robotics and Computer-Integrated Manufacturing. 22:1–16. https://doi.org/10.1016/j.rcim.2005.01.004
6. Koren Y, Shpitalni M (2010) Design of reconfigurable manufacturing systems. J Manuf Syst 29:130–141. https://doi.org/10.1016/j.jmsy.2011.01.001
7. Scholz-Reiter B, Freitag M (2007) Autonomous processes in assembly systems. CIRP Ann 56:712–729. https://doi.org/10.1016/j.cirp.2007.10.002
8. Eurostat's portal on SMEs. Retrieved on 15 Sept 2014. https://epp.eurostat.ec.europa.eu/portal/page/portal/european_business/special_sbs_topics/small_medium_sized_enterprises_SMEs
9. Tsarouchi P, Makris S, Chryssolouris G (2016) Human–robot interaction review and challenges on task planning and programming. Int J Comput Integr Manuf 29:916–931. https://doi.org/10.1080/0951192X.2015.1130251
10. Michalos G, Makris S, Spiliotopoulos J, Misios I, Tsarouchi P, Chryssolouris G (2014) ROBO-PARTNER: Seamless Human-Robot cooperation for intelligent, flexible and safe operations in the assembly factories of the future. Procedia CIRP. 23:71–76. https://doi.org/10.1016/j.procir.2014.10.079
11. URL ROBO-PARTNER EU Project: www.robo-partner.eu. Last accessed on 26 Dec 2018
12. URL LIAA EU Project: www.project-leanautomation.eu. Last accessed on 25 Dec 2018
13. URL THOMAS EU project: www.thomas-project.eu. Last accessed on 25 Dec 2018
14. URL Universal robots: www.universal-robots.com. Last accessed on 22 Nov 2015
15. International Symposium on Robotics (2010) German Conference on Robotics: Robotics (ISR), 2010 41st International Symposium on and 2010 6th German Conference on Robotics (ROBOTIK): VDE Verlag, Berlin, 7–9 June 2010

16. Makris S, Papakostas N, Chryssolouris G (2019) Productivity, Production Engineering, T.I.A. for: CIRP encyclopedia of production engineering. Springer Berlin Heidelberg, New York, NY
17. Kousi N, Michalos G, Aivaliotis S, Makris S (2018) An outlook on future assembly systems introducing robotic mobile dual arm workers. Procedia CIRP 72:33–38. https://doi.org/10.1016/j.procir.2018.03.130
18. Alexopoulos K, Makris S, Chryssolouris G (2019) Production, Production Engineering, T.I.A. for: CIRP encyclopedia of production engineering. Springer Berlin Heidelberg, New York, NY
19. Höllerer T, Feiner S (2004) Mobile augmented reality. In: Karimi HA, Hammad A (eds) Telegeoinformatics: location-based computing and services, CRC Press, Boca Raton, FL
20. Milgram P, Kishino F (1994) A taxonomy of mixed reality visual displays. IEICE Trans Inf Syst E77-D(12):1321–1329
21. Hakkarainen M, Woodward C, Billinghurst M (2008) Augmented assembly using a mobile phone. In: 2008 7th IEEE/ACM international symposium on mixed and augmented reality, IEEE, Cambridge, UK, pp 167–168. https://doi.org/10.1109/ISMAR.2008.4637349
22. CSIE 2009 (2009) 2009 WRI world congress on computer science and information engineering, March 31-April 2, 2009, Los Angeles, California. IEEE, New Jersey
23. Xin M, Sharlin E, Sousa MC (2008) Napkin sketch: handheld mixed reality 3D sketching. In: Proceedings of the 2008 ACM symposium on Virtual reality software and technology—VRST '08. ACM Press, Bordeaux, France, p 223. https://doi.org/10.1145/1450579.1450627
24. Ong SK, Pang Y, Nee AYC (2007) Augmented reality aided assembly design and planning. CIRP Ann 56:49–52. https://doi.org/10.1016/j.cirp.2007.05.014
25. Mourtzis D, Doukas M (2012) A web-based virtual and augmented reality platform for supporting the design of personalised products. 45th CIRP Conference on Manufacturing Systems (CMS 2012) 2012:234–241
26. Nee AYC, Ong SK, Chryssolouris G, Mourtzis D (2012) Augmented reality applications in design and manufacturing. CIRP Ann 61:657–679. https://doi.org/10.1016/j.cirp.2012.05.010
27. Rentzos L, Papanastasiou S, Papakostas N, Chryssolouris G (2013) Augmented reality for Human-based assembly: using product and process Semantics. IFAC Proc Vol 46:98–101. https://doi.org/10.3182/20130811-5-US-2037.00053
28. Hincapie M, Caponio A, Rios H, Gonzalez Mendivil E (2011) An introduction to augmented reality with applications in aeronautical maintenance. In: 2011 13th international conference on transparent optical networks, IEEE, Stockholm, Sweden, pp 1–4. https://doi.org/10.1109/ICTON.2011.5970856
29. Michalos G, Karagiannis P, Makris S, Tokçalar Ö, Chryssolouris G (2016) Augmented Reality (AR) applications for supporting Human-robot interactive cooperation. Procedia CIRP. 41:370–375. https://doi.org/10.1016/j.procir.2015.12.005
30. Liu H, Wang L (2017) An AR-based worker support system for Human-Robot collaboration. Procedia Manuf 11:22–30. https://doi.org/10.1016/j.promfg.2017.07.124
31. Makris S, Karagiannis P, Koukas S, Matthaiakis A-S (2016) Augmented reality system for operator support in Human–Robot collaborative assembly. CIRP Ann 65:61–64. https://doi.org/10.1016/j.cirp.2016.04.038
32. Michalos G, Kousi N, Karagiannis P, Gkournelos C, Dimoulas K, Koukas S, Mparis K, Papavasileiou A, Makris S (2018) Seamless human robot collaborative assembly—an automotive case study. Mechatronics 55:194–211. https://doi.org/10.1016/j.mechatronics.2018.08.006
33. URL Five Ways Wearable Tech Can Improve Manufacturing. https://www.magna.com/insights/article/five-ways-wearable-tech-can-improve-manufacturing. Last accessed on 6 Dec 2018
34. Kousi N, Koukas S, Michalos G, Makris S, Chryssolouris G (2016) Service oriented architecture for dynamic scheduling of mobile robots for material supply. Procedia CIRP 55:18–22. https://doi.org/10.1016/j.procir.2016.09.014
35. Kousi N, Gkournelos C, Aivaliotis S, Giannoulis C, Michalos G, Makris S (2019) Digital twin for adaptation of robots' behavior in flexible robotic assembly lines. Procedia Manuf 28:121–126. https://doi.org/10.1016/j.promfg.2018.12.020
36. URL Robot Operating System. www.ros.org Last accessed on 07 Sept 2018

37. Michalos G, Makris S, Tsarouchi P, Guasch T, Kontovrakis D, Chryssolouris G (2015) Design considerations for safe Human–Robot collaborative workplaces. Procedia CIRP 37:248–253. https://doi.org/10.1016/j.procir.2015.08.014
38. URL GAZEBO Sim. www.gazebosim.org Last accessed on 09 Nov 2018.

Chapter 16
Gesture-Based Interaction of Humans with Dual Arm Robot

16.1 Introduction

Robot programming tools are supposed to be more user-friendly with the latest trends in artificial intelligence, function blocks, universal programming languages and open source applications [1]. Via shorter training cycles, even non-expert robot programmers will be able to use those. The intuitive programming of robots requires scientific demonstration and instructional systems [2]. To order to introduce more practical robot programming systems, different methods and sensors have been used. Types include sensors of vision, speech, touch/force [3], gloves, turn-rate and acceleration sensors [4], as well as artificial markers [5]. The main challenge in this field was to establish methods for the robustness and accuracy of the sensor information to collect and reproduce data from robots [3].

An extended review on the use of human robot interaction (HRI) techniques was presented in [1]. These systems' key goal involves both designing and implementing multimodal communication structures for simplification of robot programming. Of these research studies, significant research attention has been paid to the use of gestures. Hand gestures are identified for example by the use of data gloves in [6, 7], where in [4] the sensor Wii is used. Examples using 3D cameras can be found in [8–11]. In [12], to achieve the human–robot dialog a diverse vocabulary is used. Kinect has been used as the recognition sensor in [13–16] enabling the online HRI. Kinect has also been used for ensuring safety in human robot coexistence [17]. Besides HRI, the use of Kinect can also be found in human computer interaction studies (HCI) [18]. Last but not least, HRI has also picked the leap motion sensor [19].

Despite the broad interest of research on intuitive interfaces for interaction and robot programming, challenges remain regarding the need for structured programming tools, advanced processing, integration methods, and environmental uncertainties. Natural language interactions, such as gestures and voice commands, aren't always applicable to industrial environments due to their high noise level, lighting

S. Makris, *Cooperating Robots for Flexible Manufacturing*,
Springer Series in Advanced Manufacturing,
https://doi.org/10.1007/978-3-030-51591-1_16

conditions and dynamically changing environment. The development of gesture-based language with complicated and numerous gestures, however, does not seem to be user-friendly to humans.

This research work proposes a high-level robot programming system using sensors that detect both body and hand gestures, taking into account the above challenges. The body gestures described are static and the gestures of the hands are based on the hands' dynamic movements. The user can communicate with the robot in two different ways. Special robot programming training is not required, even in this case to direct an industrial robot. During robot training, protection can be maintained using conventional emergency buttons. Microsoft Kinect and the leap motion respectively are the devices chosen to introduce body and hand gestures. For applications where accuracy of recognition is not necessary, Kinect sensor for replacing traditional 3D cameras has become substantially popular. The use of leap motion in robot programming is relatively new, allowing hand gestures to be identified to operate the robot.

With respect to programming dual arm robots, the direct interaction between humans and robots can lead to reduced production cost and programming time [1]. Human Robot Interaction (HRI) can increase the productivity and flexibility of the production lines. Currently, the industrial robot programming is mainly based on the use of teach pendants and offline programming tools. The way of editing a program is currently not user friendly and it mainly requires an expert to find the robot position to edit and replace it, without considering higher level details. Another option is the intuitive robot programming, including demonstration techniques and instructive systems [2]. The Programming by Demonstration (PbD) frameworks have traditionally used different interaction mechanisms such as voice, vision, touch sensing, motion capturing [9], data gloves, turn-rate sensors, acceleration sensors [4, 20, 21]. HRI has been achieved through gestures [4, 7, 22], voice commands [12], physical and non-physical contact [23], graphical interfaces [24, 25]. Despite the wider research interest in programming by demonstration and instructive systems, there are still open research challenges coping with the needs for easy use, extension and reconfiguration of such systems.

Different architecture schemas for robot control software have been deployed by the robotics' researchers for several years [26–28]. An extended review on robotics' architectures was presented in Amoretti et al. [29], where three different schemas, namely (a) distributed object, (b) component-based and (c) service-oriented architectures were described [30–38]. Regarding the last, the main challenges that have been faced in relation to robotics are the needs for high computing time and the lack of standardized interfaces and communication protocols that can be applied in different robotic platforms, sensors and devices [39, 40].

16.2 Approach

The dual arm robot has been improved in the proposed flexible cell with easy gripping that allows shapes of varying size and complexity to be handled. Examples of such shapes are the dashboard traverse and the fuse box. The flexible cell's main research challenges are easier robot task programming and coordination with human tasks. To this effect, a four-level hierarchical approach has been adopted, breaking down the HRC tasks in the hierarchy of program-job-task-operation. This hierarchical structure has been presented in [3, 41], enabling the structure of human and robot activities in a unique model The programming refers to the operations, where two possibilities exist. The first level concerns automated motion generation by offline programming methods and the second level relates to demonstration programming through the use of basic graphical interfaces and mechanisms of interaction.

In order to advance the traditional robot programming methods that are mainly facilitated by the use of a teach pendant for moving and teaching robot positions, this study proposes the structuring of a robotic program, around the introduced hierarchical model, as illustrated in the following Fig. 16.1.

A set of robotic positions consist an operation in the lower level. For each of them, there is a main functionality that is explained within the same table, as well as programming related parameters.

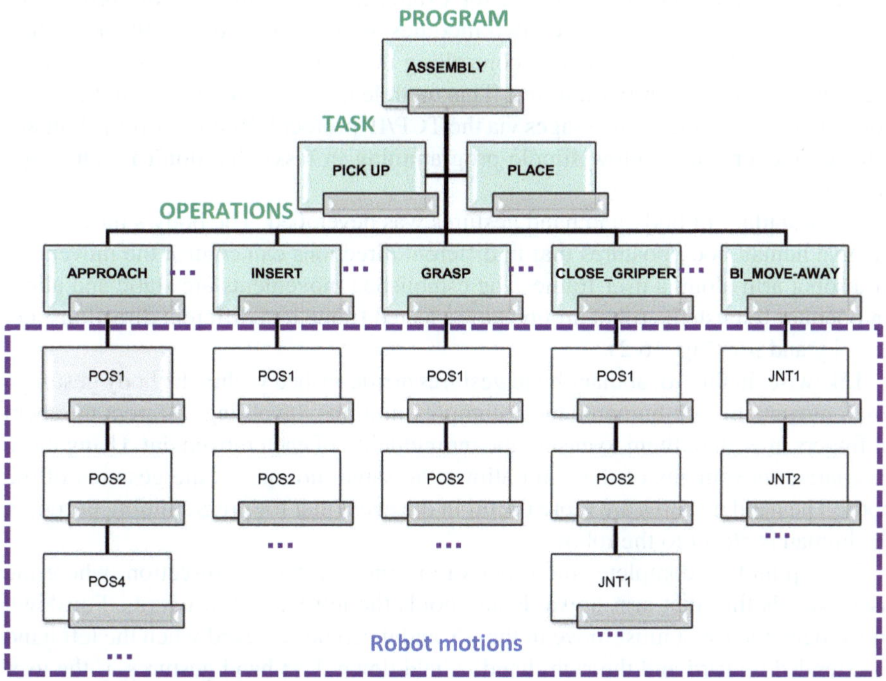

Fig. 16.1 Robot program structure [41]

The idea of the hierarchical decomposition of human–robot (HR) tasks and the use of multi-modal interfaces for interaction and programming allows to:

- Easily build a new robot program focusing on the high-level description and using multi-modal interfaces in the lower level (e.g. APPROACH, GRASP etc.) for teaching the robot a new operation. The above structure is user friendly, since the user or programmer can monitor a program by understanding the process steps rather than editing robot positions without high level description.
- Easily monitor and understand a robot program by the high-level program structure, focusing on the higher and lower level activities. The description of the different levels is meaningful, providing information about the type of activity that follows.
- Easily extend or modify an existing robot program, since the user can edit it in the different levels of the hierarchy.
- Involve one or more humans within the same program enabling their control by an external system. This external framework that controls every robot or human agent that are involved in the hybrid cell.

Within a software module, the available interaction mechanisms (sensors) are represented, enabling the detection and identification of gestures ('Gestures Vocabulary', 'Gestures Recognition'). Recognized gestures are published as messages in "rostopic," where third-party applications can subscribe via ROS middleware to listen to these messages. These messages, for example/Left gesture, are the product of recognition in each subject. Such two modules are not interacting with each other, but the external control module is controlling them. This module also controls the regulation of gestured programming. This module is connected directly to the robot controller by exchanging messages via the TCP/IP protocol. Robot starts the human-directed movement to allow simple programming in tasks that don't require high accuracy.

A vocabulary of body and hand gestures was developed. The body's movements involve human body postures that in different directions can control the movement of a robot arm along a user frame. The established movements are static and allow an operator to push the robot around any selected frame in 6 different directions, i.e. $\pm$x, $\pm$y and $\pm$z (Fig. 16.2).

Likewise, in this vocabulary hand gestures are described within the body gestures. Such movements are human hands' complex motions, involving different numbers of fingers in each of them to ensure the individuality of each movement. Using these 6 separate movements of the hand allows the same motions as the gestures of the body. The hand gestures are more useful in the event that for programming purposes the human is closer to the robot.

The up motion completes the robot's movement in the +x direction, where the user extends the right arm upwards and holds the left hand downwards. For down movement, the robot must move in the −x and it can be achieved when the left hand is extended upward and the right hand is held down. Use hand gestures, if the user uses one hand, the robot will go up and swipe it upwards, otherwise if the finger is swiped downwards, it will go down.

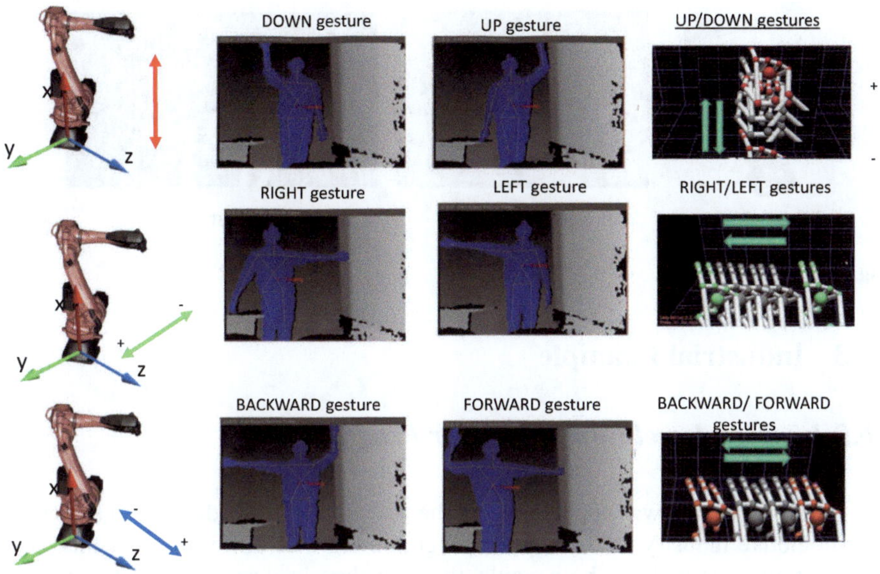

Fig. 16.2 High level commands for robot programming [42]

The movement along the +y path is done with the left gesture that involves the left hand extended at shoulder height pointing to the left direction. The robot, on the opposite, moves along the −y with the left movement involving the right hand extended in the same manner but pointing to the right direction. The signs for right and left are understood using the hand gestures, using two fingers that swip from left to right and the reverse.

The move along the +z path is performed using the forward gesture, where the right hand is extended upwards and the left hand is at shoulder height pointing to the left side. The step in the opposite direction is done with the opposite hands, the left upwards and the right at shoulder height pointing to the right (backwards gesture).

Using the same motions, where three fingers swing from left to right and the reverse, the same movements are done. Finally, two static body gestures and hand gestures are available to stop the robot during a motion. The robot motor drivers are switched off in this way and the robot won't stay active. A STOP command for the robot is illustrated in Fig. 16.3. Once the movements are understood and how they are to be performed in the robot controller, communication between the external controller module and the decoder application is created. The robot thus receives the human commands, allowing them to be performed, translated to the motions described.

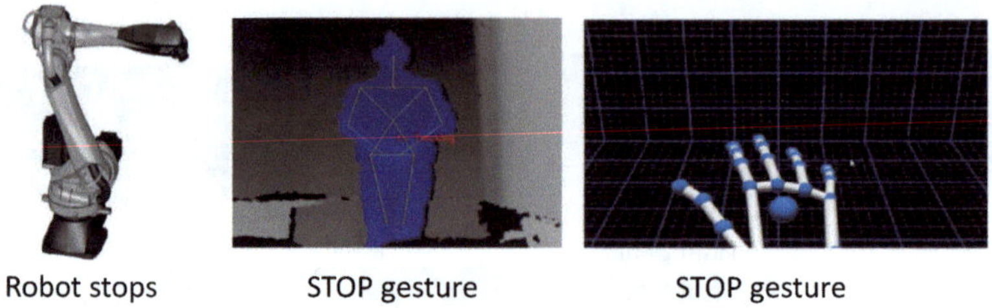

Fig. 16.3 Stop command

16.3 Industrial Example

16.3.1 High Level Commands for Programming

The proposed design was applied during the assembly of a car dashboard system to an automotive industry case study for programming dual arm robot. For this case, the robot program includes both single, time-synchronized and bi-manual assembly operations motions. Table 16.1 provides a list of high-level robot activities. The sequence of these tasks has been described in [43].

This table contains the gestures that are used in each particular robot activity. Several examples are given in Fig. 16.4 of the high-level operations and gestures.

Table 16.1 High level robot activities

High level robot task	High level robot operation	Gesture
Pick up traverse	ROTATE()	LEFT
	BI_APPROACH()	DOWN
	BI_INSERT()	BACKWARD
	BI_MOVE_AWAY()	UP
	ROTATE()	RIGHT
Place traverse	BI-APPROACH()	DOWN
	BI-EXTRACT()	FORWARD
	BI_MOVE_AWAY	UP
Pick up fuse box	ROTATE()	RIGHT
	BI_APPROACH()	DOWN
	CLOSE_GRIPPERS	–
	BI_MOVE_AWAY()	UP
	ROTATE()	LEFT
Place fuse box	BI_APPROACH()	DOWN
	OPEN_GRIPPERS	–
	BI_MOVE_AWAY()	UP

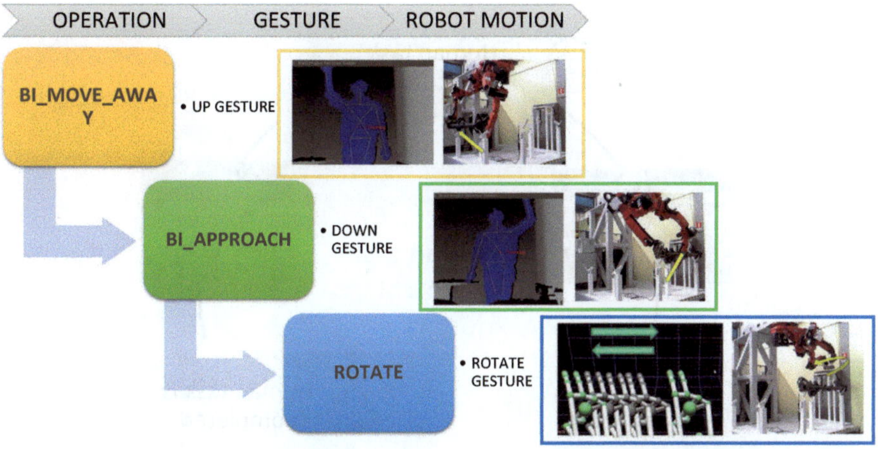

Fig. 16.4 High level programming through gestures in dual arm robot [43]

Such examples apply to the programming of four different operations, starting from the 'pick-up' task to the 'place fuse box' function.

In this situation, they both demonstrate the use of body and hand gestures. Off-line tests with the built recognition modules on body and hand gestures showed that the recognition rate was 93% and 96% respectively.

16.3.2 High Level Commands for Interaction During Execution

Installation of the wire harness is done by humans. The time available to humans is the time between the start and stop signals sent to the station's controller, either through gestures and voice commands or through the buttons on the screen (Fig. 16.5). The 'START' movement is used to indicate the beginning of a human task, while in the graphical interface the message for the human task is displayed. The human performs the task and a message is sent to the control architecture after it has been completed to signal the end of the task. According to these messages, the hierarchical module visualization is updated. The minimum safety distance between the human and the robot is measured at 1.43 m when the robot is involved, and the longest reaction time at 265 ms for this case study, according to the fenceless safety principles.

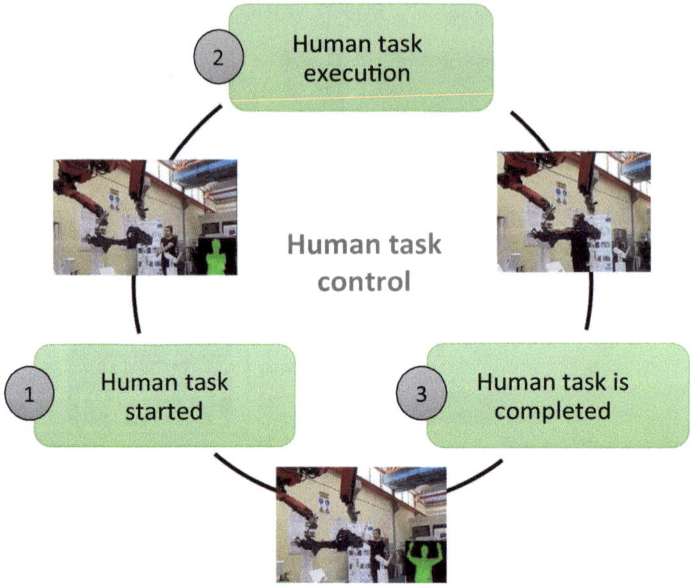

Fig. 16.5 Wire harness assembly [42]

16.4 Discussion

Robot programming requires a time, cost and robot experts, making the need for new methods of programming unavoidable. When designing intuitive robot programming techniques, the path of new programming systems is targeted. The proposed design uses two low cost sensors for robot programming to take advantage of many advantages. One path for future research is research into more robust tools to achieve better recognition performance. In addition, one direction for change is to explore the concept of more comfortable and simpler movements to rememorate. Implementing more interaction methods and testing them as to how simple they are to use and understand, could also help to develop more intuitive interfaces for robot programming.

Compared to traditional programming approaches, several benefits are anticipated due to the use of an intuitive programming structure. First the user can set up and control an industrial robot without the need of training. This helps even non-expert users to use easily implemented software to program an industrial robot. Second, these tools provide different options for the user to interact directly with a robot via a PC. Thirdly, this framework can achieve better accuracy with the use of external sensors, such as the use of visual sensors, as are already used in conventional programming.

In this direction, this chapter summarized the related challenges and contributed in the relevant research fields as follows. First, a model has been proposed for the hierarchical structuring of a HR program. This model directs the system's programmer/user to quickly create a new dual arm robot program by breaking down tasks in various levels of detail, thereby addressing 'what to do' rather than 'how to do'. The fact that the description of the different levels is meaningful is also important, providing information on the type of task, e.g. high level: Robot program for cell 1, medium level: pick up a box, low level: approach, grasp, etc.

Next, an external control system is being developed in the context of a service-oriented architecture. With regard to HR tasks teamwork, the traditional way in which their activities are coordinated will entail the programming of interlocking signals in the form of a PLC code. Any sequence changes will then entail reprogramming of the PLC. The use of the hierarchical representation of tasks removes the need to use PLCs except from the control structure introduced in this research work. In addition, the proposed framework allows the system to be easily extended, as it allows for easy integration of new agents (humans or robots), new sensors, new software applications, etc. by creating new communicating services to the existing ones. The architecture is flexible enough to allow for simple reconfiguration in different robotic platforms.

Last but not least, multi-modal interfaces for robot programming and HRI have been proposed. Despite having studied this kind of interfaces in various re-search works, this research effort contributes to research by putting together multiple interfaces that can be controlled through a specific control architecture. Such interfaces help the user concentrate on 'what' activity to teach a robot by offering programming instructions to save time and simplify the process itself.

References

1. Tsarouchi P, Makris S, Chryssolouris G (2016) Human-robot interaction review and challenges on task planning and programming. Int J Comput Integr Manuf 29:916–931. https://doi.org/10.1080/0951192X.2015.1130251
2. Biggs G, MacDonald B (2003) A survey of robot programming systems. In: proceedings of the Australasian Conference on Robotics and Automation. CSIRO p 27
3. Makris S, Tsarouchi P, Surdilovic D, Krüger J (2014) Intuitive dual arm robot programming for assembly operations. CIRP Ann 63:13–16. https://doi.org/10.1016/j.cirp.2014.03.017
4. Neto P, Norberto Pires J, Paulo Moreira A (2010) High-level programming and control for industrial robotics: using a hand-held accelerometer-based input device for gesture and posture recognition. Ind Rob 37:137–147. https://doi.org/10.1108/01439911011018911
5. Calinon S, Billard A (2004) Stochastic gesture production and recognition model for a humanoid robot. In: 2004 IEEE/RSJ international conference on intelligent robots and systems (IROaS) (IEEE Cat. No.04CH37566). IEEE, Sendai, Japan, pp 2769–2774
6. Lee C, Yangsheng X (1996) Online, interactive learning of gestures for human/robot interfaces. In: Proceedings of IEEE international conference on robotics and automation. IEEE, Minneapolis, MN, USA, pp 2982–2987
7. Neto P, Pereira D, Pires JN, Moreira AP (2013) Real-time and continuous hand gesture spotting: An approach based on artificial neural networks. 2013 IEEE international conference on robotics and automation. IEEE, Karlsruhe, Germany, pp 178–183

8. Stiefelhagen R, Fogen C, Gieselmann P, Holzapfel H, Nickel K, Waibel A (2004) Natural human-robot interaction using speech, head pose and gestures. In: 2004 IEEE/RSJ international conference on intelligent robots and systems (IROS) (IEEE Cat. No.04CH37566). IEEE, Sendai, Japan, pp 2422–2427
9. Nickel K, Stiefelhagen R (2007) Visual recognition of pointing gestures for human–robot interaction. Image Vis Comput 25:1875–1884. https://doi.org/10.1016/j.imavis.2005.12.020
10. Waldherr S, Romero R, Thrun S (2000) A gesture based interface for human-robot interaction. Auton Rob 9:151–173. https://doi.org/10.1023/A:1008918401478
11. Yang H-D, Park A-Y, Lee S-W (2007) Gesture spotting and recognition for human-robot interaction. IEEE Trans Rob 23:256–270. https://doi.org/10.1109/TRO.2006.889491
12. Norberto Pires J (2005) Robot-by-voice: experiments on commanding an industrial robot using the human voice. Ind Rob 32:505–511. https://doi.org/10.1108/01439910510629244
13. Suarez J, Murphy RR (2012) Hand gesture recognition with depth images: A review. 2012 IEEE RO-MAN: The 21st IEEE international symposium on robot and human interactive communication. IEEE, Paris, France, pp 411–417
14. Van den Bergh M, Carton D, De Nijs R, Mitsou N, Landsiedel C, Kuehnlenz K, Wollherr D, Van Gool L, Buss M (2011) Real-time 3D hand gesture interaction with a robot for understanding directions from humans. 2011 RO-MAN. IEEE, Atlanta, GA, USA, pp 357–362
15. Mead R, Atrash A, Matarić MJ (2013) Automated proxemic feature extraction and behavior recognition: applications in human-robot interaction. Int J Soc Robot 5:367–378. https://doi.org/10.1007/s12369-013-0189-8
16. Morato C, Kaipa KN, Zhao B, Gupta SK (2014) Toward safe human robot collaboration by using multiple kinects based real-time human tracking. J Comput Inf Sci Eng 14:011006. https://doi.org/10.1115/1.4025810
17. Ren Z, Meng J, Yuan J (2011) Depth camera based hand gesture recognition and its applications in Human-Computer-Interaction. 2011 8th international conference on information, communications and signal processing. IEEE, Singapore, pp 1–5
18. Rautaray SS, Agrawal A (2015) Vision based hand gesture recognition for human computer interaction: a survey. Artif Intell Rev 43:1–54. https://doi.org/10.1007/s10462-012-9356-9
19. Yalim I, Guillem A (2015) Teaching grasping points using natural movements. Front Artif Intel Appl 275–278. https://doi.org/10.3233/978-1-61499-578-4-275
20. Onda H, Suehiro T, Kitagaki K (2002) Teaching by demonstration of assembly motion in VR—non-deterministic search-type motion in the teaching stage. IEEE/RSJ international conference on intelligent robots and system. IEEE, Lausanne, Switzerland, pp 3066–3072
21. Zinn M, Roth B, Khatib O, Salisbury JK (2004) A new actuation approach for human friendly robot design. Int J Robot Res 23:379–398. https://doi.org/10.1177/0278364904042193
22. Bodiroža S, Stern HI, Edan Y (2012) Dynamic gesture vocabulary design for intuitive human-robot dialog. In: Proceedings of the seventh annual ACM/IEEE international conference on human-robot interaction—HRI '12. ACM Press, Boston, Massachusetts, USA, p 111
23. Koo S-Y, Lim JG, Kwon D-S (2008) Online touch behavior recognition of hard-cover robot using temporal decision tree classifier. RO-MAN 2008—The 17th IEEE international symposium on robot and human interactive communication. IEEE, Munich, Germany, pp 425–429
24. Koenig N, Takayama L, Matarić M (2010) Communication and knowledge sharing in human–robot interaction and learning from demonstration. Neural Netw 23:1104–1112. https://doi.org/10.1016/j.neunet.2010.06.005
25. Morioka M, Sakakibara S (2010) A new cell production assembly system with human–robot cooperation. CIRP Ann 59:9–12. https://doi.org/10.1016/j.cirp.2010.03.044
26. Bruyninckx H (2001) Open robot control software: the OROCOS project. In: Proceedings 2001 ICRA. IEEE international conference on robotics and automation (Cat. No.01CH37164). IEEE, Seoul, South Korea, pp 2523–2528
27. Fitzpatrick P, Metta G, Natale L (2008) Towards long-lived robot genes. Robot Auton Syst 56:29–45. https://doi.org/10.1016/j.robot.2007.09.014

28. Baillie J-C (2005) URBI: towards a universal robotic low-level programming language. In: 2005 IEEE/RSJ international conference on intelligent robots and systems. IEEE, Edmonton, Alta., Canada, pp 820–825
29. Amoretti M, Reggiani M (2010) Architectural paradigms for robotics applications. Adv Eng Inform 24:4–13. https://doi.org/10.1016/j.aei.2009.08.004
30. Songmin J, Hada Y, Gang Y, Takase K (2002) Distributed telecare robotic systems using CORBA as a communication architecture. In: Proceedings 2002 IEEE international conference on robotics and automation (Cat. No.02CH37292). IEEE, Washington, DC, USA, pp 2202–2207
31. Konietschke R, Hagn U, Nickl M, Jorg S, Tobergte A, Passig G, Seibold U, Le-Tien L, Kubler B, Groger M, Frohlich F, Rink C, Albu-Schaffer A, Grebenstein M, Ortmaier T, Hirzinger G (2009) The DLR MiroSurge - A robotic system for surgery. 2009 IEEE international conference on robotics and automation. IEEE, Kobe, pp 1589–1590
32. Brooks A, Kaupp T, Makarenko A, Williams S, Oreback A (2005) Towards component-based robotics. In: 2005 IEEE/RSJ international conference on intelligent robots and systems. IEEE, Edmonton, Alta., Canada, pp 163–168
33. Ando N, Suehiro T, Kitagaki K, Kotoku T, Yoon W-K (2005) RT-Component object model in RT-Middleware—distributed component middleware for RT (Robot Technology). 2005 international symposium on computational intelligence in robotics and automation. IEEE, Espoo, Finland, pp 457–462
34. Gilart-Iglesias V, Macia-Perez F, Capella-D'alton A, Gil-marti'nez-abarca J (2006) Industrial machines as a service: a model based on embedded devices and web services. In: 2006 IEEE international conference on industrial informatics. IEEE, Singapore, pp 630–635
35. Bong Keun K, Miyazaki M, Ohba K, Hirai S, Tanie K (2005) Web services based robot control platform for ubiquitous functions. In: Proceedings of the 2005 IEEE international conference on robotics and automation. IEEE, Barcelona, Spain, pp 691–696
36. Chen Y, Du Z, García-Acosta M (2010) Robot as a service in cloud computing. 2010 fifth IEEE international symposium on service oriented system engineering. IEEE, Nanjing, China, pp 151–158
37. Koubaa A (2014) A service-oriented architecture for virtualizing robots in robot-as-a-service clouds. In: Maehle E, Römer K, Karl W, Tovar E (eds) Architecture of computing systems—ARCS 2014. Springer International Publishing, Cham, pp 196–208
38. Hung M-H, Chen K-Y, Lin S-S (2004) Development of a web-services-based remote monitoring and control architecture. In: IEEE international conference on robotics and automation. Proceedings. ICRA '04. 2004. IEEE, New Orleans, LA, USA, vol 2, pp 1444–1449
39. Veiga G, Pires JN, Nilsson K (2009) Experiments with service-oriented architectures for industrial robotic cells programming. Robot Comput Integr Manuf 25:746–755. https://doi.org/10.1016/j.rcim.2008.09.001
40. Chen Y, Bai X (2008) On robotics applications in service-oriented architecture. 2008 The 28th international conference on distributed computing systems workshops. IEEE, Beijing, China, pp 551–556
41. Makris S, Tsarouchi P, Matthaiakis A-S, Athanasatos A, Chatzigeorgiou X, Stefos M, Giavridis K, Aivaliotis S (2017) Dual arm robot in cooperation with humans for flexible assembly. CIRP Ann 66:13–16. https://doi.org/10.1016/j.cirp.2017.04.097
42. Tsarouchi P, Athanasatos A, Makris S, Chatzigeorgiou X, Chryssolouris G (2016) High level robot programming using body and hand gestures. Procedia CIRP 55:1–5. https://doi.org/10.1016/j.procir.2016.09.020
43. Tsarouchi P, Makris S, Michalos G, Stefos M, Fourtakas K, Kaltsoukalas K, Kontrovrakis D, Chryssolouris G (2014) Robotized assembly process using dual arm robot. Procedia CIRP 23:47–52. https://doi.org/10.1016/j.procir.2014.10.078

Chapter 17
Synthesis of Data from Multiple Sensors and Wearables for Human–Robot Collaboration

17.1 Introduction

One of the latest trends in manufacturing foster the collaboration between human operators and robot resources [1]. Human Robot Collaboration (HRC) assembly paradigm aims to exploit the skills of humans and robots in their full extent so to optimize the production costs [2]. The combination of robots' strength, repeatability and precision with human operators' dexterity and cognition can have a positive feedback on the flexibility of the production system as well as in workers' wellbeing [3].

On one hand, focus has been given on the interaction with the human, where means for exchanging information between the robot and the operator (e.g. through gestures, voice, keyboard, screen, etc.) have been developed [2]. On the other hand, collaboration involves the cases where processes related to the execution of an activity are undertaken by the robot and the operator either jointly or in a shared workspace (e.g. carrying a load, assembling a part, etc.).

Current manufacturing practices require complete physical separation between people and active industrial robots (typically achieved using fences or similar physical barriers) as a precaution to ensuring safety [3]. However, there is an increasing demand to provide a regulatory framework capable of supporting the human–robot collaboration concept [4].

In the last years, different concepts of human robot interaction and collaboration have been investigated eliminating the use of fixed barriers, and implementing virtual safety fences where contact is not desired or allowed [5]. As a next step, the concept of shared workspace for human and a stationary robot is introduced, where only exclusive motion is allowed and the contact between them is possible [6]. The latest advance in human robot collaboration is the concept where robot and human operator share workspace and their contact is desirable in order for the human to directly interact and control the robot guiding for example its Tool Center Point (TCP) to a specific position.

S. Makris, *Cooperating Robots for Flexible Manufacturing*,
Springer Series in Advanced Manufacturing,
https://doi.org/10.1007/978-3-030-51591-1_17

In many low-volume production settings, direct physical interaction (or human-guided assembly) has several advantages when compared to "full automation". In essence, human cognitive capabilities may be sensibly joined with the robot system's force and precision [7, 8]. Such hybrid assembly systems can be divided into workplace sharing systems and workplace and time-sharing systems. Other classifications involve robot assistants as a directly interacting partner [9] collaborative robots (cobots) (mechanical devices that provide guidance using servomotors, while a human operator provides motive power) [10], portable robots where the workspace is not restricted to the place of its actual use [11]. Further research has investigated power amplifying assisting devices as well as various two-arm configurations [12, 13].

These days, as human robot collaboration becomes more efficient and easier to use, a variety of manufacturing industries aspire to introduce it in their production lines [3]. In automotive industry there have been introduced cooperative robots that work close to human operators in order to take over tasks that could cause human workers repetitive strain injury or to serve human workers as assistants passing them tools and parts during assembly processes [14].

As for robot-human-co-working [15], there have been many achievements in the fields of physical interaction, programming and teaching "by demonstration", shared workspace and safety. However, currently there are no safe industrial (high payload) robots to work in cooperative tasks or interacting with humans in a fenceless environment. There are different cooperative robots certified for safety applications (such as the UR5 and UR10 by Universal robots [16] or IWAA by KUKA [17]), but their payload is very low thus not allowing for operations like strength augmentation etc.

To achieve a safe collaboration between humans and robots it is necessary either ensure inherent safety in the utilized devices (safety certified PLCs, relays, valves etc.) [5] and/or to monitor the position of persons with respect to the robot through well designed vision systems [6] such as cameras, depth sensors, etc.

The positive experience with these seminal applications certainly opens the doors for a higher acceptance and wider usage of collaborative robots in industry. In this way, the robot becomes a production assistant in manufacturing and as such can absolve the human operator from ergonomically unfavorable work or provide assistive help during the execution of a human task.

The approach discussed in this chapter envisages the interaction with industrial grade robots that allow humans to perform assembly tasks in parallel and at the same workspace. In this sense, the role undertaken by both should be more active and provide them capabilities such as manually guiding the robot that is holding the parts, interacting with the robot by exchanging information through multi modal interfaces and so forth. The goal is to provide solutions for enabling even more direct human robot cooperation ensuring the safe cooperation with high-powered machines.

17.2 Approach

17.2.1 Requirements

Nowadays, industries are trying to find ways to automate routine tasks using robotic solutions. Nevertheless, the production operations are diverse and with high complexity. Usually, a shopfloor is organized in multiple areas, where each one of them is dedicated to specific processes and combined create the manufacturing flow. Most of the processes demand of high experience and capabilities by the human operators, who are able to adapt their actions to the specific requirements of the production scheduling.

The above apply in the white goods industries as well, that presents high revenues in millions and is expected to grow even more the following years, showing an annual growth of 8.9% during the period CAGR 2018–2022. This is the reason why this sector has attracted the attention of the research community which tries to analyze and detect the most strenuous processes for the operators. Such processes are the ones that require high levels of stress and fatigue for the humans and robotics seem the ideal solution due to the high level of precision and repetitiveness they present, improving the quality of life of workers. Such case is the pre-assembly process of the cabinet of a refrigerator. This activity includes various sealing tasks, performed by the operators, with the aim to seal the inner-liners in predefined places, before proceeding with the foaming processes.

Currently, human operators are placing sponges and tapes in order to seal any gap following the assembly of the inner-liners of the refrigerator and the freezer. After this step, a flexible panel, namely polionda, is fixed on the back of the fridge, to cover and close the inner-liner parts. The fixation takes place at the top and bottom of its edges with tape, which is not considered the best solution since it creates product quality issues such as sealant leakage of visible burrs in aesthetical parts. These activities are performed by two operators in parallel, while the product is moving on a conveyor.

Due to the delicacy of this process and in order to minimize the product defects, the operators have undergone a special training, which is repeated once a new model is introduced or when there are changes to the existing ones. Another drawback of the current approach is the high stress levels that are created to the operators due to the repetitive manual tasks they perform.

In this chapter will be described the necessary technologies to introduce robots close to human operators to perform the sealing process using a special foam, instead of tapes. In order to achieve this, multiple parameters need to be taken into consideration, such as the flexible parts that exist, the coexistence of robot with the human operator who should work in the same station, the complexity of some tasks that need to be automated, the random positioning of the components on the conveyor etc. Consequently, following this analysis, the station should be redesigned and reorganized to involve both the robot, which will execute the sealing task, and the human, who will do other delicate assembly tasks.

This station rearrangement has taken place with a human-centered perspective, allowing both resources smooth operation and ensuring human safety. As a result, two are the key pillars upon which the new cell has been created: the first is the intuitive human robot collaboration strategy and the other is the safety assurance. All the technologies developed under those pillars are described in this chapter as well as the design steps and the logic that was followed under the final demonstration.

17.2.2 Design

Following the industrial requirements that have been described above, the next step is the design of the new station paradigm. Multiple HRI technologies should be integrated together alongside with a sensorial network able to efficiently capture and fuse data under smart algorithms to enable near-real time process control.

Under this scope, from the operator's side, multimodal interfaces enhanced by sensors have been included. More specifically, force/torque sensors, microphones, cameras and wearable devices designed for difficult environments such as an industrial one. For example, voice commands and AR technology are used to interact with the robot, triggering the respective action.

From the robot's side, advanced image processing algorithms can provide in real time part detection results. Furthermore, using force sensors, the robot can move based on the vectoral force that is applied by the operator at its end effector. In other words, the operator can move the robot with his bare hands directly, without using special equipment, at any direction he/she want to freely. In Fig. 17.1 is presented

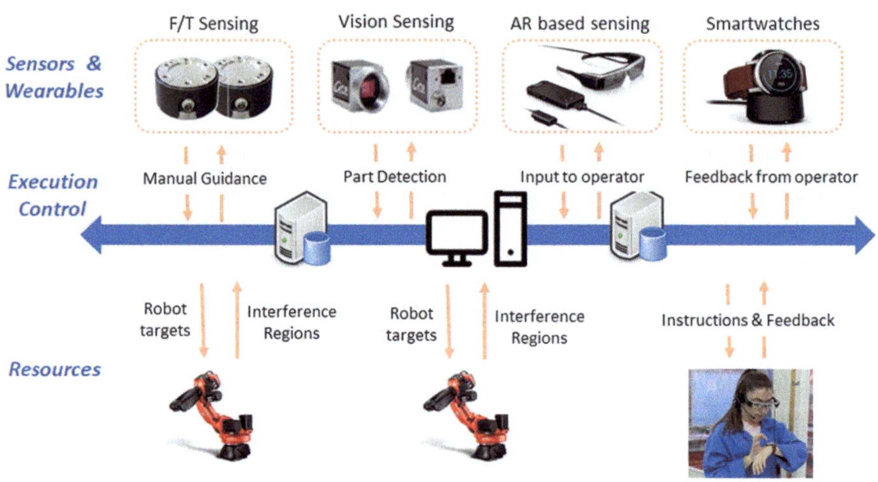

Fig. 17.1 Application architecture

the integration architecture of the whole system, showing the key components used, their interconnection and the information exchanged.

17.3 Implementation

As described in the previous section, a number of technologies is used to allow the execution control get feedback from the operator or the environment in general, adapting the process accordingly. For this reason, advanced perception, multimodal interfaces and safety algorithms have been implemented, as well as a sophisticated integration and communication platform to integrate and connect them. These components are analyzed in detail in the following sections.

17.3.1 Perception

Regarding the perception aspects of the system, a number of sensors and devices have been used and programmed to enable the integration and communication platform to receive feedback regarding the process status, human activity and robot movement. The devices that have been used are vision cameras and force/torque sensors, as well as wearable devices to get direct human feedback. The modules developed using the aforementioned hardware are described in the sections below.

17.3.1.1 Manual Guidance (Force Sensing Based Robot Motion)

The aim of this functionality is to enable the human operator to adapt robot position manually. The reason to do this can be twofold. On the one side, there is safety in which the robot should avoid collide with other objects while on the other side, a specific process needs a customized approach compared to the pre-programmed trajectory [18]. The proposed design is modular, offering the ability to install it in different robots, as well as configurability, offering the ability to adapt the control parameters based on the specific case requirements.

The key principle used for this functionality, is the fact that the human grabs robot's end effector and moves it freely in space. A force/torque sensor is attached between the robot wrist flange and the gripper as shown in pictures below (Fig. 17.2), and perceives the forces applied by the human, while the robot shows a virtual mass-damper-spring (impedance) behavior. A software running in F/T sensors controller calculates this behavior and adjusts the configuration parameters of the robots to move it accordingly in space.

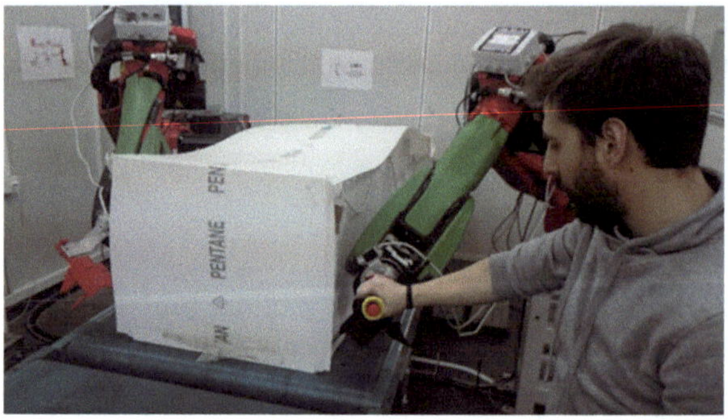

Fig. 17.2 Manual guidance for collaborative task execution

17.3.1.2 Vision System

Vision systems are very commonly used for industrial automation [19] and can be beneficial in cases where precise information needs to be instantly or repetitively extracted and used. (e.g. target tracking and robot guidance) [20]. A plethora of industrial activities have benefited from the application of vision technology on manufacturing processes. These activities include, among others, delicate electronics component manufacturing [21], metal product finishing [22], machine parts and integrated circuits manufacturing [23].

In this chapter, a vision system is described with feature tracking capabilities on specific parts. The key challenges that were faced were the high accuracy that was required for the detection, to adapt the robot motion accordingly at a later stage, the slightly different assemblies that existed in each part and should be reliable distinguished, the high speed of the whole process that was needed, the ambient conditions that may change during the process execution as well as the minor displacement of the parts under investigation, since they are not fixed on static fixtures.

In order to overcome all the aforementioned, a custom vision system has been created to track specific features of the inner liner part as shown in Fig. 17.10. This vision system detects in real time the respective features, process the detected measurement on a PC and sends the coordinates to the robot to move to a specific point and perform its activity. For this purpose, standard industrial RGB cameras have been used, since the vision algorithms are based on shape detection and color segmentation. These cameras are connected to a PC through a GigE network while the vision algorithm is executed on this PC. More specifically, Pylon SDK and an interface MVTec HALCON image processing library have been used. Pylon SDK is needed for the camera live streaming and the HALCON library for real time feature tracking and image processing. The whole setup is running continuously until a correct feature is detected. After that, the coordinates are transformed from the vision-based to robot-based coordinate system and sent to the robot.

17.3.2 Intelligent Multimodal Interfaces

Human robot collaborative systems require facilitators that would enable the intuitive communication between human operators and the robots, other resources or the main controller, during the execution of the different processes. Current practice focus on the machine connectivity through the use of PLC devices and the message exchange through signals among them, excluding the human factor, or having some static interfaces with limited capabilities.

For this reason, several interfaces have been created and will be described in the following sections, including the human in the communication loop with the rest system. The key characteristics of these interfaces are the direct input processing, timely and efficient message delivery, simplicity in their use.

17.3.2.1 Augmented Reality—Based Operator Support (AR)

AR can help close the gap between product development and manufacturing operation, mainly because of the ability to reuse and reproduce digital information and knowledge while supporting assembly operators [24]. Furthermore, AR may be used for enabling an intuitive interaction between the operator and the robot focusing on the reduction of the cognitive load of the human [24–26]. The visualization of key information in operator's field of view can increase his/her "safety feeling" or support him/her to execute his/her operations. Having the above in mind, an AR application has been created, to assist human operators in production stations using the following functionalities:

1. Provide production instructions in the form of text and virtual objects, such as 3D models, in each process step.
2. Warn the operator through visual and audio alerts when the robot is moving in the common workspace.
3. Visualization of robot's working area, in which is constrained to work, that is programmed in robot's controller.

This application has been developed using the Unity3D game engine to setup the interface and communication with the station controller through ROS, while the AR capabilities have been provided by Vuforia library. Additionally, the ROSBridge platform has been used, to enable data exchange between ROS master and non-ROS application. Lastly, the hardware where the application was running was a pair of EPSON Moverio BT-200 augmented reality glasses. The developed application as well as some of the aforementioned functionalities are visualized in the following figure, Fig. 17.3.

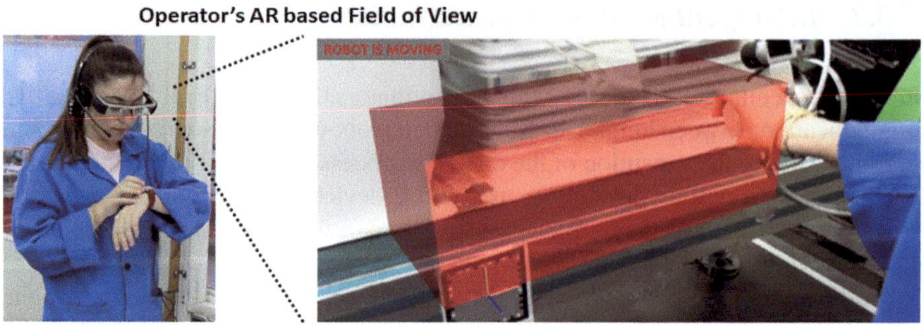

Fig. 17.3 AR based operator field of view—working areas and alerts visualization

17.3.2.2 Advanced UI on Wearable Devices

Another wearable device that has been provided to operators, apart from the AR glasses, is a smartwatch. Intuition and usability are significant factors that have been taken into consideration during the design and implementation phase of the above features [27]. This device was used to enable them provide feedback to the integration platform, including them to the execution workflow. More specifically, through the smartwatch interfaces, the operator was able to exchange information with other resources, control the overall task execution, control the robot movements using auxiliary sensors, such as microphone for audio commands or F/T sensor for manual guidance as already explained.

In terms of implementation, the application has been developed in JAVA for the Android OS, using the Android Studio IDE. This algorithm, apart from implementing the aforementioned functionalities, enables the wireless connection with the station controller and offers a set of buttons to interact with the AR application. As it can be seen in Fig. 17.4, each interface covers the whole screen of the smartwatch, while

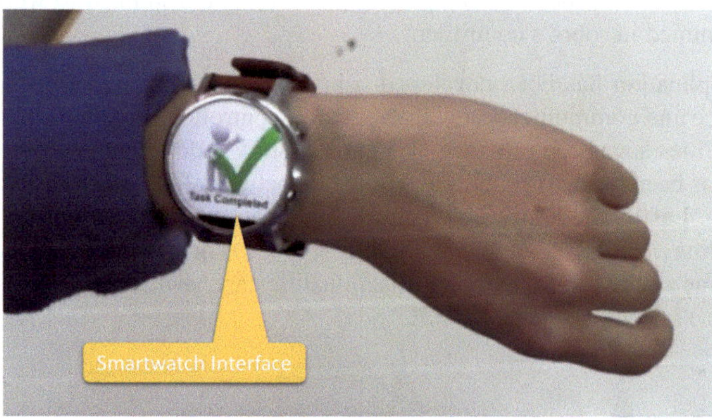

Fig. 17.4 Wearable devices/smartwatch for connected operators

the navigation among them can be easily made using swiping gestures on the screen. The interfaces that have been implemented are the following:

1. Interfaces to the AR application;
    - Show/hide textual instructions of each assigned process,
    - Show/hide robot's working areas,
    - QR code-based pairing of the watch with the AR glasses,
2. Enable/Disable manual guidance functionality of the robot;
3. Stop/Resume robot movement functionality;
4. Report task completion of a human task;
5. Audio commands functionality to move the robot.

The hardware upon which the application ran was a Motorola Moto360 2nd edition, while it can be easily transferred to other Android-based smartwatches, since the application is based on native Android wearable framework.

17.3.3 Safety

An important aspect that has been investigated and implemented in this cell is the safety assurance of the operators. Since human and robot should operate in the same workspace, sharing the same part, in parallel and without fences, it must be ensured that no harm can be done to the operators. Taking into consideration that in the application have been used industrial robots and not collaborative, therefore they cannot inherently limit the exerted forces, a risk assessment investigation was performed based on the EU safety related standards that refer to collaborative applications (ISO 10218, ISO/TS 15066). These standards, indicate the following safety methods:

- Safety—related Monitored Stop (SOS);
- Speed and Separation Monitoring (SSM);
- Power and Force Limiting (PFL);
- Hand Guiding (HG).

The specific case was oriented around SOS, PFL and HG methods and not the SSM, since the resource separation was not an option. The following subsections describe in more detail the implemented functionalities for the above methods.

17.3.3.1 Power and Force Limiting Using Robot Dynamic Modelling

Power and Force Limiting (PFL) ensures that the forces exerted upon the operator should not exceed a predefined limit, while the robot is moving, if they get in contact. Based on the requirements defined in ISO/TS 15066, a PFL application has been implemented, accomplishing online collision detection. This application is based on

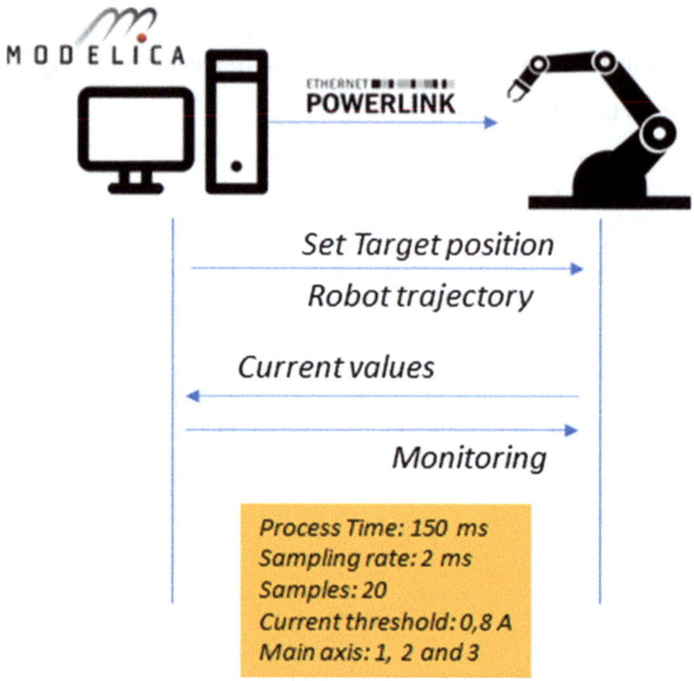

Fig. 17.5 Power and force limiting implementation

measurements on current values of the motors and the position of robot's joints. Using a time-invariant dynamic model in Open Modelica software in combination with Artificial Neural Networks, the current and torque required by each joint for a given trajectory are estimated with satisfying precision. Therefore, without using any additional sensors, for each trajectory is calculated the estimated value of current and torque and they are compared with the actual ones as the robot moves. If the above values exceed a user defined threshold, then an emergency stop command is issued. Figure 17.5 presents the implementation architecture as well as some key information of this application.

17.3.3.2 Collision Detection

Another way to detect a possible collision among the human and the robot and stop the robot movement is through the installation of a tactile safety skin on the robot body [28]. More specifically, AirSkin Safety pads from Blue Danube Robotics have been installed on the robot, which are able to detect robot contact with any object and issue an emergency stop. These pads are airtight and have embedded sensors that are able to monitor and measure air pressure variations inside the pad. In Fig. 17.6

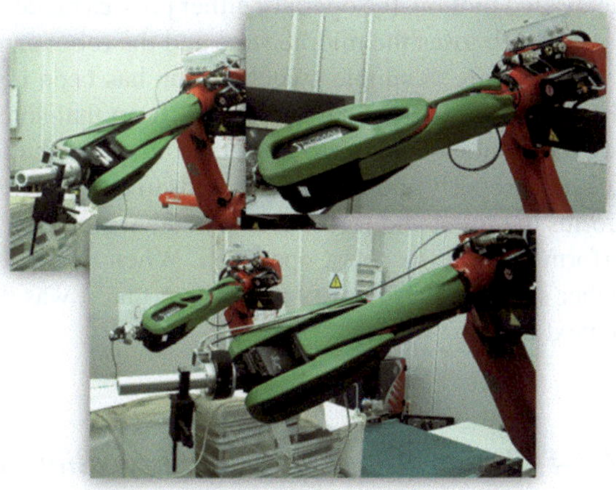

Fig. 17.6 Safety skin for emergency stop upon contact

are visualized how these pads have been installed in the last three axes of the robot, which are close to the operator, covering almost all their surface.

17.3.3.3 Interference Regions

As already described in a previous section, the COMAU controller offers the possibility to design specific 3D cartesian volumes, in which robot's movement is constrained. This function is in accordance with the requirements of the Safety—rated monitored stop safety method as indicated in ISO/TS 15066 Clause 5.5.2. The programmer is able to define different types of these volumes, based on the requirements of the applications, such as monitoring, constraining, stop on violation etc. Additionally, the programmer is able to define the stop category these zones may issue, which in this application the category 1 stop has been selected. The size and position of these zones is visualized to the operator through the AR application as shown in Fig. 17.3.

17.3.3.4 Enabling Devices and Intervention Buttons

Several buttons have been also used in the cell for safety reasons. Apart from the emergency stop buttons, enabling devices have been used in the case of manual guidance functionality. This allowed the operator to move the robot, when the robot was in manual guidance mode, only once they are pressed, reducing the risk of injuries. This enable device consists of a three-state switch, which is active only in the middle state. When the operator presses the switch to the middle state, a signal is sent to the robot and it starts to move based on the force sensor data. In case of an unexpected motion, the operator either presses harder the switch or complete releases it, the switch loses the middle state and the robot cannot move.

Furthermore, a simple switch button has been installed in the cell. This button, when pressed, informs the control platform that something unexpected occurred in the cell, but not so serious to press the emergency button, for example a process is not executed properly and the human needs to intervene. As a result, the control platform informs the robot and forces it into manual guidance mode, allowing the operator to perform manual the robot operation. When the operation finishes, the human uses his/her smartwatch to notify the system that it was done and the cell proceeds with the next task.

17.3.4 Integration and Communication Architecture

All the aforementioned functionalities that have been implemented, need a strong and robust integration platform to coordinate them. This would enable the communication and would control the different resources achieving an autonomous and robust Human Robot Collaboration (HRC) [29, 30].

Both the frontend and backend have been developed in Java EE environment. The frontend of this platform is visualized in Fig. 17.7 and was a web-based application, offering an easy to use interface for data access and production execution, operation per operation.

Additionally, like most components described in this chapter, it was based on ROS topics and services to ensure proper message exchange among the resources. For this reason, in the backend platform a ROS-JAVA framework has been implemented, while the non-ROS applications, such as the AR and smartwatch applications, used the Rosbridge server that provides a JSON API to acquire ROS functionalities and connect directly via ROS and the WebSocket protocol.

Lastly, in the final scenario, a number of topics such as status, command, emergency and others have been issued and used for XML file exchange, based on predefined XSD schemas, dedicated per topic.

Fig. 17.7 Service based reporting and operation sequencing (execution control system)

TASK	RESOURCE	STATUS
Recognize path 1	Camera	Completed
Guide robot	Human	
Apply sealant2	Human	In progress
Reset position	Robot	Completed
Apply sealant1	Robot	Completed
Recognize path 2	Human	
Stop guide robot	Robot	

17.4 Industrial Example

This section discusses, the implementation of the presented HRC system in a case study for the white goods sector. This industrial example involves the pre-assembly of a refrigerator's cabinet including the sealing process in specific inner points as visualized in Fig. 17.9. In the current production system, this assembly process is performed manually with product variability of around thirteen different models throughout the annual production. The cycle time of line is thirty-eight seconds achieving a production throughput of eighty pieces per hour. The movement of the part from one workstation to the next is performed through a conveyor belt system moving with a speed of three and a half meters per hour.

In the introduced human robot collaborative assembly system, the human operators perform only the taping and cabling operations on the refrigerator cabinet. The industrial robots, working in parallel with the operators, undertook the sealing process, avoiding any leakages and faults in the aesthetical part that was a possible outcome from the manual process.

The 3 pieces of the inner liner (in upper, central and lower position) are currently fixed by adhesive tapes which are also used for the fixing of the "polionda" panel at the two ends. Referring on the main component, as depicted in Fig. 17.8 there are three different sealing paths, which require high accuracy because the dimensions and clearances of the paths are very tight.

The design of the implemented HRC cell is presented in Fig. 17.9. This workstation employs two human operators and two industrial COMAY Racer robots. The latter use a sealing gun as a tool and are able to automatically detect the sealing points of the refrigerator that moves in the conveyor and dynamically approach them for performing the process. The accurate detection of the sealing point along with the generation of the robot trajectories to reach this point is quite critical in order to avoid collision with the humans working in proximity.

The physical set up of the HRC cell is presented in Fig. 17.9. In the beginning of the shift the human operators enter the cell equipped with the wearable devices (AR

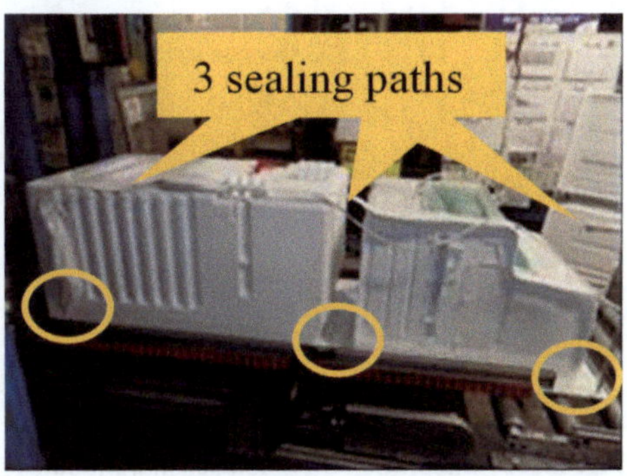

Fig. 17.8 Sealing path

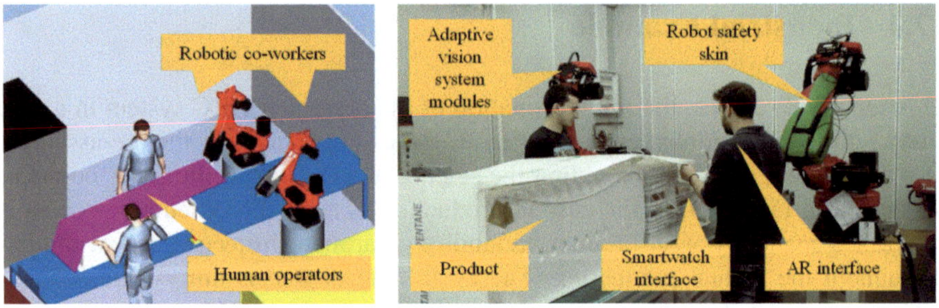

Fig. 17.9 Physical installation of the HRC cell for the white goods case study

glasses, smartwatch and headset). These devices ensure the intuitive communication of the operators with the central execution system.

At the beginning of each cycle, a refrigerator arrives on the cell through the moving conveyor. The first step is for the operators to load the upper panel, called polionda, on the product. Then, they add cables on the product and use tapes to fix the polionda on the refrigerator. The operators receive textual instructions on the tasks they have to perform, sent from the central execution control system, through their AR glasses.

In parallel with human operators' assembly tasks, the robots are able to operate autonomously. The vision system integrated is trigger by the central execution system in order to detect the sealing point coordinates as shown in Fig. 17.10. Then the robot controller generates the robot arms trajectories considering the conveyor movement, in order to ensure that the robot will reach the sealing point efficiently in terms of timing. The robot motions are monitored in terms for cartesian position and speed through C5G controller provided safety functions in order to avoid unexpected robot movements.

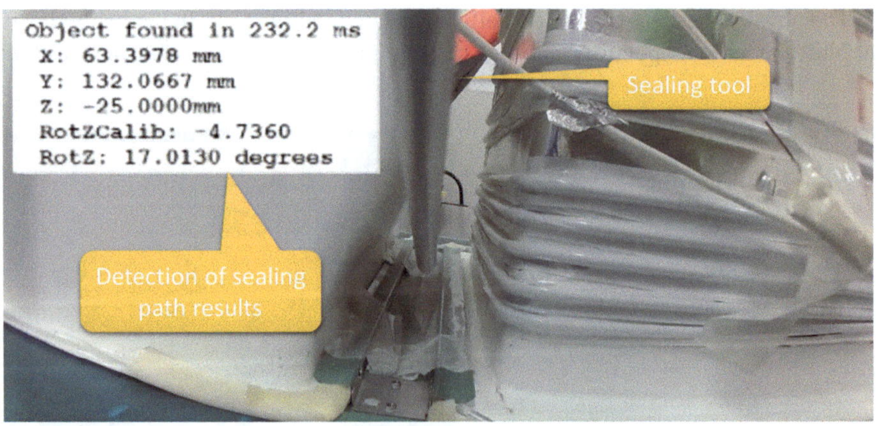

Fig. 17.10 Integration of vision/sensing modules

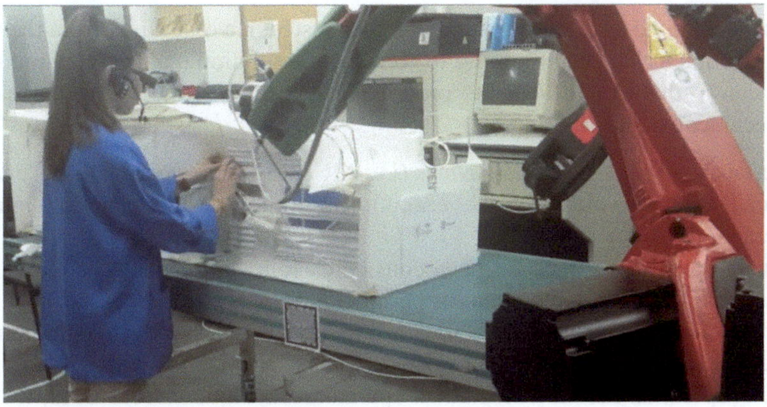

Fig. 17.11 H-R collaboration and workspace sharing

The visual and audio alerts are activated and sent to the operators when any of the two robots is moving, even if the latter is not in their field of view. The operators can still visualize the instructions for their running assembly task. By combining these features, the workers can increase their "safety feeling" while reducing assembly errors. The operators are using their smartwatch to send a signal in the execution system when they finish their running tasks. In that way, the execution system monitors the execution and coordinates the tasks that need to be performed by the operators and the robots, respective the required sequence. The collaborative assembly continues in the same way for the middle and rear part of the refrigerator as shown in Fig. 17.11.

The human operators, apart from their assembly tasks, are also inspecting the quality of the sealing operations performed by the robots. In case the robot does not perform efficiently the process, the worker can manually guide the robot again in the sealing point for repeating the process. During this manual guidance operation and in order to avoid accidents, the robot tool position and orientation are constrained inside the Interference Regions programmed in the C5G controller. The operators can see these virtual zones through the AR glasses and in that way, they are aware of the safe and allowed robot working volumes. Another important part of the deployed safety system is the deployed safety skin, which upon contact triggers an emergency stop of the robot.

17.5 Discussion

This chapter introduced a human robot collaborative assembly cell enabled by the integration of multiple sensing and wearable devices. The target of this cell was to provide a safe workplace for humans and robots where the operators are able to intuitive interact with their robot co-workers during the collaborative execution. The performance of the developed system has been tested in a case study from the white goods manufacturing sector aiming to increase the solution maturity and industrial relevance. To this extend under to proposed HRC solution the following features have been deployed:

- Interfaces in smart wearable devices (i.e. AR glasses and smartwatches) enabled human active involved in the execution workflow while enabling intuitive interaction;
- Human operators' safety has been considered following the EU standards indications through the deployment of peripheral devices (i.e. safety skin) and software applications (Power and Force Liming functionality) as well as programmed safety functions in robot's controller (i.e. C5G Interference Regions) to ensure safe cooperation between the human operators and the robot resources;
- The robot perception has been enhanced through the integration of:
  - The vision system for dynamic sealing point detection,
  - The force sensing system enabling the robot manual guidance by the worker.
- The coordination of the execution of the manufacturing process, performed by multiple resources has been achieved by integrating of the hardware devices and software components in the central execution system.

The system has been validated through the deployment in the refrigerators' assembly case. The recorded benefits by implementing the suggested solution are summarized below:

- Increased flexibility and productivity in sealing process. The joint operability of the human–robot team and the cooperative control of the correctness of the process may assure high level of quality and repetitiveness.
- A better-quality performance of a reproducible process (sealing) without requiring high relevant investment. The preliminary calculation of ROI has demonstrated that the potential investments are viable when compared to the manual labor case.
- Opportunity for employees to skip repetitive work that could not previously be automated. In the demonstrator, the OCRA analysis is under evaluation. Also considering the sealing material is consider dangerous for the operators if they come in contact, the suggested solutions reduces this danger since the robot is performing the sealing task.
- A good technological base to reduce risks of injuries. The safety package developed for the pilot cell (technologies for robot as impedance control, force limitation, safety skin) and the auxiliary devices to support the human operability and the process control (such as AR glasses, robot-connectable objectives etc.) are a good starting point to ensure operator's safety.

Overall, the results indicate that the collaboration between humans and industrial robots can bring tangible benefits in current manual production lines. However, there is still a long way to be covered for ensuring a seamless collaboration and safe solution. Future work should focus on developing methods for better immersing the human in the new safety measures that are becoming available as well as on integrating the safety induced restrictions inside design and planning tools that can efficiently simulate their effect on the manufacturing process. Last but not least, the introduction of standardized services that can integrate all the heterogeneous sensing and interaction equipment would greatly simplify the process of deploying such cells.

References

1. Chryssolouris G (2006) Manufacturing Systems: Theory And Practice. Springer, New York
2. Tsarouchi P, Makris S, Chryssolouris G (2016) Human–robot interaction review and challenges on task planning and programming. Int J Comput Integr Manuf 29:916–931. https://doi.org/10.1080/0951192X.2015.1130251
3. Michalos G, Makris S, Tsarouchi P, Guasch T, Kontovrakis D, Chryssolouris G (2015) Design considerations for safe human-robot collaborative workplaces. Procedia CIRP 37:248–253. https://doi.org/10.1016/j.procir.2015.08.014
4. Kousi N, Michalos G, Aivaliotis S, Makris S (2018) An outlook on future assembly systems introducing robotic mobile dual arm workers. Procedia CIRP 72:33–38. https://doi.org/10.1016/j.procir.2018.03.130
5. Michalos G, Kousi N, Karagiannis P, Gkournelos C, Dimoulas K, Koukas S, Mparis K, Papavasileiou A, Makris S (2018) Seamless human robot collaborative assembly—an automotive case study. Mechatronics 55:194–211. https://doi.org/10.1016/j.mechatronics.2018.08.006
6. Papanastasiou S, Kousi N, Karagiannis P, Gkournelos C, Papavasileiou A, Dimoulas K, Baris K, Koukas S, Michalos G, Makris S (2019) Towards seamless human robot collaboration: integrating multimodal interaction. Int J Adv Manuf Technol 105:3881–3897. https://doi.org/10.1007/s00170-019-03790-3
7. Bley H, Reinhart G, Seliger G, Bernardi M, Korne T (2004) Appropriate human involvement in assembly and disassembly. CIRP Ann Manuf Technol 53:487–509. https://doi.org/10.1016/S0007-8506(07)60026-2
8. Kruger J, Lien TK, Verl A (2009) Cooperation of human and machines in assembly lines. CIRP Ann Manuf Technol 58:628–646. https://doi.org/10.1016/j.cirp.2009.09.009
9. Helms E, Schraft RD, Hagele M (2002) rob@work: robot assistant in industrial environments. IEEE, pp 399–404
10. Akella M, Peshkin M, Colgate E, Wannasuphoprasit W (1998) Cobots: a novel material handling technology
11. Gkournelos C, Kousi N, Christos Bavelos A, Aivaliotis S, Giannoulis C, Michalos G, Makris S (2019) Model based reconfiguration of flexible production systems. Procedia CIRP 86:80–85. https://doi.org/10.1016/j.procir.2020.01.042
12. Bernhardt R, Surdilovic D, Katschinski V, Schreck G, Schröer K (2008) Next generation of flexible assembly systems. In: Azevedo A (ed) Innovation in manufacturing networks. Springer, US, Boston, MA, pp 279–288
13. Bernhardt R, Surdilovic D, Katschinski V, Schröer K (2007) Flexible assembly systems through workplace-sharing and time-sharing human-machine cooperation—PISA. IFAC Proc 40:247–251. https://doi.org/10.3182/20070523-3-ES-4908.00041
14. Michalos G, Makris S, Spiliotopoulos J, Misios I, Tsarouchi P, Chryssolouris G (2014) ROBO-PARTNER: seamless human-robot cooperation for intelligent, flexible and safe operations in the assembly factories of the future. Procedia CIRP 23:71–76. https://doi.org/10.1016/j.procir.2014.10.079
15. Goodrich MA, Schultz AC (2007) Human-robot interaction: a survey. Found Trends® Hum-Comput Interact 1:203–275. https://doi.org/10.1561/1100000005

16. URL Universal Robots. https://www.universal-robots.com/
17. URL KUKA Robots. https://www.kuka.com/
18. Kang M-K, Lee S, Kim J-H (2014) Shape optimization of a mechanically decoupled six-axis force/torque sensor. Sens Actuators Phys 209:41–51. https://doi.org/10.1016/j.sna.2014.01.001
19. Chen S, Li Y, Kwok NM (2011) Active vision in robotic systems: a survey of recent developments. Int J Robot Res 30:1343–1377. https://doi.org/10.1177/0278364911410755
20. Davies ER (1990) Automated visual inspection. In: Machine vision. Elsevier, pp 411–434
21. Sanz JLC, Petkovic D (1988) Machine vision algorithms for automated inspection thin-film disk heads. IEEE Trans Pattern Anal Mach Intell 10:830–848. https://doi.org/10.1109/34.9106
22. Malamas EN, Petrakis EGM, Zervakis M, Petit L, Legat J-D (2003) A survey on industrial vision systems, applications and tools. Image Vis Comput 21:171–188. https://doi.org/10.1016/S0262-8856(02)00152-X
23. Li H, Lin JC (1994) Using fuzzy logic to detect dimple defects of polished wafer surfaces. IEEE Trans Ind Appl 30:317–323. https://doi.org/10.1109/28.287528
24. Rentzos L, Papanastasiou S, Papakostas N, Chryssolouris G (2013) Augmented reality for human-based assembly: using product and process semantics. IFAC Proc 46:98–101. https://doi.org/10.3182/20130811-5-US-2037.00053
25. Michalos G, Karagiannis P, Makris S, Tokçalar Ö, Chryssolouris G (2016) Augmented reality (AR) applications for supporting human-robot interactive cooperation. Procedia CIRP 41:370–375. https://doi.org/10.1016/j.procir.2015.12.005
26. Makris S, Karagiannis P, Koukas S, Matthaiakis A-S (2016) Augmented reality system for operator support in human–robot collaborative
27. Gkournelos C, Karagiannis P, Kousi N, Michalos G, Koukas S, Makris S (2018) Application of wearable devices for supporting operators in human-robot cooperative assembly tasks. Procedia CIRP 76:177–182. https://doi.org/10.1016/j.procir.2018.01.019
28. Blue Danube Robotics. https://www.bluedanuberobotics.com
29. Karagiannis P, Giannoulis C, Michalos G, Makris S (2018) Configuration and control approach for flexible production stations. Procedia CIRP 78:166–171. https://doi.org/10.1016/j.procir.2018.09.053
30. Argyrou A, Giannoulis C, Sardelis A, Karagiannis P, Michalos G, Makris S (2018) A data fusion system for controlling the execution status in human-robot collaborative cells. Procedia CIRP 76:193–198. https://doi.org/10.1016/j.procir.2018.01.012

Chapter 18
Virtual Reality for Programming Cooperating Robots Based on Human Motion Mimicking

18.1 Introduction

Programming of hybrid production systems requires to handle the robot and the human behavior as well as their interaction, which makes the programming effort and time rather high [1]. Current research has been investigating more intuitive programming methods [2] for the reduction of programming effort and time in dual arm robots. Dual arm robots have been recently employed for a number of assembly tasks, aiming to increase the flexibility and re-configurability of the manufacturing systems [3], exploiting the advantages of dexterity, flexibility, space saving and a human-like structure.

This chapter discusses a method for the programming of an industrial dual arm robot by imitating human motions. An hierarchical structure for robot programming is proposed for this purpose based on task oriented programming principles. Human motion data are captured in a virtual environment and are further processed. The output of this process is the human motion identification that is finally imitated by a dual arm robot. The proposed processing approach merges the different human motion data, representing them with mathematical models. The development of a global algorithm that automatically identifies the human motion under a higher-level hierarchical model is also an advantage. The proposed framework was implemented as a user-friendly offline programming tool for dual arm robots. It is demonstrated in a case study from the automotive final assembly, for the pick-and-place of a cable in a repository using human captured data from a CAVE environment.

S. Makris, *Cooperating Robots for Flexible Manufacturing*,
Springer Series in Advanced Manufacturing,
https://doi.org/10.1007/978-3-030-51591-1_18

18.2 State-of-the-Art Solutions

Human Robot Interaction (HRI) could eventually increase the productivity and flexibility of the production lines and therefore research attention has been paid to in [2, 4–6]. Currently, the industrial robot programming is mainly based on the use of teach pendants and offline programming tools. The way of editing a program is currently not user friendly and it mainly requires low level programming, without considering higher level details. Another option is the intuitive robot programming, including demonstration techniques and instructive systems [7]. The Programming by Demonstration (PbD) frameworks have traditionally used different interaction mechanisms such as voice, vision, touch sensing, motion capturing, data gloves, turn-rate sensors, acceleration sensors [8–10]. HRI has been achieved through gestures [8, 11–13], voice commands [14], physical and non-physical contact [15], graphical interfaces [2, 16, 17]. Despite the wider research interest in programming by demonstration and instructive systems, there are still open research challenges coping with the needs for easy use, extension and reconfiguration of such systems.

Imitation learning is a user-friendly instruction technique to transfer new skills to robots. The steps of a mimesis framework include the robot observing an instructor, generalizing and reproducing the observed tasks [18]. Numerous research attempts have focused on the development of imitation learning frameworks for robot manipulators and humanoid robots [19–24].

Research work has also been done on human motion recognition using virtual reality [25, 26] and imitation, both for online and offline programming frameworks [27–34]. Some of these studies focused on task oriented programming and automatic task recognition in a virtual reality environment [9, 35]. Motion primitives were evaluated in [36, 37], while approaches for motions classification were discussed in [2]. In the same research, an intuitive programming framework for industrial dual arm robots was presented, introducing the use of multimodal interfaces such as gestures, vocal commands and graphical interfaces.

18.3 Approach

18.3.1 Hierarchical Model for Programming

An hierarchical structure of the robot program has been proposed focusing on 'what' to do rather than 'how', as opposed to motion-oriented methods. The robot program model is decomposed in three levels, namely, program, task and operation (Fig. 15.1). The program level is a general-purpose complex activity and includes a number of robot and human tasks. The task level is of a general purpose however, it is a simple activity, comprising a set of operations.

A set of robotic positions consist an operation in the lower level. The list of these operations is summarized in Table 18.1. For each of them, there is a main functionality

Table 18.1 List of operations

Operation	Main functionality	Parameters						
		Arm No	TF	BF	UF	COOP	SYNC	SINGLE
CLOSE_GRIPPER	It enables the current input of an electrical gripper for the closing of clamps	x						
OPEN_GRIPPER	It disables the current input of an electrical gripper for the opening of clamps	x						
MOVE_AWAY	It allows single arm movement away from a position, with or without carrying a load	x	x	x	x			x
APPROACH	It allows single arm movement towards a position, with or without carrying a load	x	x	x	x			x
BI_ MOVE_AWAY	It allows the cooperative motion of two robot arms away from a position, while carrying an object. One arm plays the role of the master	x	x	x	x	x		

(continued)

Table 18.1 (continued)

Operation	Main functionality	Parameters						
		Arm No	TF	BF	UF	COOP	SYNC	SINGLE
BI_APPROACH	It allows the cooperative motion of two robot arms towards a position, while carrying a load. One arm plays the role of the master	x	x	x	x	x		
SYNC_ MOVE_AWAY	It allows the synchronized motion of two robot arms away from one or more positions without carrying a load. Each arm acts independently	x	x	x	x		x	
SYNC_APPROACH	It allows the synchronized motion of two robot arms towards one or more positions, without carrying an object. Each arm acts independently	x	x	x	x		x	
CUSTOM*	It allows the user to prepare a customized operation, depending on the application (e.g. operation related to a vision system)	Depending on application						

that is explained within the same table, as well as programming related parameters. These parameters refer to the robot arm number (Arm No), the tool frame (TF), the robot base frame (BF), the user frame (UF), as well as the mode of motion among cooperative (COOP), synchronized in time only (SYNC) and single.

The idea of the hierarchical decomposition of human–robot (HR) tasks and the use of multi-modal interfaces for interaction and programming allows to:

- Easily build a new robot program focusing on the high level description and using multi-modal interfaces in the lower level (e.g. APPROACH, GRASP etc.) for teaching the robot a new operation. The above structure is user friendly, since the user or programmer can monitor a program by understanding the process steps rather than editing robot positions without high level description.
- Easily monitor and understand a robot program by high-level program structure, focusing on the higher and lower level activities. The description of the different levels is meaningful, providing information about the type of activity that follows.
- Easily extend or modify an existing robot program, since the user can edit it in the different levels of the hierarchy.
- Involve one or more humans within the same program enabling their control by an external system. This external framework that controls every robot or human agent that are involved in the hybrid cell.

18.3.2 Human Motion Data Capturing and Processing

In this approach, the human motion has been captured through the CAVE technology in order to be imitated by a dual arm robot. A set of physical objects is used for capturing the human two hands' motion. These objects are constructed by combining three or more infrared light reflectors in a unique geometry. Each one of these objects is tracked by the cameras of the CAVE system. A virtual object is created in the virtual environment, corresponding to each physical object (Fig. 18.1).

The virtual object data constitute the human motions inside the virtual environment and are finally stored in text file. When a human holds a part with the two hands, the physical object reflectors are not always visible from at least three cameras, since they are covered by the human's hand. This absence of visibility leads into errors and noise in the captured data. Therefore, the cartesian position data include the

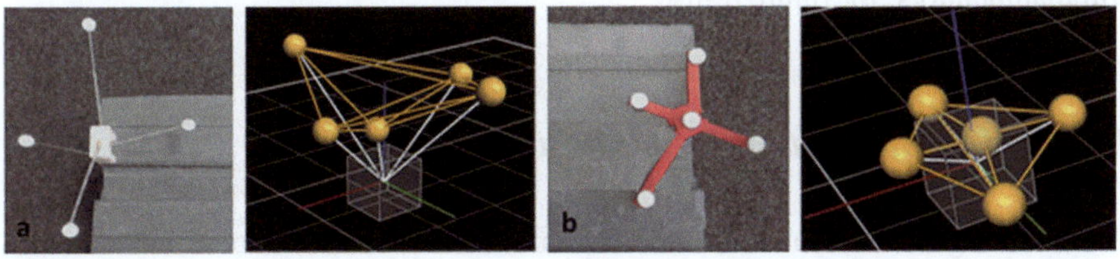

Fig. 18.1 **a** Physical and virtual object of Arm1; **b** physical and virtual object of Arm2

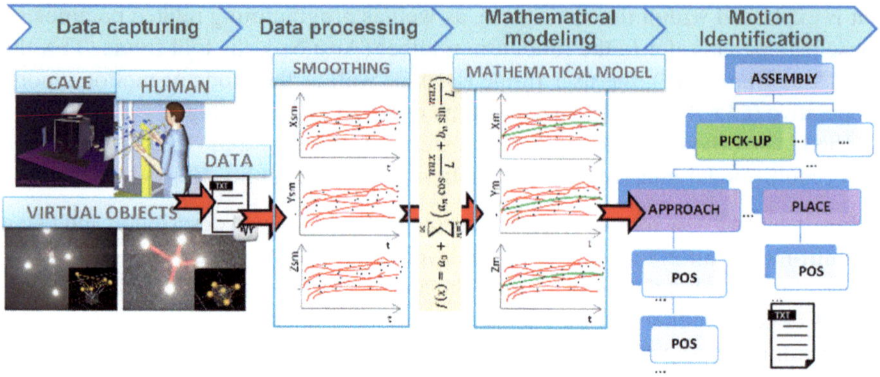

Fig. 18.2 Overall approach from human motion to robot motion generation

noise that should be eliminated in order to allow the generation of a smoother robot motion. As a result, the use of smoothing methods is important for the exclusion of this malfunction.

In order to take advantage of the numerous experiments during the data capturing in the CAVE environment, a smoothing approach was adapted for achieving a satisfying result. As illustrated in Fig. 18.2, the data processing of multiple experiments for the creation of smoothed data was the second step of the proposed method. The local regression using weighted linear least squares and a 2nd degree polynomial model was evaluated for smoothing the captured data via Eq. (18.1). The estimation of the coefficients ao, a1 is done so that Eq. (18.2) will be minimized [38].

$$\mu(x_i) \approx a_o + a_1(x_i - x) + \frac{a_o}{2}(x_i - x)^2, x - h \leq x_i \leq x + h \tag{18.1}$$

$$\sum_{i=1}^{n} W\left(\frac{x_i - x}{h}\right)(Y_i - (a_o + a_o(x_i - x)))^2 \tag{18.2}$$

After the generation of smoother motions, a final curve that reproduces the motion according to the majority of the smoothed data is generated. The smoothed data are used in order to fit a curve that mathematically represent the human motion. This will allow the description of the motion that can be eventually reproduced by dual arm manipulators, for future re-use.

18.3.3 Human Motion Data Fitting

The final target of the method is to fit human data in the form of a mathematical models based on different curve fitting parameters. For the curve fitting process, different kinds of curves, such as high order polynomials, Fourier and power series,

Gaussians etc. were evaluated. Observations shown that the Fourier of 6th and even higher order polynomials can satisfactorily fit the smoothed data, since it had the minimum relative errors compared to the other models. Additionally, the choice of the Fourier series can be explained from physical point of view. The human motion, similar with most of the motions on physical world, can be approached in an extremely satisfying degree with the use of cosine and sine functions. Fourier series take advantage of these trigonometrical functions and enable the data fitting according to a smoothed curve. The form of this model is presented in Eq. (18.3), where ao is a constant term in the data and is associated with the i = 0 cosine term, nπ/L is the fundamental frequency of the signal, n is the number of harmonics in the series, and $1 \leq n \leq 6$.

$$f(x) = a_0 + \sum_{n=1}^{6}\left(a_n \cos\frac{n\pi x}{L} + b_n \sin\frac{n\pi x}{L}\right) \tag{18.3}$$

18.3.4 Motion Identification and Classification

The final step is to use the model fitting values, f(x), and develop an algorithm for identifying the motions and classifying them in a sequence of operations under a structured hierarchy (Fig. 18.1). The first and second levels, called order and task respectively, are high level activities. The third level includes the operations that are low level activities, such as APPROACH, MOVE_AWAY, GRASP, OPEN/CLOSE_GRIPPER etc. The proposed algorithm for the automatic identification of these operations is based on the evaluation of the coordinates' sign, the gradient values and the sign of the 2nd order differential.

The algorithm for identifying an operation includes two stages (Fig. 18.3). In the first stage, the analysis of the z axis data helps with the initial estimation of the APPROACH and MOVE_AWAY operations as well as the estimation of the minimum position, signaling their beginning and ending. The point where the APPROACH and MOVE_AWAY operations are crossed is the potential GRASP/RELASE and OPEN/CLOSE_GRIPPER operations. The second stage is based on the analysis of the x axis coordinates, where the differentials of 1st and 2nd order are calculated for deciding the operations. The evaluation of the gradient sign, the sum and average values of the data were taken into account for ensuring the result.

18.3.5 Human Robot Frames Transformation

In order for the captured human motion to be replicated into a robot cell, transformation between the frames between the robot and CAVE environment is required.

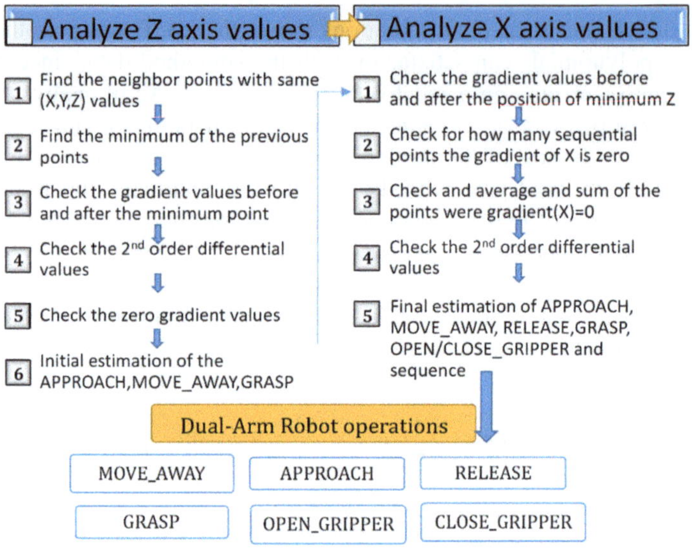

Fig. 18.3 Motion identification method

The model describing the human motion determines the way that the robot TCP (Tool Center Point) coordinates will be updated. Figure 18.4 illustrates an example of the paths according to the CAVE and robot reference frames. In the CAVE, the two objects have Acave and Bcave frames respectively. The equations describing the relation between the positions of virtual objects in the CAVE reference systems are Eq. (18.4). A transformation matrix T transforms the CAVE frames into robot frames. In the specific case, the CAVE reference system rotates 90° around the x axis and the robot reference frames are estimated [Eqs. (18.6) and (18.7)].

The result is a text file including the vectors w'1i and w'2i for Arm 1 and Arm2 respectively. The robot path is identical to the data from the fitted curve and the

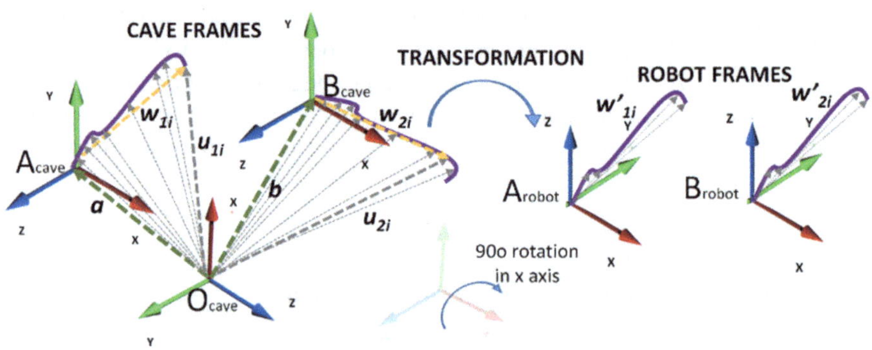

Fig. 18.4 CAVE to robot frames transformation

transformation ensures the execution of the paths according to the initial human motions.

$$\mathbf{w}_{1i} = \mathbf{u}_{1i} - \mathbf{a} \tag{18.4}$$

$$\mathbf{w}_{2i} = \mathbf{u}_{2i} - \mathbf{b} \tag{18.5}$$

$$\mathbf{w'}_{1i} = \boldsymbol{T}\mathbf{w}_{1i} \tag{18.6}$$

$$\mathbf{w'}_{2i} = \boldsymbol{T}\mathbf{w}_{2i} \tag{18.7}$$

18.3.6 System Implementation

The CAVE system used for data capturing has three DLP Projectors Barco RLM W12, each one connected to a HP Z820 workstation and combined with an optical tracking system. The system is controlled by a VR engine software called Virtools. The Vicon software monitors the movement of the virtual objects and publishes all the information using the VRPN protocol to Virtools. The metric system was used in the CAVE environment and therefore, the matching between the virtual reality and the robot cell workspace was one to one conversion.

The implementation architecture (Fig. 18.5) focused on the connection between the CAVE and the robot work cell. The fitting values, f(x), were sent to the robot using the TCP-IP communication protocol given the fact that the frames matching process

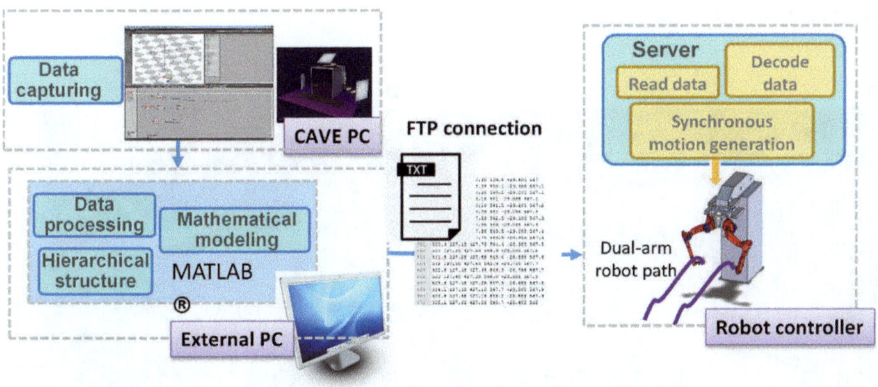

Fig. 18.5 Implementation architecture

between the CAVE and the robot has been completed. The processing, modelling and motion identification algorithms have been developed in the form of a Graphical User Interface (GUI) in MATLAB.

A server program resides in the dual arm robot controller. This program reads the communicated data and decodes step by step the coordinates, with the synchronous movement of both arms. The coordinates in the text file are in the form of a six-column table. The first and last three columns are the coordinates for the x, y and z axis of the robot's frame for Arm1 and Arm2 respectively. The program then matches these coordinates with the ones for the position of both robot arms.

18.4 Industrial Example

18.4.1 Cable Handling Use Case

The proposed programming method is applied in a case study for the pick-and-place task of a cable in two repositories. Two layouts have been created; one in the CAVE and the other in a dual-arm robot cell. The layout of the CAVE cell (Fig. 18.6a) has two repositories with a slot where the cable will be placed. The physical objects allowed the human motion path to be recorded during the experiment execution. The layout of the robot's cell (Fig. 18.6b) includes a smart dual-arm robot (COMAU) and two repositories as well.

Two different tasks were obtained. Firstly, the human approached with both arms, the two ends of the cable, which was placed in the slot of the repository 1 in front of the human. Following that, the human grasped the two ends of the cable, while still holding the physical objects and transferred them to repository 2. Finally, since the human had placed the cable in repository 2, both human arms moved away from the repository 2.

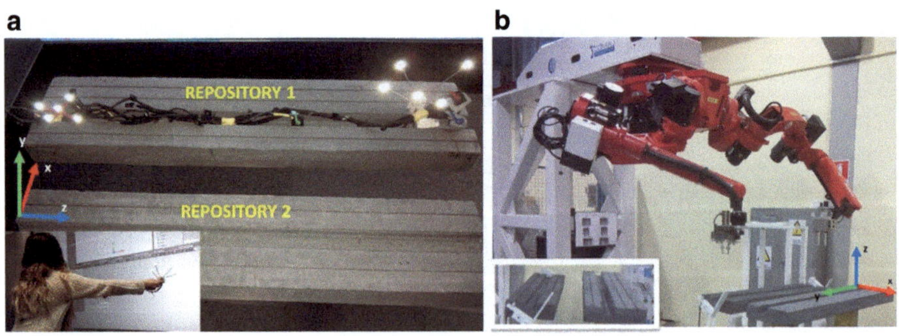

Fig. 18.6 **a** CAVE cell layout and **b** dual-arm robot cell layout

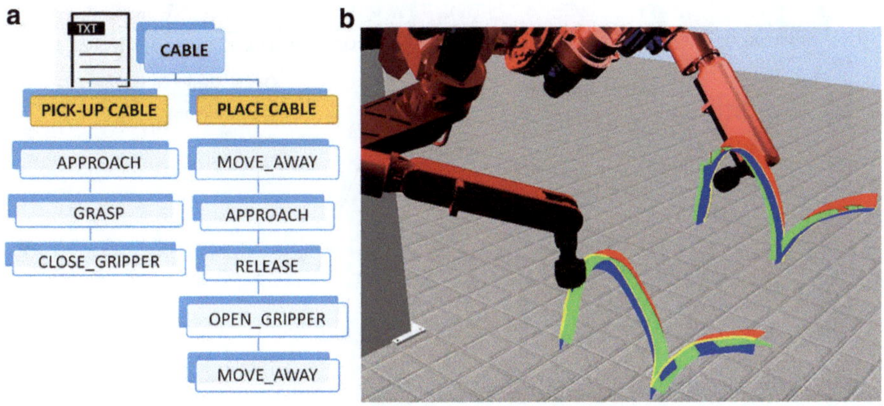

Fig. 18.7 **a** Hierarchical structure of identified motions for cable placement; **b** dual-arm robot path

The repository 1 was placed in front of the robot, while repository 2 was placed in a distance equal to the one that was used in the CAVE environment. The two separate tasks to be performed by the human and imitated by the dual-arm robot are illustrated in Fig. 18.7a. The generated robot trajectory is illustrated in Fig. 18.7b in a 3D simulation environment. This robot path has been generated from the 6th order of the Fourier series fitting values, as explained in Eq. (18.3). The coefficients of the three different dimensions x, y, z are used in order to generate the robot path.

18.4.2 Results

The data of the pick-and-place tasks for Arm1 are illustrated in Fig. 18.8 and are similar to Arm2. Firstly, the smoothing process is applied to the human motion data for the reduction of noise. All x, y, z data follow a smoother path according to the initial ones. The next step is the generation of the mathematical model that fits the smoothed data. The fitting values are the data used for the robot motions. The motion is identified and the recognized operations are visualized. The following sequence of operations was recognised: APPROACH, GRASP, CLOSE_GRIPPER, MOVE_AWAY, APPROACH, RELEASE, OPEN_GRIPPER, MOVE_AWAY.

A closer look in the results of Fig. 18.8 is given in Fig. 18.9, where the different data sets are more clearly distinguishable. The sequence of operations RELEASE, OPEN_GRIPPER and MOVE_AWAY is automatically recognized. The difference between the captured and the smoothed data is depicted here clearly, as well as the identified operations.

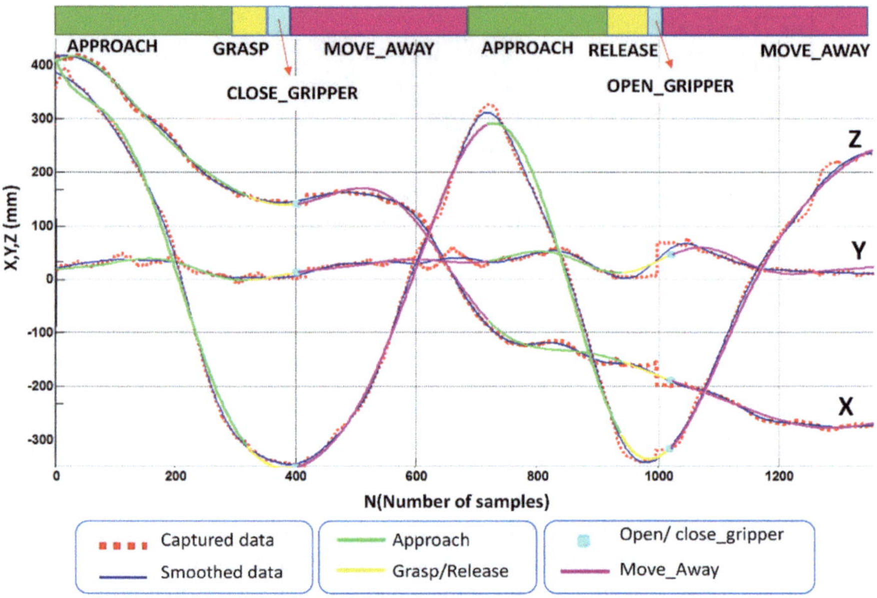

Fig. 18.8 Motion capturing and identification results-Pick-and-place task

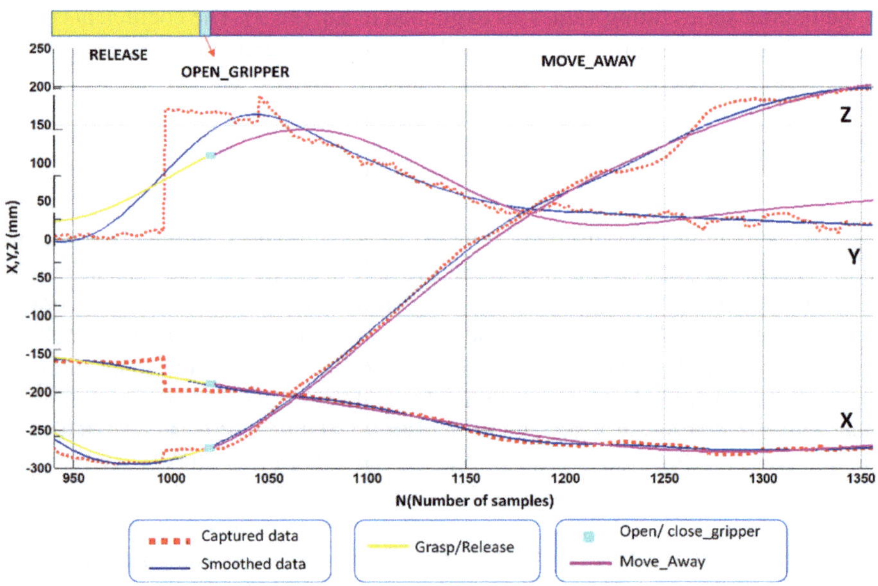

Fig. 18.9 Motion capturing and identification for the operations—RELEASE, OPEN_GRIPPER, MOVE_AWAY

18.5 Discussion

The advantage of the proposed method is the simplification of robot programming by capturing human motion data for robot paths' generation. The introduction of the hierarchical representation enables the better understanding of the robot program and allows the easy programming of new tasks by reusing the same methods. In this way, the proposed framework offers the user the possibility to construct and program new robot tasks without advanced training in robot programming.

The feasibility and reliability evaluation of this approach in a laboratory environment had a good impact on the industrial application. The possibility of easy programming dual-arm robots was valid without considering the synchronization or cooperation of both arms. The following aspects can be considered as the system's advantages:

- Since it is a learning method by imitating human motions, it allows reduction in the programming time. Additionally, it allows to reuse the human data in the form of a mathematical model for new robot paths generation in the future.
- It is a user-friendly offline programming system allowing to check the robot motion generation path before executing it to the robot.
- The data capturing and the robot motion generation is carried out in cartesian space coordinates and thus allows the imitation of data, without requiring a rather complex inverse kinematics solver.
- The automated recognition of the human operations in an hierarchy in another advantage towards the direction of simplifying the robot programming approach.

References

1. Tsarouchi P, Makris S, Chryssolouris G (2016) Human-robot interaction review and challenges on task planning and programming. Int J Comput Integr Manuf 29:916–931. https://doi.org/10.1080/0951192X.2015.1130251
2. Makris S, Tsarouchi P, Surdilovic D, Krüger J (2014) Intuitive dual arm robot programming for assembly operations. CIRP Ann 63:13–16. https://doi.org/10.1016/j.cirp.2014.03.017
3. Chryssolouris G (2006) Manufacturing systems: theory and practice, 2nd edn. Springer, New York
4. Moniz AB (2013) Organizational concepts and interaction between humans and robots in industrial environments 6
5. Ameri Ekhtiarabadi A, Akan B, Çürüklu B, Asplund L (2011) A general framework for incremental processing of multimodal inputs. In: Proceedings of the 13th international conference on multimodal interfaces—ICMI '11. ACM Press, Alicante, Spain, p 225
6. Mayer MPh, Odenthal B, Faber M, Winkelholz C, Schlick CM (2014) Cognitive engineering of automated assembly processes: cognitive engineering of automated assembly processes. Hum Factors Ergon Manuf Serv Ind 24:348–368. https://doi.org/10.1002/hfm.20390
7. Biggs G, MacDonald B A survey of robot programming systems, 10
8. Neto P, Norberto Pires J, Paulo Moreira A (2010) High-level programming and control for industrial robotics: using a hand-held accelerometer-based input device for gesture and posture recognition. Ind Robot Int J 37:137–147. https://doi.org/10.1108/01439911011018911

9. Onda H, Suehiro T, Kitagaki K (2002) Teaching by demonstration of assembly motion in VR—non-deterministic search-type motion in the teaching stage. IEEE/RSJ international conference on intelligent robots and system. IEEE, Lausanne, Switzerland, pp 3066–3072
10. Zinn M, Roth B, Khatib O, Salisbury JK (2004) A new actuation approach for human friendly robot design. Int J Robot Res 23:379–398. https://doi.org/10.1177/0278364904042193
11. Neto P, Pereira D, Pires JN, Moreira AP (2013) Real-time and continuous hand gesture spotting: an approach based on artificial neural networks. 2013 IEEE international conference on robotics and automation. IEEE, Karlsruhe, Germany, pp 178–183
12. Bodiroža S, Stern HI, Edan Y (2012) Dynamic gesture vocabulary design for intuitive human-robot dialog. In: Proceedings of the seventh annual ACM/IEEE international conference on human-robot interaction—HRI '12. ACM Press, Boston, Massachusetts, USA, p 111
13. Tsarouchi P, Athanasatos A, Makris S, Chatzigeorgiou X, Chryssolouris G (2016) High level robot programming using body and hand gestures. Procedia CIRP 55:1–5. https://doi.org/10.1016/j.procir.2016.09.020
14. Norberto Pires J (2005) Robot-by-voice: experiments on commanding an industrial robot using the human voice. Ind Robot Int J 32:505–511. https://doi.org/10.1108/01439910510629244
15. Koo S-Y, Lim JG, Kwon D-S (2008) Online touch behavior recognition of hard-cover robot using temporal decision tree classifier. RO-MAN 2008—the 17th IEEE international symposium on robot and human interactive communication. IEEE, Munich, Germany, pp 425–429
16. Koenig N, Takayama L, Matarić M (2010) Communication and knowledge sharing in human–robot interaction and learning from demonstration. Neural Netw 23:1104–1112. https://doi.org/10.1016/j.neunet.2010.06.005
17. Morioka M, Sakakibara S (2010) A new cell production assembly system with human–robot cooperation. CIRP Ann 59:9–12. https://doi.org/10.1016/j.cirp.2010.03.044
18. Eppner C, Sturm J, Bennewitz M, Stachniss C, Burgard W (2009) Imitation learning with generalized task descriptions. 2009 IEEE international conference on robotics and automation. IEEE, Kobe, pp 3968–3974
19. Yamane K, Hodgins JK, Brown HB (2003) Controlling a marionette with human motion capture data. In: 2003 IEEE international conference on robotics and automation (Cat. No.03CH37422). IEEE, Taipei, Taiwan, pp 3834–3841
20. Riley M, Ude A, Wade K, Atkeson CG (2003) Enabling real-time full-body imitation: a natural way of transferring human movement to humanoids. In: 2003 IEEE international conference on robotics and automation (Cat. No.03CH37422). IEEE, Taipei, Taiwan, pp 2368–2374
21. Muhlig M, Gienger M, Hellbach S, Steil JJ, Goerick C (2009) Task-level imitation learning using variance-based movement optimization. 2009 IEEE international conference on robotics and automation. IEEE, Kobe, pp 1177–1184
22. Inamura T, Nakamura Y, Ezaki H, Toshima I (2001) Imitation and primitive symbol acquisition of humanoids by the integrated mimesis loop. In: Proceedings 2001 ICRA. IEEE international conference on robotics and automation (Cat. No.01CH37164). IEEE, Seoul, South Korea, pp 4208–4213
23. Englert P, Paraschos A, Peters J, Deisenroth MP (2013) Model-based imitation learning by probabilistic trajectory matching. 2013 IEEE international conference on robotics and automation. IEEE, Karlsruhe, Germany, pp 1922–1927
24. Calinon S, Guenter F, Billard A (2005) Goal-directed Imitation in a Humanoid Robot. In: Proceedings of the 2005 IEEE international conference on robotics and automation. IEEE, Barcelona, Spain, pp 299–304
25. Moeslund TB, Hilton A, Krüger V (2006) A survey of advances in vision-based human motion capture and analysis. Comput Vis Image Underst 104:90–126. https://doi.org/10.1016/j.cviu.2006.08.002
26. Chryssolouris G, Mavrikios D, Fragos D, Karabatsou V (2000) A virtual reality-based experimentation environment for the verification of human-related factors in assembly processes. Robot Comput-Integr Manuf 16:267–276. https://doi.org/10.1016/S0736-5845(00)00013-2

27. Hein B, Worn H (2009) Intuitive and model-based on-line programming of industrial robots: new input devices. 2009 IEEE/RSJ international conference on intelligent robots and systems. IEEE, St. Louis, MO, pp 3064–3069
28. Jenkins OC, Mataric MJ (2002) Deriving action and behavior primitives from human motion data. IEEE/RSJ international conference on intelligent robots and system. IEEE, Lausanne, Switzerland, pp 2551–2556
29. Ude A, Atkeson CG (2003) Online tracking and mimicking of human movements by a humanoid robot. Adv Robot 17:165–178. https://doi.org/10.1163/156855303321165114
30. Luo RC, Shih B-H, Lin T-W (2013) Real time human motion imitation of anthropomorphic dual arm robot based on Cartesian impedance control. 2013 IEEE international symposium on robotic and sensors environments (ROSE). IEEE, Washington, DC, USA, pp 25–30
31. Calinon S, Billard A (2007) Active Teaching in Robot Programming by Demonstration. RO-MAN 2007—the 16th IEEE international symposium on robot and human interactive communication. IEEE, Jeju, South Korea, pp 702–707
32. Libera FD, Minato T, Fasel I, Ishiguro H, Menegatti E, Pagello E (2007) Teaching by touching: an intuitive method for development of humanoid robot motions. 2007 7th IEEE-RAS international conference on humanoid robots. IEEE, Pittsburgh, PA, USA, pp 352–359
33. Pan Z, Polden J, Larkin N, Van Duin S, Norrish J (2012) Recent progress on programming methods for industrial robots. Robot Comput-Integr Manuf 28:87–94. https://doi.org/10.1016/j.rcim.2011.08.004
34. Gaschler A, Springer M, Rickert M, Knoll A (2014) Intuitive robot tasks with augmented reality and virtual obstacles. 2014 IEEE international conference on robotics and automation (ICRA). IEEE, Hong Kong, China, pp 6026–6031
35. Takahashi T, Ogata H (1992) Robotic assembly operation based on task-level teaching in virtual reality. Proceedings 1992 IEEE international conference on robotics and automation. IEEE Comput. Soc. Press, Nice, France, pp 1083–1088
36. Ott C, Lee D, Nakamura Y (2008) Motion capture based human motion recognition and imitation by direct marker control. Humanoids 2008–8th IEEE-RAS international conference on humanoid robots. IEEE, Daejeon, Korea (South), pp 399–405
37. Maeda Y, Nakamura T (2015) View-based teaching/playback for robotic manipulation. ROBOMECH J 2:2. https://doi.org/10.1186/s40648-014-0025-4
38. Loader C (1999) Local regression and likelihood. Springer, New York

Chapter 19
Mobile Dual Arm Robots in Cooperation with Humans

19.1 Introduction

The latest trend in today's globalized market expresses an evident desire towards a greater level of product personalization [1]. As a result, the customer, needs to be treated as an individual and not as a market segment, his/her requirements must be efficiently identified and sufficiently satisfied. However, in current production systems the transition to mass customization generates complexity and high costs in the design and production [2] that manufacturers need to address in order to maintain their competitiveness and sustainability. The capability of offering more variants per model, and introducing new models faster, is constrained by the current technologies and the equipment of mass production operations [3].

Achieving flexibility [2] and adaptability that can be defined as the production system's sensitivity to internal and external changes is regarded as one of the most promising solutions over the last years 43. Robots have been considered as a major enabler for autonomous assembly systems. However, in current robot-based production systems, flexibility and reconfiguration are still constrained [4] due to [5]: (a) the rigidity of the used stationary robotic units, (b) the use of fixed and product model dedicated equipment, (c) the use of fixed robot control logic and (d) the absence of perception abilities that would allow the robots to dynamically adapt their behavior on the production needs. Overcoming these limitations may be realized through the introduction of flexible robot workers enabling autonomy and collaboration between all production resources (including human operators and robot resources). Mobility both in resources and product level can play a vital role towards the realization of such production concepts as discussed in [3]. To this end, a hybrid and dynamically reconfigurable shopfloor is suggested employing mobile dual arm workers, namely Mobile Robot Platforms—MRP, and human operators as well as Mobile Product Platforms (MPP).

The last decades, extensive research has been made in the field of mobile robotic systems, either in the field of mobile robot manipulators or simple mobile platforms

S. Makris, *Cooperating Robots for Flexible Manufacturing*,
Springer Series in Advanced Manufacturing,
https://doi.org/10.1007/978-3-030-51591-1_19

[6]. However, existing applications have limited perception capabilities not allowing real time adaptation of the system and robot behavior to dynamic environments [7]. Most of the manipulators are restricted to performing off-line programmed tasks only when they are in fixed positions, thus not fully exploiting their mobility [8]. On the other hand, digital representation and simulation of the production environment and process have emerged over the last decades as a means of partially handling the optimization of the production system performance [9]. In this era of digitalization in manufacturing, the Digital Twin concept has gained a lot of attention given the advantages that it may offer in terms of system autonomy [10]. The main principle of this concept relies on the digital representation of the physical world using multiple data input formats such as CAD files or other unified formats [11,12] as well as real time update of the virtual world based on real time data (e.g. shopfloor/resource sensors, process related data etc.) [13]. This is a very promising approach for providing perception and cognition abilities towards more autonomous and intelligent robotic systems [14].

Existing applications of dynamic robot control based on digital modelling and sensor data for ensuring collision free paths are based on the functionalities provided by Robot Operating System (ROS) framework [15]. The latter provides a rich content of data types and formats for virtual representing various hardware devices and multi-sensor data as well as a network of services and topics for broadcasting the captured knowledge. However, existing infrastructure are not mature enough to support the representation of the discussed hybrid production paradigm given the complexity of the various automated devices used such as multiple mobile dual arm workers and products as well as human operators. To overcome the existing limitations, this chapter introduces a Digital World Model infrastructure able to support the registration of multiple resources as well as the shopfloor scene real time reconstruction based on multiple sensor data and CAD models. A unified semantic representation of the geometrical and the workload state on top of the ROS provided data structures is proposed so for the model to be able to support real time task planning decision based on the shopfloor status.

The following sections are organized as follows: The second section provides the description of the hybrid production paradigm and the approach on the enabling technologies. In the third section the implementation of the Digital World Model based task planning components is described. The performance of the system is analyzed on an automotive case study in Sect. 19.4 and evaluated in the fifth section. The last section is dedicated to drawing the conclusions and providing an outlook towards future work.

19.2 Approach

The latest trends in European manufacturing research deal with the deployment of mobile robot teams for serving multiple operations in shopfloor level. This work discusses a hybrid production paradigm, as shown in Fig. 19.1, aiming to enable dynamically reconfigurable shopfloors using autonomous mobile resources. There

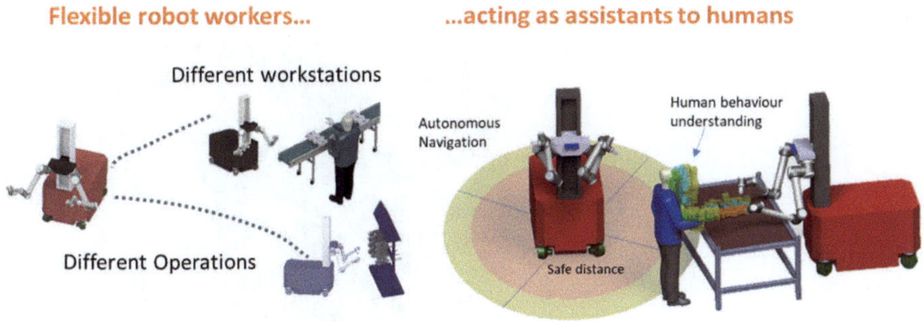

Fig. 19.1 Hybrid production paradigm

resources are mobile dual arm workers able to perceive their environment and cooperate with each other and with human operators.

In order to address the re-configurability aspect, a dual mobility approach is proposed: (1) at the resource level and (2) at the product level. Both robots and human operators are considered and the supporting technologies to bring these entities together are presented in this section.

19.2.1 Mobility in Resource Level

The resources considered in the discussed paradigm are Mobile Robot Platforms (MRPs) considering mobile and versatile dual—arm manipulators (Fig. 19.2). These manipulators are able to perform various processes and navigate across the shopfloor, easily changing their role and tasks. In particular, the MRPs are:

- able to navigate freely and in a safe way across the shop floor avoiding humans and obstacles,
- able perform tasks on products that are being transported by autonomous robotic units as well,
- able to position themselves inside workstations and execute tasks using the available tooling,
- able to be dynamically re-allocated to different tasks without human intervention,
- able to perceive their environment and adjust their behavior to collaborate with humans,
- able to communicate with each other through a network of services and reason over their course of actions in order to achieve the production goals.

This allows the system to be able to flexibly adapt to changes in production by quickly and efficiently dispatching resources wherever they are most needed.

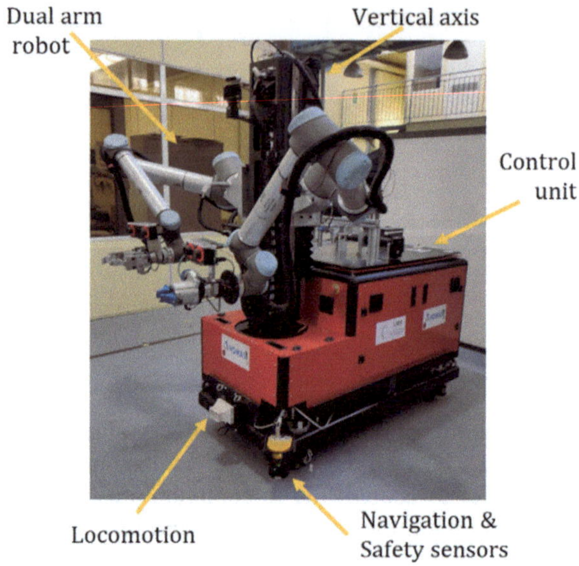

Fig. 19.2 Mobile robot platform—MRP

19.2.2 Mobility in Product Level

In product level, the proposed paradigm employs Mobile Product Platforms (MPPs) which are "assistant" vehicles responsible for carrying parts and products (Fig. 19.3). These will act as a moving "workstation" for human operators or the MRPs based

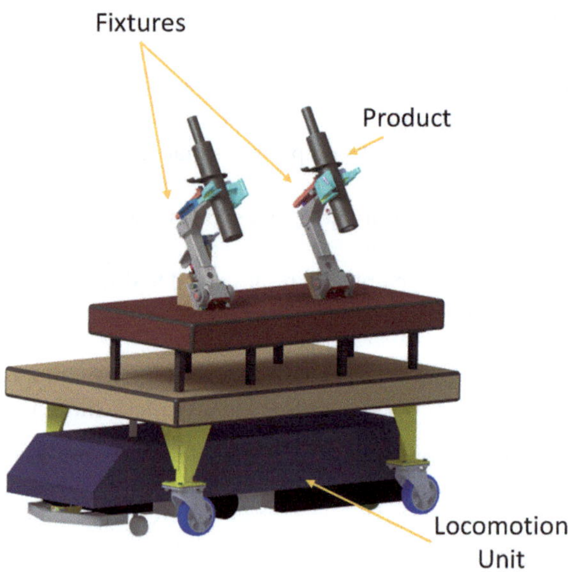

Fig. 19.3 Mobile product platform—MPP

on the Automated Guided Vehicles (AGVs) principles. Nowadays, AGVs are extensively used on industrial settings for multiple operations such as kitting, part transfer etc. However, such systems lack of flexibility since they follow strictly defined paths without considering the dynamic changes of the working environment [16]. Thus, novel methods for enhancing AGVs' perception as well as autonomy are required so to enable their dynamic operation. The specific challenge is to enable the coordination and synchronization movements between such vehicles and the proposed MRPs allowing them to follow different routes based on the process requirements. Modular integration architectures may enable the communication of the MPPs with the rest of the production entities to achieve such purposes.

19.2.3 Shopfloor Virtual Representation—Digital World Model

The success of the proposed paradigm that aspires to achieve autonomy at different levels heavily relies on the ability to efficiently organize the production entities (both humans and robots). This means that a common integration and communication network where these resources can share the data from their sensors or update their status is required. To enable the dynamic behavior and communication among these MRPs, MPPs and human operators, this chapter introduces a Digital World model enabling dynamic task planning. This model aims to provide the infrastructure for enabling the shopfloor data acquisition as well as combine them in a common representation to be consumed by the different decision-making mechanisms involved in the execution. A continuous feedback from the actual shopfloor (using resource and sensor data) enables the dynamic update of digital twin involving two main functionalities:

- Virtual representation of the shopfloor using multiple sensor data combination and CAD models. The digital shopfloor will be rendered in the 3D environment exploiting the related capabilities provided by ROS framework.
- Storing Semantic Information in the World repository. A unified data model will be implemented in order to semantically represent the geometrical as well as the workload state. This data model should be generic enough in order to be able to address multiple cases as well as to be consumed by multiple components inside execution system.

The virtual representation of the shopfloor is achieved through the deployment of four sub-components, namely (a) *Resource Manager*, (b) *Sensor Manager*, (c) *Layout Manager* and (d) *3D environment constructor.*

19.2.3.1 Resource Manager

The Resource Manager is responsible for registering in the Digital Twin any new resource introduced in the shop floor. Based on a unified Resource Data model, the Resource Manager stores in the Digital Twin repository all the attributes of each resource entity. Attributes such as maximum payload, minimum velocity etc. compose the resource model. Two sub-components, namely the *Resource location monitoring* and *Resource status monitoring* are responsible for real-time monitoring the status and location of each mobile resource and update online the actual values in Digital Twin. As visualized in Fig. 19.4 a number of topics and services are initiated for each subscribed resource. In order to efficiently handle all entities, the resource manager applies a unified naming convention as follows.

/(USER_NAMESPACE)/($RESOURCE_TYPE)_($RESOURCE_ID)/…

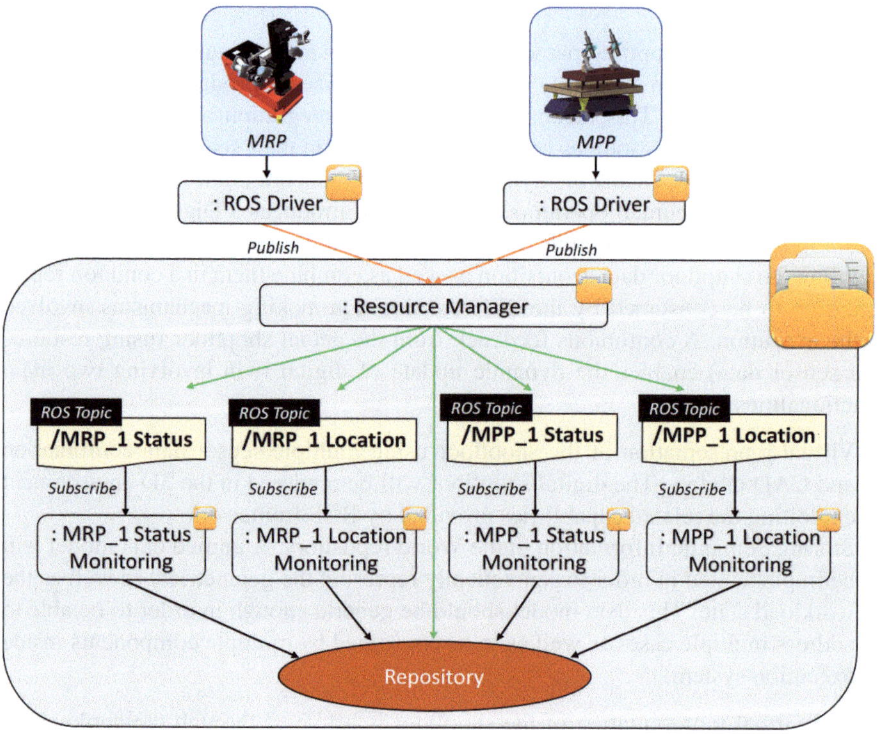

Fig. 19.4 Resource manager

19.2.3.2 Sensor Manager

The Sensor Manager is responsible for interfacing with the existing sensors' ROS drivers and registering their configuration data in a common world repository using a unified data model format. All sensorial data are made available to a dedicated communication bus, using a publish/subscribe pattern as a communication mechanism. Alongside with the Sensor Manager, the Digital World Model contains a discovery module for the available sensors to facilitate the integration of various sensing devices. More specifically, this module is running as daemon on the ROS-based communication bus and periodically checks for new unregistered topics.

The values of each sensor and measurement is published on a specific topic. Sensors and measurements are uniquely identified by their IDs and a specific naming convention needs to be followed by every module that intends to publish or subscribe to a sensor data topic for the following requirements:

- Given a topic name the unique id of the sensor and the unique id of the measurement should be identifiable.
- Given a sensor id and measurement id a unique topic name should be produced.
- The topic naming is compliant with the ROS naming conventions.
- The topic naming can be used in conjunction with the conventions used to provide ids in the Digital World model.
- The naming convention has an efficient implementation.

The following convention complies with all the above requirements:

```
(RESOURCE_NAMESPACE)/($SENSOR_CATEGORY)_($SENSOR_ID)/($SENS
OR_DESCRIPTION)/...
```

The Sensor Manager module maintains the list of topics that each sensor uses to publish data and also offers some additional functionality, such as controllable logging of the sensor data and querying of the logged sensor data. In order to facilitate the use of action planning algorithms like gmapping [17], amcl [18], ompl [19], for motion and path planning the 2D-3D sensor data fusion module allows to easily and dynamically merge multiple, sensor data into a single one topic. The latter feature allows the re-construction of the scene based on real time sensor data.

19.2.3.3 Layout Manager

In order to represent the entire shopfloor, the static layout needs to be described inside the world model. The Layout Manager is responsible for the control and storage of all CADs files related to static fixtures, parts and products. Similar to a resource or a sensor, a product needs to be registered inside the digital twin and to be described through a unified data model. This component allows the user to upload the CAD file and configure various parameters concerning (a) the parts involved in the

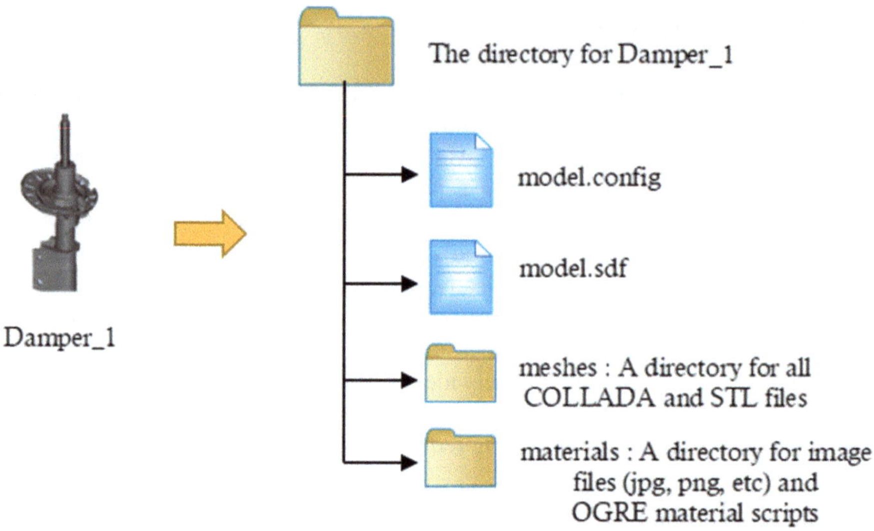

Fig. 19.5 Object model files structures

assembly process, (b) the stationary fixtures included in the shopfloor. The supported file structures are visualized in Fig. 19.5.

19.2.3.4 3D Environment Constructor

The final component in the process chain is the 3D environment constructor. This component retrieves the locations of all parts, fixtures, sensors and resources in order to construct an environment with a global world frame. Apart from the static parts which their position is defined at the configuration phase from the Resource, Sensor, or Layout manager as visualized in Fig. 19.6, there are also moving objects and obstacles that their position is not well fixed and need to be identified during the execution. For this reason, the 3D environment constructor provides an interface with ROS Services and Topic to track and update the position of all parts inside shop floor.

19.2.3.5 Unified Semantic Data Model

To harmonize the information captured under the Digital World Model a unified data model, represented the process and the environment following the principles of hierarchical modelling is suggested as shown in Fig. 19.7. This data model contains the data structures that hold the semantic information relevant to the execution system. These structures are used by all the involved modules ensuring the seamless communication and exchange of data.

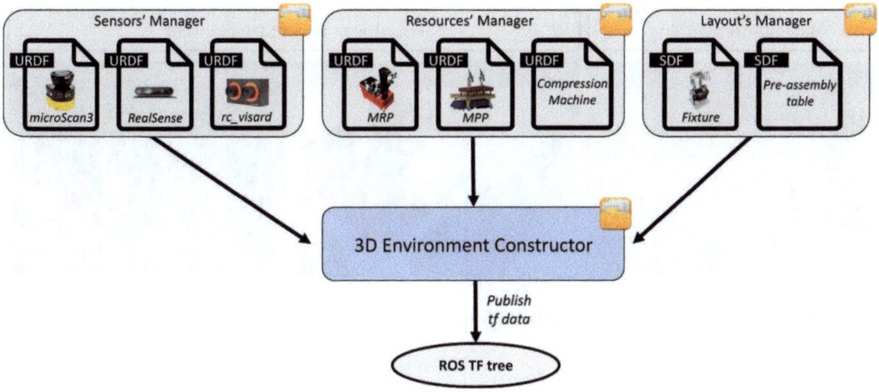

Fig. 19.6 TF tree construction

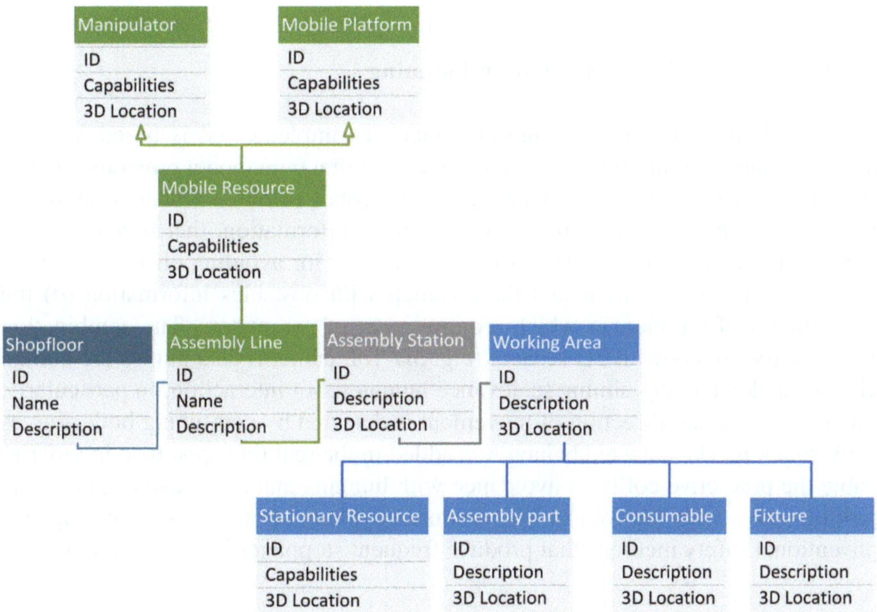

Fig. 19.7 Digital world unified data model

19.2.4 Real Time Robot Behavior Adaptation

It is essential demand the quick adaptation of a robot in an unknown environment. For this reason, lot of work can be found in literature related to the avoidance of potential collisions with other resources and unmapped obstacles inside the shopfloor

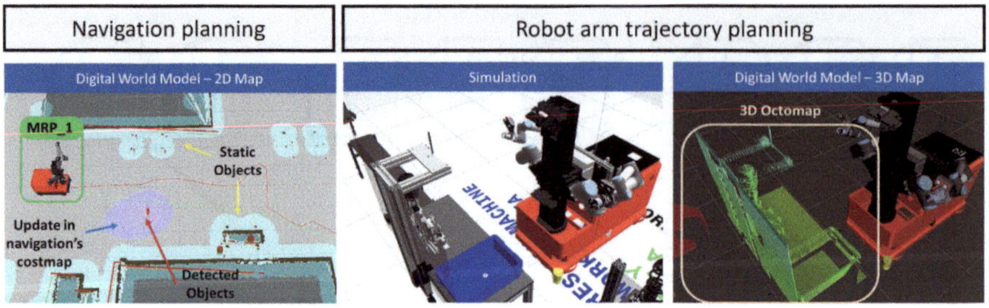

Fig. 19.8 Two-level robot behavior adaptation

environment. Digital twin model provides interfaces with robot's path and trajectory planners, in order to achieve online re-planning based on real-time information coming from the shopfloor as visualized in Fig. 19.8.

19.2.4.1 Human Aware Navigation Planning

The path planning component interface that was implemented is based on ROS navigation stack for mobile robots. In particular digital twin model provides: (a) The transformation for every coordinate frame of a robot resource which is described in a.urdf file, (b) distance sensors and odometry information, that is required for mapping and localization of the resource as long as for avoiding obstacles, (c) the global map of the shopfloor and the costmap with obstacles information (d) the configuration of the planner which requires a set of parameters. The combination of 2D laser scanners with 3D sensors (e.g. 3D ToF camera) data allows the human behavior understanding aiming to advance human robot interaction. In particularly, human presence and direction of movement is detected by combining both sources of information. The detected behavior is added in the real time cost map in order to enable the predictive collision avoidance with humans that are present in the same working space with the robots. This feature supplements workers' safety against conventional safety methods that produce frequent stoppages in robot motion.

19.2.4.2 Collision Free Robot Arm Trajectory Planning

For the motion planning of robotic arms, the digital twin model facilitates an interface with the MoveIt! framework. Three kinds of information are needed for the setup the motion planning: (a) robot's urdf file [20], (b) robot's srdf file [21] and (c) MoveIt! configuration files [22] for joint limits, kinematics, motion planning, perception input and other information. For ensuring a collision free trajectory of the robot, digital twin model provides the global planning scene with all the objects and resources. This planning scene is published as an occupancy map. The construction of this occupancy map is done by the 3D environment constructor. A standard procedure needs to be followed for the creation of an object's octree. Afterwards, the occupancy map updated online using the available sensor data.

19.3 Implementation

The software components of the proposed approach were developed based on ROS principles, enabling the communication among network nodes and quick integration with existing robotic applications. In particular, the described Digital Twin infrastructure has been deployed and tested on a PC running on Ubuntu 14.04 with ROS indigo version. Each of the aforementioned components are implemented as C++ nodes providing ROS interfaces (topics, services, actions) for exchanging the appropriate information, following the proposed naming convention. The configuration files that are handled and generated from digital twin model are following the standard convention of ROS community (Table 19.1).

The Digital World Model is integrated to an execution controller that has been developed to coordinate the execution of tasks from multiple mobile robot resources. Figure 19.9 provides an overview of the established architecture. Digital World Model after the configuration phase, communicates with robot controller and sensor nodes receiving data and generates the semantic information that is needed from execution controller.

In the next step the execution controller receives this information and based on the available resources dispatches the commands that need to be executed for a task. The communication between the different nodes is established over the ROS transport network. This network apart from Ethernet connections, contains also connections over Wi-Fi, due to the mobility that is required.

Table 19.1 List of file formats used by the Digital Twin's components

File format	Description
.yaml [23]	YAML syntax used for query and set parameters from ROS parameter server
.launch	An XML file used for easily launching multiple ROS nodes
.urdf	A standard format for descripting robot models, sensors and scenes
.urdf.xacro [24]	An advanced syntax of urdf, useful for describing large files
.sdf [25]	File format for describing objects and environments for robot simulators like Gazebo
.srdf	Semantic robot description format is the file format used for MoveIt! motion planner configuration
.dae [26]	File format used for digital assets based on the COLLADA XML schema

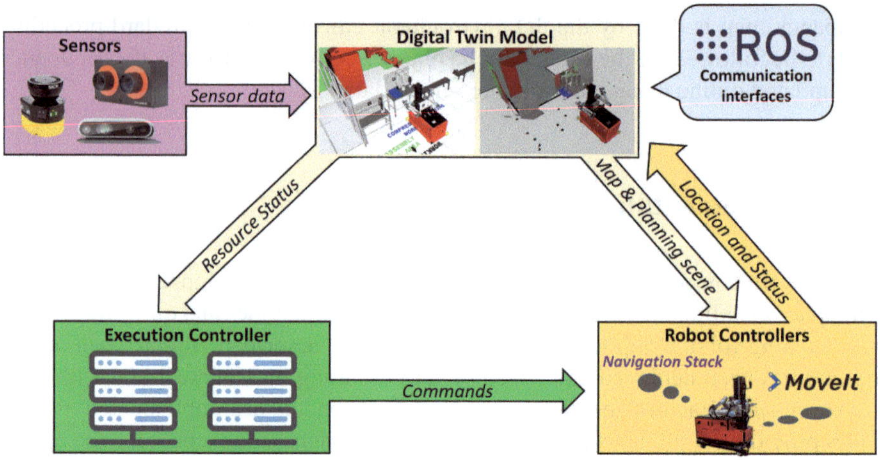

Fig. 19.9 Overall integration architecture

19.4 Industrial Example

19.4.1 Current State—Manual Assembly

The proposed production paradigm has been applied in an assembly line inspired from the automotive industry. The final product of the line is the front axle of a passenger vehicle. Although the line consists of nine working stations, this study focuses on the first four (Fig. 19.10) that involve the assembly of the dampers on the disks: (1) S1—Right Damper Assembly (RDS), (2) S2—Left Damper Assembly

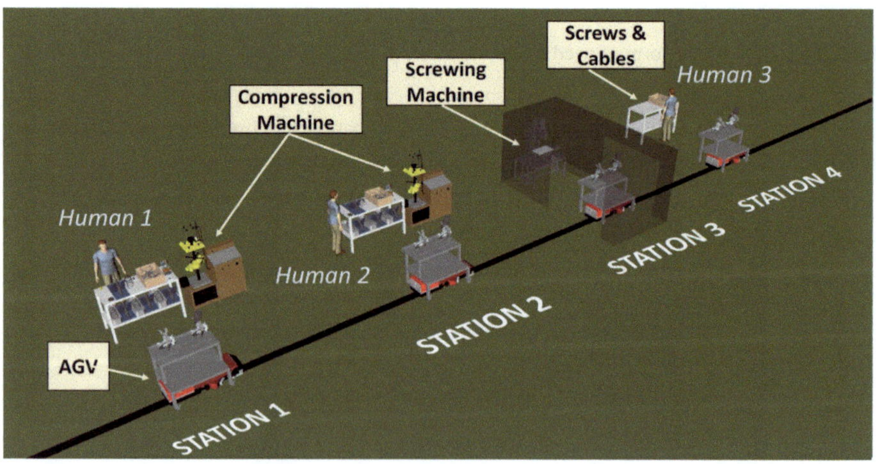

Fig. 19.10 Front axle assembly line—manual assembly

Table 19.2 Current state—assembly stations workload

Station	Task	Resource	Duration
S1	Pre-assembly of right damper	Human	12 s
S1	Load pre-assembled damper on the compression machine, compression of right damper and load compressed damper on the AGV	Human	30 s
S2	Pre-assembly of left damper	Human	12 s
S2	Load pre-assembled damper on the compression machine, compression of left damper and load compressed damper on the AGV	Human	30 s
S3	Screwing machine connect each damper with one disk	Screwing machine	15 s
S4	Cables/screws insertion	Human	28 s

(LDS), (3) S3—Screwing and (4) S4—Cabling stations. In these stations various assembly tasks are performed such as handling, insertion, screwing etc. and the majority of them is performed manually. Table 19.2 lists the tasks performed in each station.

The main challenges that the current set up faces concern:

- Ergonomy issues: The workers in S1 and S2 lift 480 compressed dampers during an 8-h shift. The weight of these parts is up to 6 kg, causing considerable physical strain to the operators.
- Flexibility issues: The automated screwing machine can only handle the screwing for specific front axle's models (2 out of the 3 models produced). For the third model, the screwing is performed in S4 by the human creating a delay on the cycle. When new models need to be produced major changes are performed in the current set up requiring considerable time, effort and space.

19.4.2 Hybrid Production Paradigm

To resolve these ergonomic issues a hybrid production paradigm is applied. In this human robot collaboration environment, tasks with manipulation of heavy parts and uncertainty succeed are being assigned to robot assistants. Process time cycle is being decreased because of mobile robot's capability to perform parallel screwing tasks for both dampers by the time it is docked with the MPP. Perfect synchronization of human and robot tasks reduces the idle time between human tasks. This hybrid production system in conjunction with high robot's flexibility and accuracy, contribute to the limitation of working areas from four to two. Table 19.3 presents the workload distribution under the new production paradigm employing two MRPs and one human operator for both stations.

Table 19.3 Hybrid production paradigm—assembly stations workload

Station	Task	Resource	Duration
S1	Pre-assembly of right damper	Human	12 s
S1	Load right pre-assembly in the compression machine	MRP 1	13 s
S1	Right damper's compression using the compression machine and loading on the AGV	MRP 1—Human	33 s
S1	Cables/screws insertion	Human	15 s
S2	Pre-assembly of left damper	Human	12 s
S2	Load right pre-assembly in the compression machine	MRP 1	13 s
S2	Left damper's compression using the compression machine and loading on the AGV	MRP 2—Human	33 s
S2	Cables/screws insertion	Human	15 s
S2	Tightening of cables on both dampers	MRP 2	10 s
S2	Docking of MPR and MPP and screwing tasks on both dampers	MRP 2—MPP	15 s

The model-based task planning framework has been validated on the right damper assembly and disk cabling/screwing tasks in the industrially relevant environment presented in Fig. 19.11 comprising three working areas: (a) the right Pre-assembly area, (b) the right damper compression area and (c) the MPP working area where both the left and the right dampers assembled on the respective disks.

Figure 19.12 visualizes the Digital World Model of the assembly environment based on the sensor data: (a) two laser scanners located in the mobile platform of the MRP, (b) one depth sensor located on its torso and (c) two stereo cameras located in the two robot arms. The Station Controller mechanism is responsible for dispatching the assigned tasks to the MRP, as well as the human operator and monitoring their progress so to coordinate the execution. For the efficient execution of the scenario, the MRP needs to perceive the: (a) damper and working tables to compensate the

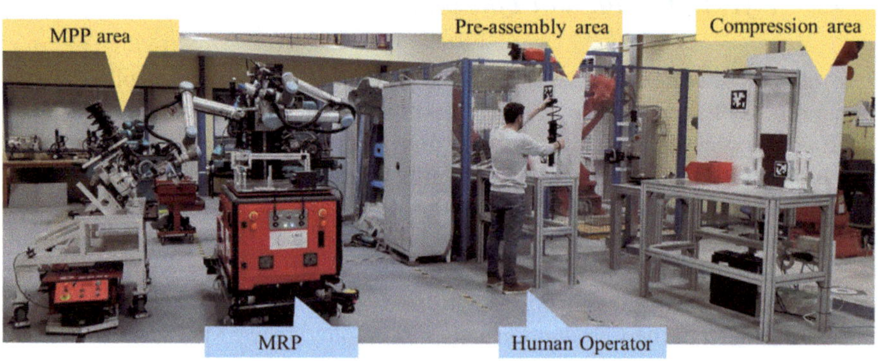

Fig. 19.11 Automotive case study layout

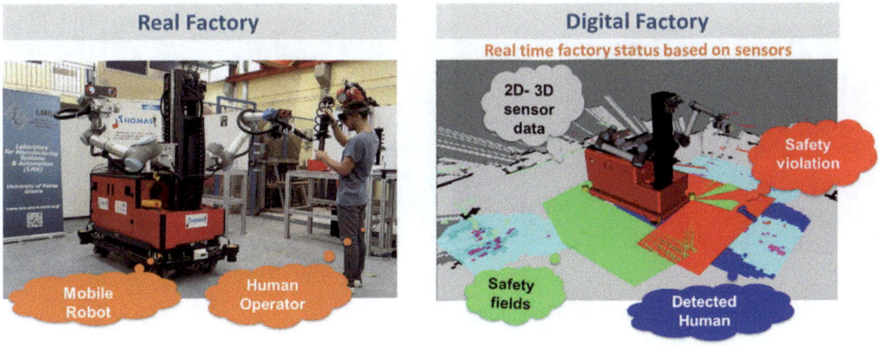

Fig. 19.12 Digital world model sensor-based scene re-construction

localization errors that cannot be foreseen offline, using the depth sensor, (b) static obstacles and moving humans/obstacles for ensuring collision free navigation, using 2D laser scanner data combined with the 3D data.

Each time the robot is re-located by the Station Controller to a different workstation it needs to autonomously navigate from its current location to the new one. 2D SLAM based navigation is used achieving accuracy of 5–10 cm as shown in Fig. 19.13. However, this error creates constraints to the robot arms when required to perform assembly operations to the respective working area. Current industrial practice exploits mechanisms for physical docking of the mobile platforms to the assembly areas. In that way, the error can be reduced up to millimeters. However, this creates high dependency of the process to the hardware setup and therefore it takes considerable time to re-structure the setup when changes are dictated in the layout. Virtual docking may significantly increase the flexibility of the system to adapt to such changes eliminating the dependency among the hardware components [8]. To achieve the required accuracy when the MRP is virtually docked to a working area, 3D in cell localization is used based on AprilTag detection. In this case the navigation error is decreased by to 1 cm.

The dynamic nature of the Digital World model allows the real time update of the planning scene, so for the navigation planners to consider in the local plan generation

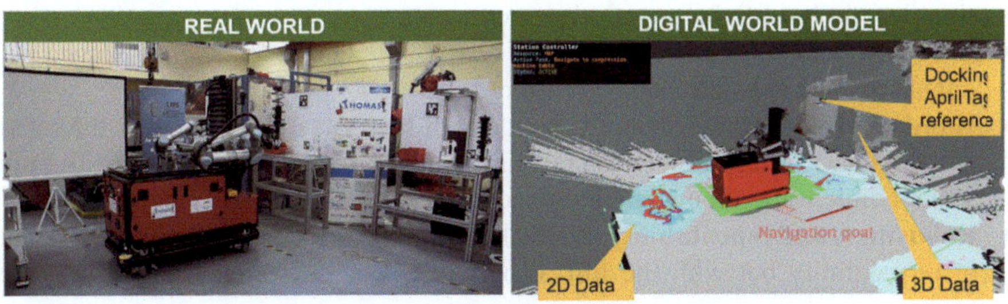

Fig. 19.13 MRP cell to cell navigation and in cell accurate localization

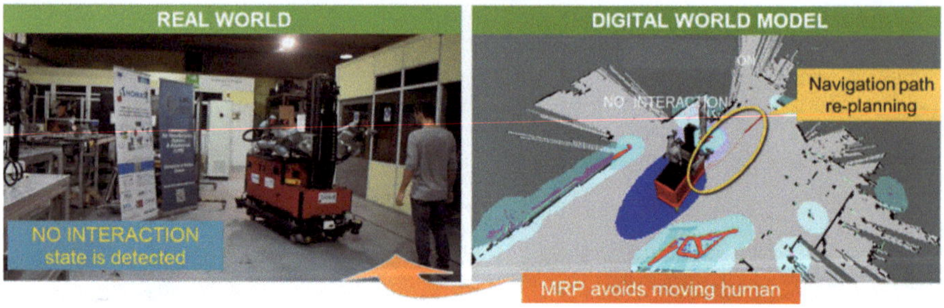

Fig. 19.14 Human aware navigation planning

the new, unmapped obstacles. Figure 19.14 presents an instance where the human operator interferes to MRP's planned path. The respective visualization of the Digital World model instance during this case is also presented. In that way, the MRP may avoid collision with humans in a dynamic way while both are in motion.

19.5 Discussion

To evaluate the performance and efficiency of the developments several key performance indicators need to be considered for capturing the benefits of the hybrid approach:

- **Maximum weight handled by operator**: Denotes the weight of the parts handled by operator without any support, accounting for the ergonomy issues in all stations. The target is to minimize it, leaving to operators only the handling of light weight parts and the execution of complex dexterous operations.
- **Number of models/diversity**: Indicating the total number of different car models/variants that the assembly line can accommodate. This heavily depends on the ability of resources to perform multiple tasks.
- **Operator activity**: The KPI refers to the percentage of time that the operator is occupied within a cycle. Minimization of waiting times (e.g. due to compression machine cycle) and inclusion of more tasks that require dexterity but induce significantly less strain are to be pursued.
- **Production throughput**: Target must be to maintain the vehicles produced per minute to be at least the same as current number.
- **Number of operators**: Since the majority of heavy tasks are done by the robot in S3, S4 and S6 the number can be reduced to 1 with a proper redesign. To fit in a small dimension as compared to the current setup.
- **Robustness and repeatability**: Manual operations can guarantee a certain amount of repeatability but with the full or partial robotization enhancement is to be expected.

- **Quality**: With the manual assembly meeting a maximum of 95% quality standards, a fully or partially robotized assembly must improve this number up to a 99%.
- **Safety**: Current assembly line has achieved a TF1 value of 1.37 which is targeted to be safer in a robotized assembly with a value less than 1 by the end of this project.
- **Return of investment**: Referring to the relation of profits against the capital invested.

The baseline and target values for these KPIs are shown in the following table:

KPI	Baseline	Target
Weight handled by operators (Kg)	5	1
No. of models	3	6
Operator activity (%)	60	70
Production throughput (parts/h)	60	60
No. of operators	3	1
Repeatability	95	99
Quality (%)	95	99
Safety	1.37	<1
RoI	–	<1 year

19.6 Conclusions and Ongoing Research

This work has discussed the requirements and challenges for introducing autonomous mobile dual arm robots allowing them to cooperate with other robots and human operators inside a production system. The analysis of applying this hybrid paradigm in the case of vehicle suspension assembly has been used to highlight the required transformations in terms of (a) production system layout, (b) process and environment sensing and (c) integration and control. The analysis indicates that with the proper re-assignment of roles between robotic and human resources can lead to significant enhancement in the operator working conditions, maintaining at least the same production levels as the current practice. Future work will involve the validation of the proposed system: (a) in a larger scale set up, involving more than one MRPs (b) in an actual manufacturing environment starting from the automotive sector.

References

1. Kousi N, Gkournelos C, Aivaliotis S, Giannoulis C, Michalos G, Makris S (2019) Digital twin for adaptation of robots' behavior in flexible robotic assembly lines. Proc Manuf 28:121–126. https://doi.org/10.1016/j.promfg.2018.12.020
2. Chryssolouris G (2006) Manufacturing systems: theory and practice. Springer, New York
3. Kousi N, Michalos G, Aivaliotis S, Makris S (2018) An outlook on future assembly systems introducing robotic mobile dual arm workers. Proc CIRP 72:33–38. https://doi.org/10.1016/j.procir.2018.03.130
4. Michalos G, Kousi N, Makris S, Chryssolouris G (2016) performance assessment of production systems with mobile robots. Proc CIRP 41:195–200. https://doi.org/10.1016/j.procir.2015.12.097
5. Michalos G, Makris S, Chryssolouris G (2015) The new assembly system paradigm. Int J Comput Integr Manuf 28:1252–1261. https://doi.org/10.1080/0951192X.2014.964323
6. Kousi N, Koukas S, Michalos G, Makris S (2019) Scheduling of smart intra—factory material supply operations using mobile robots. Int J Prod Res 57:801–814. https://doi.org/10.1080/00207543.2018.1483587
7. Michalos G, Sipsas P, Makris S, Chryssolouris G (2016) Decision making logic for flexible assembly lines reconfiguration. Robot Comput-Integr Manuf 37:233–250. https://doi.org/10.1016/j.rcim.2015.04.006
8. Gkournelos C, Kousi N, Christos Bavelos A, Aivaliotis S, Giannoulis C, Michalos G, Makris S (2019) Model based reconfiguration of flexible production systems. Proc CIRP 86:80–85. https://doi.org/10.1016/j.procir.2020.01.042
9. Váncza J, Monostori L (2017) Cyber-physical manufacturing in the light of professor Kanji Ueda's legacy. Proc CIRP 63:631–638. https://doi.org/10.1016/j.procir.2017.04.059
10. Michalos G, Makris S, Papakostas N, Mourtzis D, Chryssolouris G (2010) Automotive assembly technologies review: challenges and outlook for a flexible and adaptive approach. CIRP J Manuf Sci Technol 2:81–91. https://doi.org/10.1016/j.cirpj.2009.12.001
11. Makris S, Michalos G, Chryssolouris G (2012) Virtual commissioning of an assembly cell with cooperating robots. Adv Decis Sci 2012:428060. https://doi.org/10.1155/2012/428060
12. Caggiano A, Teti R (2012) Digital manufacturing cell design for performance increase. Proc CIRP 2:64–69. https://doi.org/10.1016/j.procir.2012.05.041
13. Qi Q, Tao F (2018) Digital twin and big data towards smart manufacturing and industry 4.0: 360 degree comparison. IEEE Access 6:3585–3593. https://doi.org/10.1109/ACCESS.2018.2793265
14. Uhlemann TH-J, Lehmann C, Steinhilper R (2017) The digital twin: realizing the cyber-physical production system for industry 4.0. Proc CIRP 61:335–340. https://doi.org/10.1016/j.procir.2016.11.152
15. Robot Operating System (ROS). https://www.ros.org/
16. Kousi N, Dimosthenopoulos D, Matthaiakis A-S, Michalos G, Makris S (2019) AI based combined scheduling and motion planning in flexible robotic assembly lines. Proc CIRP 86:74–79. https://doi.org/10.1016/j.procir.2020.01.041
17. URL ROS GMAPPING. https://wiki.ros.org/gmapping. Accessed 24 Aug 2019
18. URL ROS AMCL. https://wiki.ros.org/amcl. Accessed 17 Aug 2019
19. URL ROS OMPL. https://wiki.ros.org/ompl. Accessed 22 Oct 2019
20. URL ROS URDF. https://wiki.ros.org/urdf. Accessed 14 Sep 2019
21. URL ROS SRDF. https://wiki.ros.org/srdf. Accessed 11 Aug 2019
22. URL ROS MoveIt. https://moveit.ros.org/. Accessed 11 Dec 2019
23. URL ROS YAML. https://wiki.ros.org/YAML%20Overview. Accessed 12 Jun 2019
24. URL ROS URDF XACRO. https://wiki.ros.org/urdf/Tutorials/Using%20Xacro%20to%20Clean%20Up%20a%20URDF%20File. Accessed 7 Jul 2019
25. URL ROS SDF. https://gazebosim.org/tutorials?tut=build_model
26. URL ROS COLLADA URDF. https://wiki.ros.org/collada_urdf

Chapter 20
Allocation of Manufacturing Tasks to Humans and Robots

20.1 Introduction

Over the last decade, extensive research has been done on designing flexible production systems with adaptive capabilities, able to react and adapt to the fluctuating market's needs. The manufacturing of products in the modern industrial world requires the combined and coordinated efforts of people, machines and equipment [1]. HR interaction for shared task execution improves the capacity of robots to respond to unexpected events with human flexibility. This holds the promise of major production efficiency improvements. One of the main challenges is to have a decision-making mechanism capable of either assigning tasks to specific resources or for cooperation dynamically. This would include adequate modeling and preparation of the assembly processes as well as the means to evaluate runtime working conditions.

Complex processes can be separated into smaller processes with minimal complexity. Advanced factory automation can help solve these problems by sufficient manufacturing and production process management and optimization. There are multiple operations stemming from manual processes that require more flexibility and robustness [2, 3]. Even with primitive intelligence, robots have shown their ability to increase the production, performance and quality of a factory. The combination of human workers with robots in a manufacturing cell may allow greater flexibility with shorter production cycles resulting in better adaptation to changes in the demand profile of the markets [4, 5]. A detailed study regarding human–robot collaboration may be reviewed in [6]. The introduction of robots to manual assembly lines is a research area that has attracted great interest in order to enhance the competitiveness and efficiency of modern production systems [7].

Despite the fact that industrial robots are used in a wide range of production lines, it is not deemed sufficiently advanced to be in close cooperation with humans. This is consistent with the fact that human capabilities cannot be completely replaced by

S. Makris, *Cooperating Robots for Flexible Manufacturing*,
Springer Series in Advanced Manufacturing,
https://doi.org/10.1007/978-3-030-51591-1_20

those of the robots; however, by combining the advantages of both sides, a solution can be achieved. Unconstrained environmental conditions in production lines, including variable lighting conditions, crossing staff, and autonomous mobile units, such as AGVs, are some important reasons for working together in complex assembly cells.

In this chapter, there is a proposed structure for Human–Robot Collaboration (HRC), which allows sequential execution of human and robot tasks. It is part of the shared human–robot relationship (HRI) tasks and workspace, when the human and robot tasks are sequential. The suggested approach focuses on the task allocation of a hybrid assembly cell with multiple human and robot tools. A smart decision-making algorithm is the basis for scheduling tasks by testing various criteria. The resource suitability, quality and processing time are included in an overview of the decision steps in task allocation.

20.2 Related Work

An extended review of the challenges faced by human robot collaboration (HRC) on a shared environment is discussed in [2], while an extended research on HR task scheduling methods is presented in [3]. Artificial intelligence (AI) techniques for scheduling are considered in [4]. A multi-objective optimization based on genetic programing is presented in [5]. HR task planning methods have been considered as well in [6] and [7], with their main criterion being the reduction of the non-ergonomic human tasks. Additionally, the evaluation of HR teams' productivity and human satisfaction through a rule-based decision-making framework is discussed in [8]. The criterion for scheduling was the completion time of the tasks in this work. In addition to scheduling, movement planning is considered in [9] to minimize the cycle time of production tasks by selecting the appropriate trajectory.

In order to achieve high levels of adaptability and robustness in the cell, HRI for collaborative task execution and scheduling aims to improve the efficiency of manufacturing and assembly processes [8, 9]. Therefore, in a hybrid assembly cell, a safe and versatile co-operation between a robot and a human will lead to increased productivity. The specifications of the hybrid assembly lines are related to ensuring the coordination and timing of the activities, so all agents operating in a cell should be coordinated by a centralized control.

HRI is separated into four main categories. The coexistence of a human and a robot consists of hybrid cells divided into shared workspace and tasks, common workspace and common tasks and separate workspace [10]. The tasks and working spaces of the human and robot resources are concurrent in the case of 'shared tasks and workspace,' and are allocated accordingly. For assembly and production cells, there are various research studies on HRI.

A hybrid assembly cell, as described in [9,11] consists of an industrial robot capable of learning from a human new tasks and executing them in a shared workspace. Multiple sensors are used to monitor the area and detect objects and humans where multi-modal channels are available for communication purposes (speech, gaze, haptic). The reason for using multi-modal devices was to help the operator undertake and perform his tasks.

Kwon et al. [12] presents a study that is based on the prediction of behaviors for human–robot cooperative tasks in a probabilistic way. An approach based on semantic maps was described in [13] and has been evaluated in the case of HRI and robot task planning.

The aim of this approach was to enhance the use of common information and to increase the autonomy of agents. A determining framework for the performance of the HRI task is presented in [14], where the human being is aware of the performance of the task.

Cooperation between human operators and robots and social interaction are new concepts as robots' anthropomorphism increases [10, 15]. In addition, human and robot communication is described in [16] and is based on analyzing the behavior and collaboration of operators in assembly lines. This study focused on gestures as a normal way to interact. In addition, in the case of a robot's programming, where abundant research was done [3, 17–20], gestures were also considered.

20.3 Approach

The allocation and scheduling of human robot tasks is achieved by means of a decision-making algorithm based on a hierarchical sequence of tasks and multiple criteria [1, 3, 7, 21]. A universal model is proposed for the simultaneous representation of hybrid cells, including both humans and robots. As far as human operators are concerned, each individual human is considered a re-source, while each arm is also considered a re-source in the case of single arm robots. As far as the dual-arm robot system is concerned, each single arm of the dual-arm robot is considered a specific resource that can be assigned to a task. At the lowest level, they are considered as part of the grippers and equipment used by the human resources.

The HR workload model represents the total work to be done in the hybrid production cell. The workload can be considered as a production plan in this context. The workload is made up of various processes, each of which is divided into tasks (Fig. 20.1).

A Human–Robot task allocation is accomplished by determining such assignment constraints in this methodology. For this effect, a series of decision steps has been considered allowing for evaluate the effectiveness of the particular task assigned to a unit. The metrics used for deciding the potential resources being assigned to a task are the resource suitability, availability and processing time. Suitability of resources is the metric used to ensure that a re-source is appropriate for the success of a task. For example, before the decision on the assignment of the tasks, it takes into account the payload capabilities of the resources. Availability of resources is the metric used to determine if a re-source is available in the hybrid assembly cell to perform a task. Processing time is the time that should be minimized to allow a resource to perform

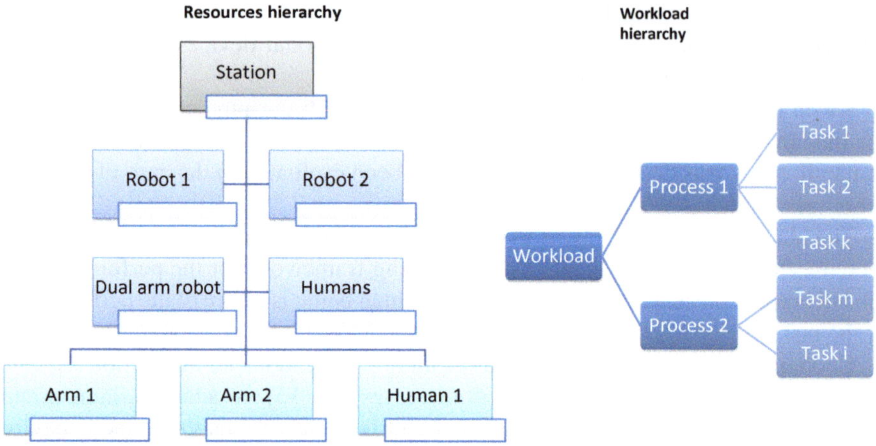

Fig. 20.1 Resources and workload models [22]

a task. In this direction, a task with minimum processing time will be assigned to the resource. The above metrics are described in [21].

Based on the principle of depth search, smart scheduling algorithms take the previous steps to evaluate a schedule based on multiple parameters between the assigned tasks. This algorithm is defined in [23], where DH (decision horizon), sampling rate (SR) and maximum number of alternatives (MNA) are the main control parameters used. Many performance measures used in scheduling issues are based on the times of completion, such as the mean flow time and the cost of utilization [1]. Examples of criteria for choosing a good alternative are the optimal use of resources (ARU), the average flow time and are described in [21, 24].

Based on the above criteria, the next step is to select the best alternative through a matrix that restricts the various design alternatives based on the value of these criteria. The ultimate decision involves a minimization or maximization of these values. A resource selection incorporates the total utility factor score or each alternative, taking into account the weighted factor of each criterion.

20.4 Industrial Example

The suggested method was applied in the automotive industry to a hybrid assembly cell for the assembly of a vehicle dashboard [22]. The cell is made up of a dual arm robot and a human. Many functions can be done by a human for this job [20, 25] and others by the dual arm robot. The loading and assembly area are used for the picking up and the placement of the traverse.

The fuse box, with the support of three screws for stability reasons, is the next element to be installed in its final position in the traverse slot. The last part to be

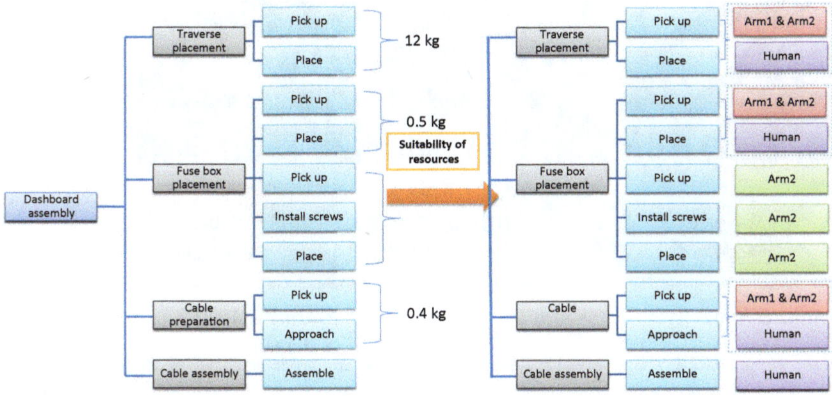

Fig. 20.2 Dashboard workload and suitable resources [22]

assembled is the cable pack and is very versatile. This job is a typical case of co-operation with HR. As an input to the decision-making process, the hybrid cell model and the workload model are used. The workload involves four procedures, including traverse placement, device body placement, as well as the planning and installation of wire harness.

Such procedures require a number of tasks and a number of suitable resources for each task (Fig. 20.2). Taking into account the weight of the components relative to the re-sources' payload capacities, the appropriateness of the resources for a task is determined. Therefore, the versatility of the parts relative to the handling capabilities of the resource is taken into account.

The indication 'ARM 1' or 'ARM 2' means that single-arm robots will complete the tasks while the indication 'ARM1 & ARM2′ means that the cooperation/synchronization of both arms is considered for a single task. The 'ARM1 & ARM2' and 'HUMAN' can perform both the pick-up and placement of the traverse as well as the computer's body. 'ARM2′ can perform the screwing tasks in the computer's body, while 'ARM1' & 'ARM2' and 'HUMAN' can perform the pick-up of the cable and its placement in the assembly bank. The cable installation can be executed by the 'HUMAN', as it is a difficult task, requiring advanced sensing capabilities.

Two parameters are used to produce and test the alternative work plans. These two parameters are the average utilization time (ARU) and the mean flow time (F). The first four tasks are eventually allocated to the 'ARM1 & ARM2' while the 'ARM2' is assigned to the next eight tasks. The last three tasks contribute to the planning and deployment of the cable and are delegated to the 'HUMAN'. The execution of the dashboard workload is visualized in a, while the Gantt chart of the best alternative is depicted in Fig. 20.3b. The pie charts in the same figure, show the effect of different assignments to the overall performance of the manufacturing system. In one alternative, the robot system takes the 47% of the time and the human the remaining 53%. In the second one, robot is performing for 93% while the human

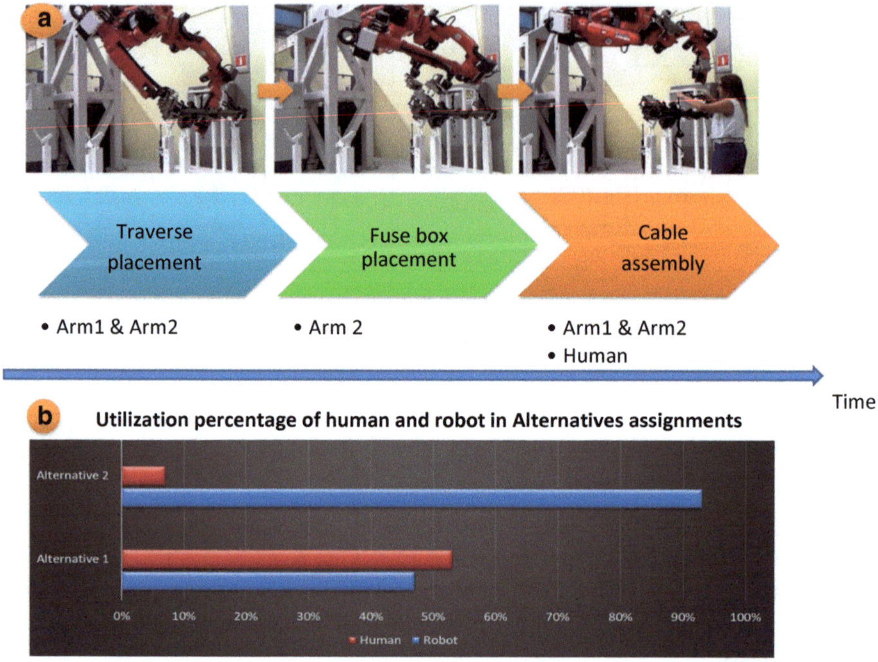

Fig. 20.3 (**a**) Dashboard assembly case; (**b**) Gantt chart of the best alternative [22, 25]

is performing 7%. Therefore, in large manufacturing systems, comprising a large number of humans and robots that can perform interchangeably a large number of tasks, employing such a decision making logic can help to achieve a proper balancing of the time and better utilization of resources.

20.5 Discussion

The proposed HR task planning approach allows single, dual-arm robots and humans to be integrated into the same unified model. The suggested modelling approach can also be implemented in the simple case of single-arm robots, cooperating with humans but also with multi-arm robots and humans in more complex cells.

The main contribution of this research effort is summarized as follows. First, it proposes a framework for complex scheduling between human and robot tasks to be introduced. This is a double innovation, as it also applies to the design-built hybrid assembly lines field by leveraging resource suitability. Second, humans and robots can cooperate in the same fenceless cell, simplify the cell design and setup process by enabling the use of separate workspaces. This research work proposes a framework that is applied to a case study stemming from a manual assembly line in automotive final assembly stations and permits the aforementioned functionalities.

The proposed HR task planner allows a hybrid assembly cell to be quickly and efficiently reconfigured. Human involvement in the assembly cell has been significantly reduced, allowing for the potential for human involvement as a supervisor or operator in nearby stations. Through introducing a new criterion, it is possible to address the ergonomics problems for human activities. The multi-arm robot system design also provides the potential to reduce the space required in the shop floor compared to single-arm robot systems.

Using such a structure for decision-making allowed the HR task planning to be carried out in a short time and can also be used for re-planning. The evaluation criteria can be modified if a new product or a new cell introduces new conditions and specifications. In a shared structure, the HR tools and their tasks are concurrently modelled. This enables good alternative approaches to be created, taking into account HR skills at the same time.

The key findings found in this research work are summarized as follows:

- conceptual analysis of separate and shared HR tasks and resources;
- short-term evaluation of various alternative solutions, taking into account current requirements and specifications;
- re-scheduling taking into account unexpected events during execution of tasks. The rescheduling is focused on the evaluation of events during the execution of assigned tasks either by human operators or robots.

The above findings have given rise to some potential improvement. Work on determining parameters that are ideal for mutual HR roles is still underway. Also an important issue for future research is the representation of knowledge for humans, robots, and HR teams. In the future, integration with sensors will be considered for analysis of online feedback from humans and robots. It will also be a challenge to integrate with current manufacturing execution systems (MES). These facts will make it easy to integrate HR collaboration concepts into hybrid work cells by assigning tasks and reassigning them to HR resources.

References

1. Chryssolouris G (2006) Manufacturing systems: theory and practice, 2nd edn. Springer, New York
2. Hermawati S, Lawson G, D'Cruz M, Arlt F, Apold J, Andersson L, Lövgren MG, Malmsköld L (2015) Understanding the complex needs of automotive training at final assembly lines. Appl Ergon 46:144–157. https://doi.org/10.1016/j.apergo.2014.07.014
3. Makris S, Tsarouchi P, Surdilovic D, Krüger J (2014) Intuitive dual arm robot programming for assembly operations. CIRP Ann 63:13–16. https://doi.org/10.1016/j.cirp.2014.03.017
4. Alexopoulos K, Papakostas N, Mourtzis D, Gogos P, Chryssolouris G (2007) Quantifying the flexibility of a manufacturing system by applying the transfer function. Int J Comput Integr Manuf 20:538–547. https://doi.org/10.1080/09511920600930046
5. Georgoulias K, Papakostas N, Mourtzis D, Chryssolouris G (2009) Flexibility evaluation: a toolbox approach. Int J Comput Integr Manuf 22:428–442. https://doi.org/10.1080/09511920802527582

6. Krüger J, Lien TK, Verl A (2009) Cooperation of human and machines in assembly lines. CIRP Ann 58:628–646. https://doi.org/10.1016/j.cirp.2009.09.009
7. Nikolakis N, Kousi N, Michalos G, Makris S (2018) Dynamic scheduling of shared human-robot manufacturing operations. Proc CIRP 72:9–14. https://doi.org/10.1016/j.procir.2018.04.007
8. Tsarouchi P, Makris S, Chryssolouris G (2016a) Human—robot interaction review and challenges on task planning and programming. Int J Comput Integr Manuf 29:916–931. https://doi.org/10.1080/0951192X.2015.1130251
9. Wilcox R, Shah J (2012) Optimization of multi-agent workflow for human-robot collaboration in assembly manufacturing. In: Infotech@Aerospace 2012. American Institute of Aeronautics and Astronautics, Garden Grove, California
10. Mayer MPh, Odenthal B, Faber M, Winkelholz C, Schlick CM (2014) Cognitive engineering of automated assembly processes: cognitive engineering of automated assembly processes. Hum Factors Man 24:348–368. https://doi.org/10.1002/hfm.20390
11. Wallhoff F, Blume J, Bannat A, Rösel W, Lenz C, Knoll A (2010) A skill-based approach towards hybrid assembly. Adv Eng Inform 24:329–339. https://doi.org/10.1016/j.aei.2010.05.013
12. Kwon WY, Suh IH (2014) Planning of proactive behaviors for human–robot cooperative tasks under uncertainty. Knowl-Based Syst 72:81–95. https://doi.org/10.1016/j.knosys.2014.08.021
13. Galindo C, Fernández-Madrigal J-A, González J, Saffiotti A (2008) Robot task planning using semantic maps. Robot Auton Syst 56:955–966. https://doi.org/10.1016/j.robot.2008.08.007
14. Alami R, Clodic A, Montreuil V, Sisbot EA, Chatila R (2005) Task planning for human-robot interaction. In: Proceedings of the 2005 joint conference on smart objects and ambient intelligence innovative context-aware services: usages and technologies—sOc-EUSAI '05. ACM Press, Grenoble, France, p 81
15. Moniz AB (2013) Organizational concepts and interaction between humans and robots in industrial environments, p 6
16. Gleeson B, MacLean K, Haddadi A, Croft E, Alcazar J (2013) Gestures for industry Intuitive human-robot communication from human observation. 2013 8th ACM/IEEE international conference on human-robot interaction (HRI). IEEE, Tokyo, Japan, pp 349–356
17. Neto P, Norberto Pires J, Paulo Moreira A (2010) High-level programming and control for industrial robotics: using a hand-held accelerometer-based input device for gesture and posture recognition. Industr Robot 37:137–147. https://doi.org/10.1108/01439911011018911
18. Bodiroža S, Stern HI, Edan Y (2012) Dynamic gesture vocabulary design for intuitive human-robot dialog. In: Proceedings of the seventh annual ACM/IEEE international conference on Human-Robot Interaction—HRI'12. ACM Press, Boston, Massachusetts, USA, p 111
19. Neto P, Pereira D, Pires JN, Moreira AP (2013) Real-time and continuous hand gesture spotting: an approach based on artificial neural networks. 2013 IEEE international conference on robotics and automation. IEEE, Karlsruhe, Germany, pp 178–183
20. Makris S, Tsarouchi P, Matthaiakis A-S, Athanasatos A, Chatzigeorgiou X, Stefos M, Giavridis K, Aivaliotis S (2017) Dual arm robot in cooperation with humans for flexible assembly. CIRP Ann 66:13–16. https://doi.org/10.1016/j.cirp.2017.04.097
21. Tsarouchi P, Matthaiakis A-S, Makris S, Chryssolouris G (2017) On a human-robot collaboration in an assembly cell. Int J Comput Integr Manuf 30:580–589. https://doi.org/10.1080/0951192X.2016.1187297
22. Tsarouchi P, Makris S, Chryssolouris G (2016b) On a human and dual-arm robot task planning method. Proc CIRP 57:551–555. https://doi.org/10.1016/j.procir.2016.11.095
23. Michalos G, Sipsas P, Makris S, Chryssolouris G (2016) Decision making logic for flexible assembly lines reconfiguration. Robot Comput-Integr Manuf 37:233–250. https://doi.org/10.1016/j.rcim.2015.04.006
24. Michalos G, Kousi N, Makris S, Chryssolouris G (2016) Performance assessment of production systems with mobile robots. In: Procedia CIRP. Naples, pp 195–200
25. Tsarouchi P, Makris S, Michalos G, Stefos M, Fourtakas K, Kaltsoukalas K, Kontrovrakis D, Chryssolouris G (2014) Robotized assembly process using dual arm robot. Proc CIRP 23:47–52. https://doi.org/10.1016/j.procir.2014.10.078

Chapter 21
Sensoreless Detection of Robot Collision with Humans

21.1 Introduction

Modern manufacturing systems are characterized by characteristics such as flexibility, reusability, agility and reconfigurability [1–3]. Robots are powerful assets in production lines due to their ability to be adapted for supporting in a number of tasks such as machining tasks [4], assembly activities and recently, human robot cooperative activities [5, 12]. Nowadays, human robot collaborative applications necessitates the development of safety technologies for human robot interaction (HRI) (Fig. 21.1). The human intelligence and skills should be combined with robots' velocity, strength and repeatability under a safe working environment [6]. The ability of robot to be reallocated in the working space (mobile robot), requires robots to be more safe for humans working in the same environment [11]. A variety of control, safety and operator support techniques are required to ensure the human safety [7, 8]. In the terms of human-robot collaboration (HRC), human operators and robots can be work together in a common workspace if they are equipped with safety elements according the pre-defined standards.

30 EU Directives and around 600 standards related to safety are already be created. Indicative examples of them are presented in Table 21.1.

A number of researchers have investigated the topic of potential collision detection among a robotic arm inside its dynamic environment. Power and Force limiting scope is two-fold; the prediction of collision (proactive stage) and the detection of the collision (reactive stage).

Nowadays, research focuses mainly in the section of sensitivity of the collision detection aiming to detect very light collisions for preventing possible failures or accidents inside robot's envelope [10].

This research presents an approach for online collision detection for human-robot contact able to pause the robot's motion before a certain amount of force is applied by the robot. The novelty arises from the online detection of unexpected collisions in very short time based on the physics-based simulation models. A control

S. Makris, *Cooperating Robots for Flexible Manufacturing*,
Springer Series in Advanced Manufacturing,
https://doi.org/10.1007/978-3-030-51591-1_21

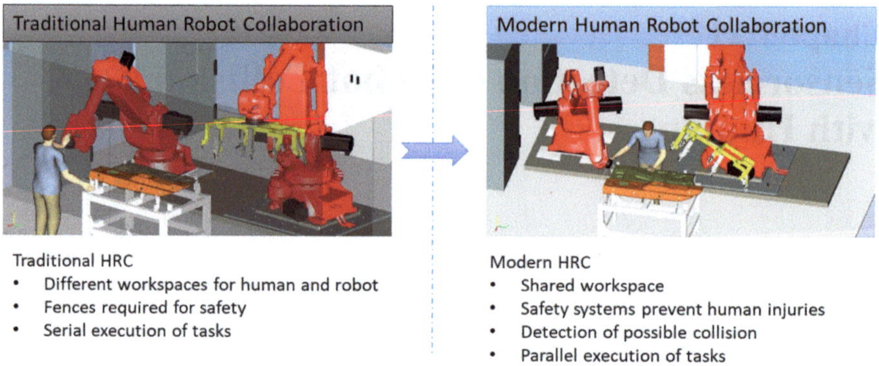

Fig. 21.1 Traditional and modern human robot collaboration

Table 21.1 EU Directives

Title	Description
2006/42/EC	Machinery Directive (MD)
2009/104/EC	Use of Work Equipment Directive
89/654/EC	Workplace Directive
2001/95/EC	Product Safety Directive
2006/95/EC	Low Voltage Directive (LVD)
2004/108 EC	Electromagnetic Compatibility Directive (EMC)

function that uses robot's signals gathered by controlled without the need for extra hardware. Supervised learning Neural Networks and signal filtering techniques are used to improve the sensitivity of the system and to reduce the computational time. The discussed approach hereinafter called Power and Force Limiting (PFL), can be implemented not only on lightweight robot arms but also on industrial robots.

21.2 Approach

The objective of this work is the development of a safety system able to prevent the human from possible accidents caused by unexpected and undefined motion of an industrial robot during their collaboration. Power and Force Limiting bases on the detection of unexpected external forces which applied to the structure of a robotic arm. The external forces could be applied by two main reasons. The first one, which is based on the third Newton's law of motion, includes all the cases that a robot comes in collision with a body (e.g. human or other parts). In these cases, the robot exerts a force on the body and the body simultaneously exerts a force equal in magnitude and opposite in direction on the robot. The second reason is the directly applying of an external force to the robot's structure by a human. In both cases, the industrial robot's control loop recognizes the external forces due to the position error between the calculated target position of a robot's motor and the real one which gathered by the encoder of each motor. Taking under consideration that the robot's control loop is closed, it tries to eliminate the position error by sending more current to the motors in order to reach to their target positions. Accordingly, the detection of deviations between the nominal current signal of each robot's motor and the corresponding real one can be translated in a detection of applied external forces to the structure of the robot. As nominal is characterized the current signal which applied to a robot's motor in order to execute an ideal motion without collisions or contacts. In contrast, the real current signal is the motor's current signal which measured during the execution of a robot's industrial tasks. Of course, the nominal and the real motor signals are very close and possible deviation arise only in the case of external forces applied.

During the execution of a robot motion along a certain path, current signals are applied to the robot's motors by the controller in order to actuate the corresponding links. For example, in Fig. 21.2a, the motion of the robot from a start point to a target point requires the rotation of robot axis 1 which is realized by applying an indicative current signal as shown Fig. 21.2b. Figure 21.2c depicts an unexpected contact point with a human during the robot's motion. In the diagram of Fig. 21.2d, the blue line

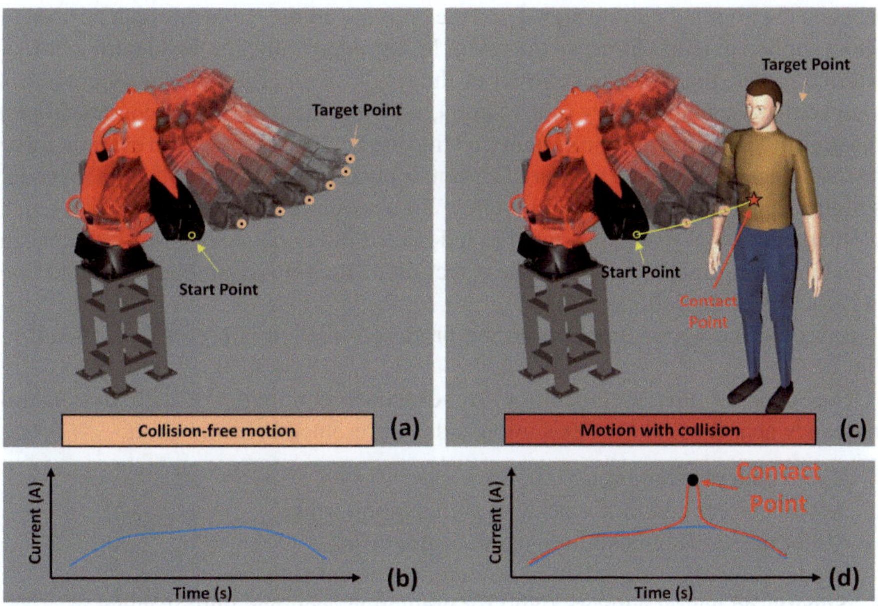

Fig. 21.2 Collision detection approach

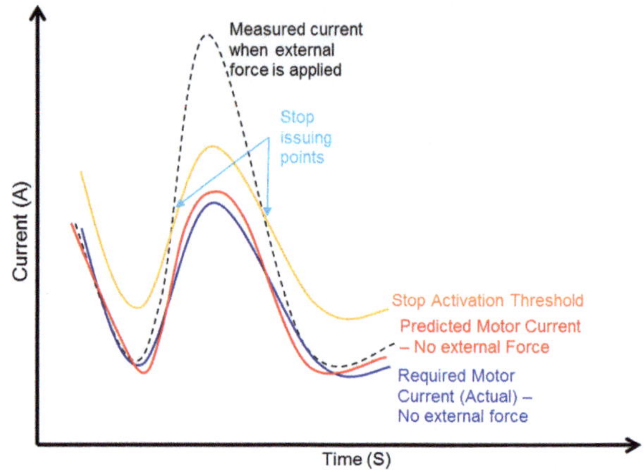

Fig. 21.3 Excessive force detection strategy

corresponds to the nominal current signal which will be applied (and thus measured at the motor input) if no collision event occurred (same as Fig. 21.2b), while the red line diagram is the actual current signal which is monitored during the robot motion. The peak of the current in Fig. 21.2d is generated by the robot's controller which tries to eliminate the position error by sending more current to the motor in order to reach the target position. Accordingly, the detection of such deviations between the nominal current signal and the corresponding actual one, can be used to detect applied external forces to the robot. More specifically, for the identification of external forces, the measured current of the robot motors (blue line in Fig. 21.3) is compared at the end of every time interval, with a computed current coming from a dynamic model of the robotic system (red line in Fig. 21.3). If due to an external force the difference between the measured current (dotted black line) and that of the model is higher than a specific threshold (yellow line) a simple post-impact reaction (such as emergency stop), is implemented as it depicted in Fig. 21.3. This work extents an already existing methodology in the direction of the online simulation, prediction and reaction of the robotic system.

Based on the above description, the proposed method is presented in detail in Fig. 21.4.

The steps of this methodology are executed continuously with a standard frequency near the real time execution and are presented below in short:

1. Robot path programming.
2. Actual current and motor position signal monitoring.
3. Motor position signals filtering and restructuring.
4. Model simulation—prediction of nominal current signals.
5. Supervised neural network (NN) for prediction accuracy improvement.
6. Signal processing to cut off noise.

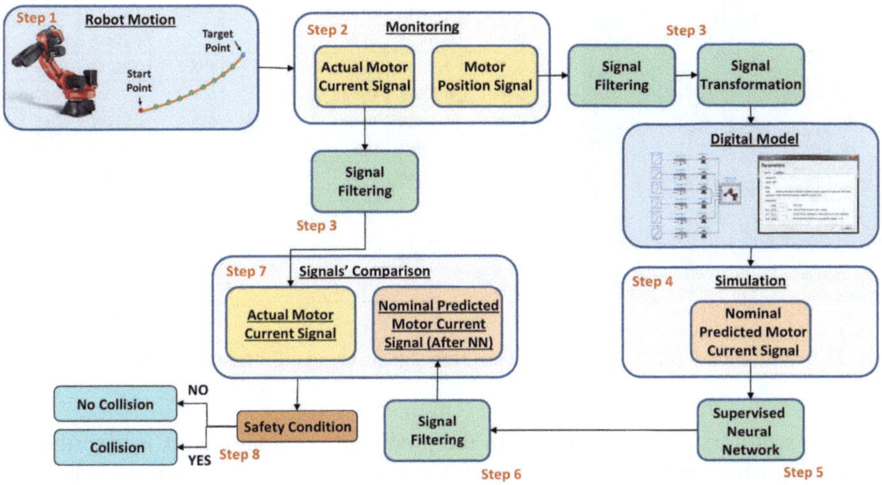

Fig. 21.4 Overall concept

7. Signal correlation.
8. Collision detection check.

In addition to the aforementioned steps which are executed online, another set of offline preparatory activities are required before the method can be applied:

1. Inverse dynamic digital model creation.
2. Neural Network training.
3. Signal processing—filter creation.

All the activities involved in the preparatory phase as well as in the online execution phase are presented and described below in detail.

21.2.1 Task 1: Monitoring and Filtering of Robot's Current and Position Signals

Initially, the monitoring of the needed data for the implementation of the presented approach takes place. These data gathered from the controller of the robot online in real time. An intelligent communication algorithm allows the monitoring and gathering of robot's current and position signals each 2 ms. This algorithm runs continuously and saves the monitored data in a local database. The monitored position signals are used as input for the prediction of the nominal motors' current signals as it described in Task 2. The current signals are represent the real time current signals

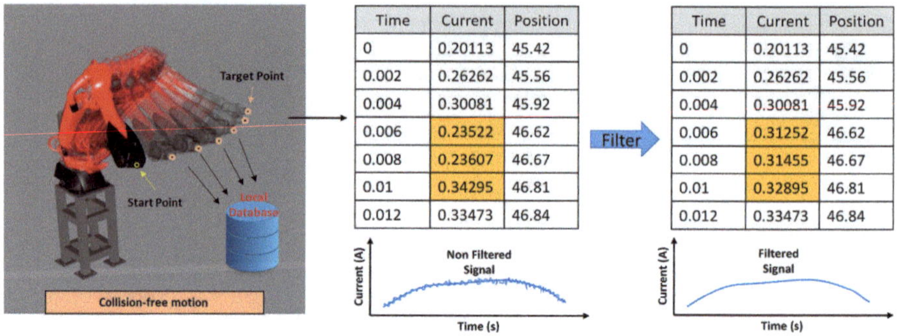

Time	Current	Position
0	0.20113	45.42
0.002	0.26262	45.56
0.004	0.30081	45.92
0.006	0.23522	46.62
0.008	0.23607	46.67
0.01	0.34295	46.81
0.012	0.33473	46.84

Time	Current	Position
0	0.20113	45.42
0.002	0.26262	45.56
0.004	0.30081	45.92
0.006	0.31252	46.62
0.008	0.31455	46.67
0.01	0.32895	46.81
0.012	0.33473	46.84

Fig. 21.5 Filtering of robot's current and position signals using a Butterworth filter

of robot's motors and compared with the predicted nominal one as it described in Task 3. Last but not least, a Butterworth filter is employed to cut off possible noise of the signals which presented during monitoring of them (Fig. 21.5).

21.2.2 Task 2: Prediction of Nominal Industrial Robot's Current Signals

In order to detect the deviations between the nominal motor's current signals of the robot and the corresponding real one, the nominal motor's current signals should be obtained. A simple solution is to execute a robot's path initially (predefined motion of the robot without external forces applied) and to monitor the current signals of each motor. Eventually, they can be saved in a local database and to be used as nominal signals for the detection of possible deviations each time the robot executes the same path. This maybe is a simple and quickly solution but it is limited in only already defined paths of the robot. Due to the need of the increasing of flexibility in the new production lines, the above method is not efficiency. The only way to obtain the nominal current signals in real time and for not predefined robot's motion is to predict of them. Therefore, nominal current signals are the nominal current signals which obtained as it described above while predicted nominal current signals are the nominal current signals which be obtained using the following predictive procedure. Both signals are almost equal in a non-collision motion of the robot. Although, each of them is used in different phases of PFL.

Table 21.2 List of components and sub-components

Component	Sub-component	Function
Mechanical structure	Link	Represent the rigid bodies of a robot
	Joint	Represent the rotation of a rigid body around one axis
	Mass	Represent the tool of the robot at its TCP
	World frame	Represent the base frame of the robot
Gearbox	Ideal gear	Represent the gear ratio of the axis
	Bearing friction	Represent the Coulomb friction of the axis
	Inertia	Represent the mass and the inertia tensors of the gears

21.2.3 Task 3: Inverse Dynamic Modelling

Firstly, an inverse rigid-body dynamic simulation model of the robot is created using an appropriate modelling software and language. The simulation of inverse rigid-body dynamic robot model allows the estimation of the forces and torques based on the kinematic and structural/inertia properties of a robot. Every component is characterized by a number of parameters which correspond to a specific robot module. The components are analysed as following:

- *Mechanical Structure* represents the kinematic model of the robot and the mass properties definition.
- *Gearbox* represents the robot's gearbox.
- *Actuator* has the role of robot' motor.
- *Input signals* is used to provide the monitored position signals to the digital model.

The Mechanical Structure and Gearbox components consist of several subcomponents which are presented in Table 21.2.

21.2.4 Task 4: Simulation and Prediction of Nominal Current Signals

The *Input signals* component will be used to import the monitored position signals for the model simulation. The monitored data are converted in order to be used as input in the simulation software. A variety of linear and non-linear solvers are available to solve multi-equations systems like this one. Parameters such as the simulation interval, the tolerance of the solver and others are defined in the properties of the simulation software. The simulation output is the predicted nominal torque signals of the robot's motors. Assuming a linear relation among the torque and current of a motor, the following expression allows the calculation of the nominal current signals (Fig. 21.6).

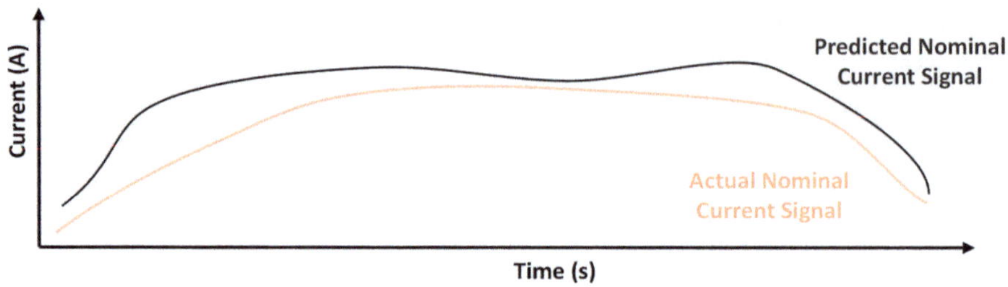

Fig. 21.6 Diagram of actual and predicted nominal current signals

$$I = \frac{T_m}{K_T} \tag{21.1}$$

where,

I is motor current,
T_m is motor torque and,
K_T is Motor Torque constant and is known for each motor.

21.2.4.1 Supervised Learning for Final Adjustments and Prediction

As it mentioned above, the inverse dynamic simulation model contains a number of assumptions for two reasons. The first reason is that some of the robot's component have been modelled as grey and some as black boxes due to their unknown parameters. According to this, the following assumptions take place:

- At the component of Mechanical Structure, the elasticity of the joints and the links are thought to be negligible.
- At the component of Gearbox, the transmission's inertia as well as the effect of backlash are thought to be negligible.
- At the whole inverse dynamic model, the effect of lost motion is thought to be negligible.

The second reason is the assumption of the linear behavior of the model in order to manage the modelling of it as well as to reduce the computational time of the simulation. Accordingly, a huge number of uncertainties have been created especially duo to the non-linear behavior of some phenomena which be modelled above (e.g. the phenomenon of friction and others which is relative at robot's dynamic features).

Supervised learning are used to compensate the non-linearity of the simulation models aiming to improve the robot's nominal current signals prediction. The survived learning algorithm which suits more to this approach is a nonlinear regression one for continues-response values [9] (Fig. 21.7).

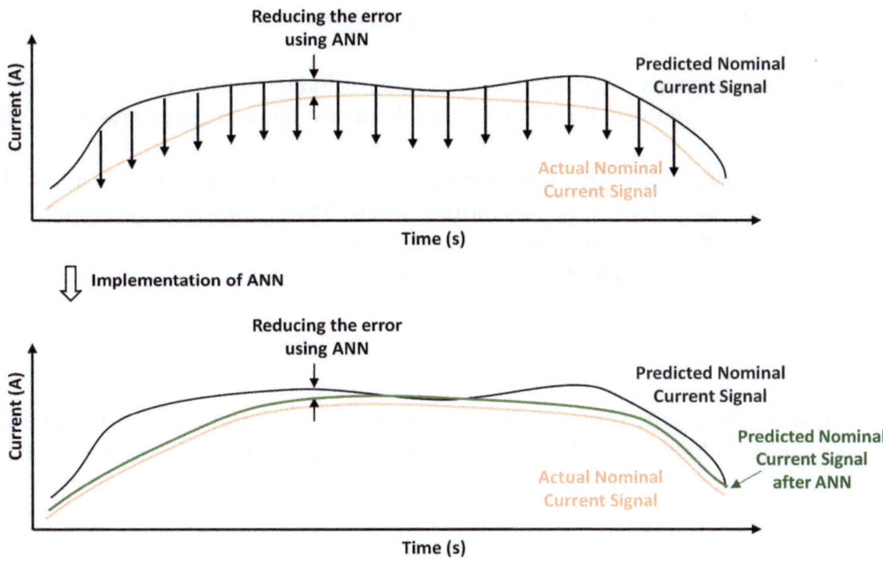

Fig. 21.7 NN for nominal current signal prediction

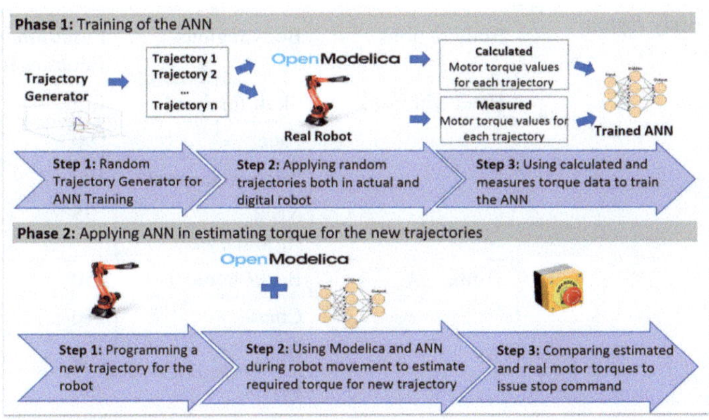

Fig. 21.8 Power and force limiting overview

The proposed scheme for the identification of uncertainties is shown in Fig. 21.8.

21.2.4.2 Safety Concept—Conditions and Actions

Based on the European directives in the section of Safety and Health, the Biomechanical limits to minimize the potential for a human's injury are presented in Table 21.3.

Based on the transfer energy equation which is analyzed in the maximum allowed velocity of the manipulator in collaborative cells is 0.25 m/s. Based on the aforementioned standards the maximum allowed contact force value is defined to 60 N. The velocity should also be under 0.25 m/s, which is determined as a velocity threshold for safe human–robot collaboration (Fig. 21.9).

The actions which take place after a violation of the safety condition are the following:

- Actuators power off: It means that the current is stopped to be provided to the robot and eventually, the robot system stays a stable configuration using its brakes.
- Robot motion paused: It means that the robot is still enabled for motion, but the initial programmed motion will be paused.
- Robot manual guidance: It means tha the operator drives the robot using a passive motion mechanism (Fig. 21.10).

Table 21.3 Body region forces' values

Body regions	Body regions	Limit/minimum values of forces (N)
Head with neck	Skull/forehead	175
	Face	90
	Neck (sides/neck)	190
	Neck (front/larynx)	35
Trunk	Back/shoulders	250
	Chest	210
	Belly	160
	Pelvis	250
	Buttocks	250
Upper extremities	Upper arm/elbow joint	190
	Lower arm/hand joint	220
	Hand/finger	180
Lower extremities	Thigh/knee	250
	Lower leg	170
	Feet/toes/joint	160

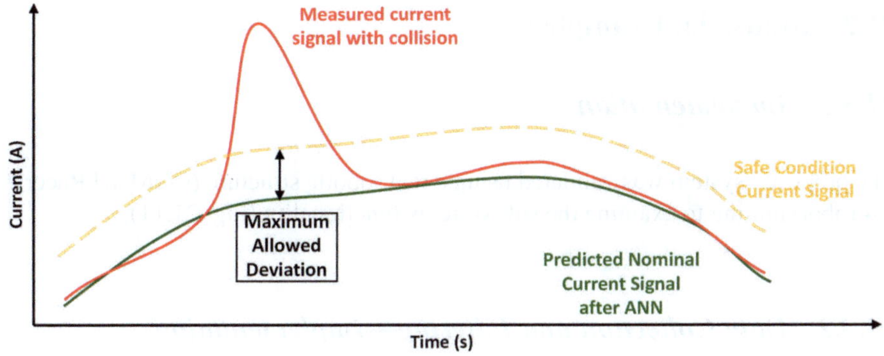

Fig. 21.9 Safety condition—max. allowed deviation

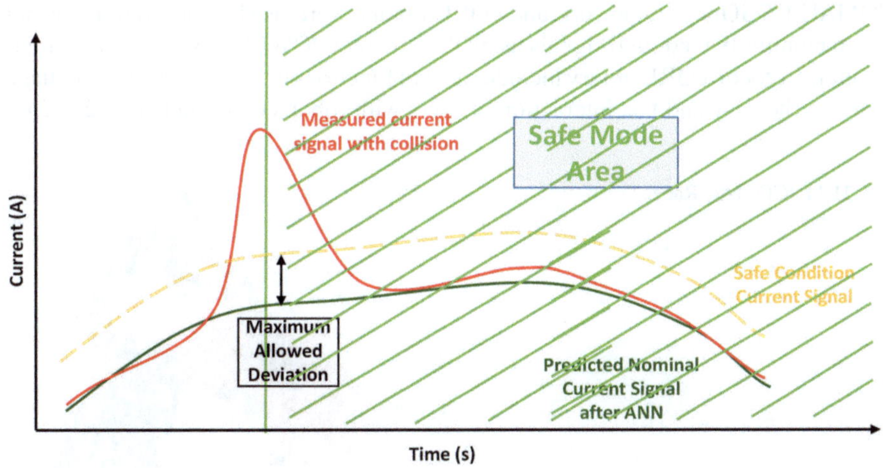

Fig. 21.10 Safe mode area

To sum up, the aforementioned procedure presents a potential system to predict the collision of a robotic arm within its workspace and more specifically, with human operators. PFL safety system is capable to be utilized kind of the robot included the industrial robots.

21.3 Industrial Example

21.3.1 Implementation

The proposed system was validated using a real robotic structure (COMAU Racer 7 1.4 robot) aiming to examine the sub-systems functionality (Fig. 21.11).

21.3.2 Data Collection and Filtering—Implementation

The objective of this Implementation Phase is to monitor and gather the required position and current signals of robot's controller as well as the filtering of them. COMAU C5GOpen framework and eORL library were used to program the robot and to monitor the required signals using the Ethernet POWERLINK (EPL) communication protocol. eORL library includes a set of functions implemented at C language enabling the easy and fast interaction of users with robot's controller (Fig. 21.12).

Fig. 21.11 COMAU Racer 7 1.4 robot

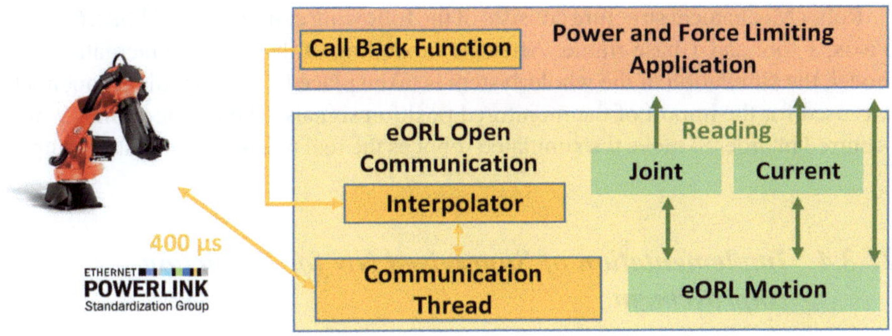

Fig. 21.12 Data gathering architecture

21.3.3 Robot Modelling and Simulation

OpenModelica was used for the creation of COMAU Racer 7 1.4 robot inverse dynamic model, an object-oriented language that is used for component-oriented modelling (Fig. 21.13).

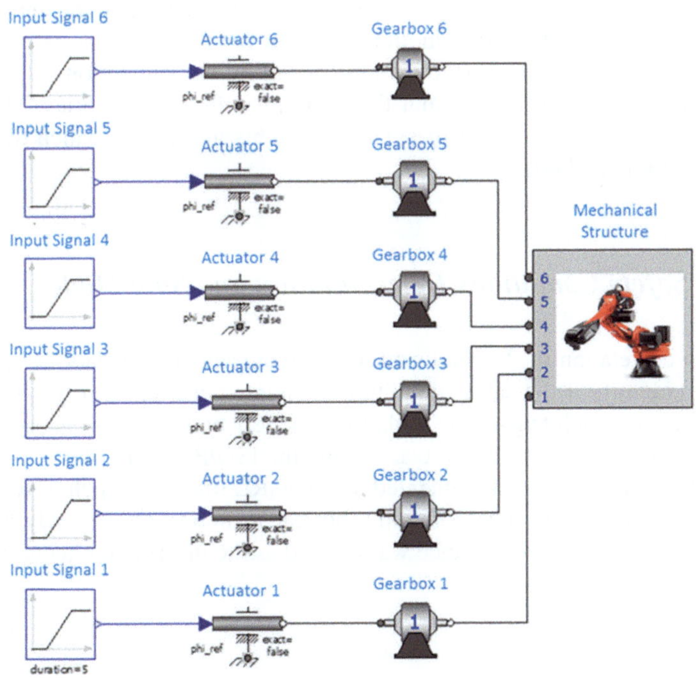

Fig. 21.13 Modelling of the robotic structure

Robot Mechanical structure consists of the following components; 7 links, 6 joints, 6 axis, 1 tool and 1 base frame. After the modelling of the inverse kinematic robot model, the simulation of the whole system is taken placed. Input signals components was used for the import of the monitored position signals. The torque signals which occurred in order to move the simulated robot as the real one is the simulation output.

21.3.4 Implementation of Supervised NN for Prediction Improvement

Due to the uncertainties and complex phenomena which can not be modelled, deviations were occurred between the predicted nominal current signals and the real one. To compensate these deviation, supervised feedforward input-delay neural networks are created and trained in MATLAB software. Two neural networks are used for the motion of each robot's axis, one for the positive motion and one for the negative motion. The training dataset consists of predicted nominal current signals (input) as well as the corresponding real one (output) for a number of predefined trajectories. These trajectories have been designed in a way to lead the robot to critical points during its motion e.g. singularities, joints' limits and other dynamic phenomena.

COMAU robotics function which allows the user to monitor all the motor signals during the execution of a robot motion are used for the real data gathering. The predicted nominal current signals were estimated by the simulation of the inverse dynamic robot model using as input the same predefined trajectories. The output of the training phase was a set of synaptic weights and biases which arise using backpropagation methods.

21.3.5 Safety Conditions Determination and Safe Mode

The online correlation of the predicted nominal current signals and the real one is perfume to identify if the safety condition is violated or not. As it was mentioned above, the error which trigger the violation of safety condition has been defined at 0.6 A or approx. 60 N contact force according the ISO/PDTS 15,066.

Safe Mode of the robot implemented using eORL function which have provided by COMAU. The motors power off and robot pause motion have already defined as function while the manual guidance was created as the inverse architecture of PFL.

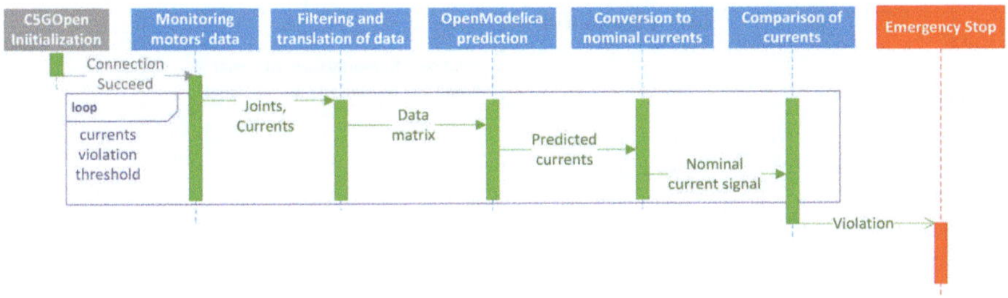

Fig. 21.14 Software sequence diagram

Table 21.4 Duration of each activity

Operation phase	Duration in each cycle (ms)
Data transition and filtering—implementation	11.7
Robot modelling and simulation	102.6
Implementation of supervised NN for prediction improvement	24.7
Safety conditions determination and safe mode	10.2
Total	**149.2**

21.3.6 Integration and Communication Architecture

All the above activities are integrated in a common framework aiming to interact each other is the minimum time (Fig. 21.14).

The time cycle of the whole procedure is about 150 ms, including the monitoring, the filtering, the simulation and prediction as well as the comparing phases (Table 21.4).

21.3.7 Experiments and Results

A simply industrial pilot case including a COMAU Racer 7 1.4 robot was selected for the execution of the experiments. The predicted nominal current and real-time monitored current signals for robot Axis 1, Axis 2 and Axis 3 are depicted below. In the same way, the diagrams of the Axis 4, Axis 5 and Axis 6 are arise (Figs. 21.15, 21.16, 21.17, 21.18, 21.19 and 21.20).

The successful rate of the presented safety system depends mainly in the accurate prediction of the nominal current signals (Figs. 21.21 and 21.22). As the safety conditions have been defined in a standard limit, possible inaccurate predictions of the current signals may lead to deviations higher of the safety condition without any external force applied. In this case, the safety system blocks the motion of robot

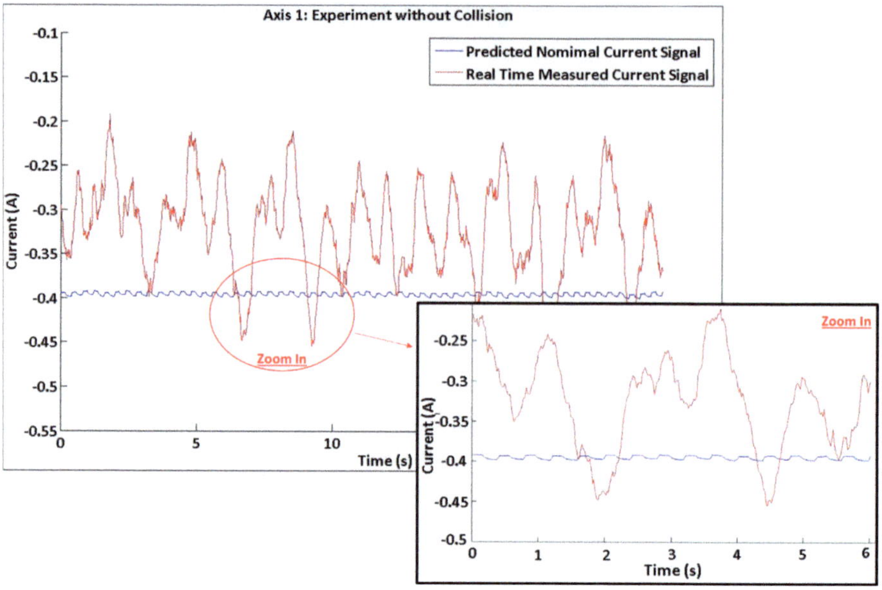

Fig. 21.15 Axis 1—Motion without collision

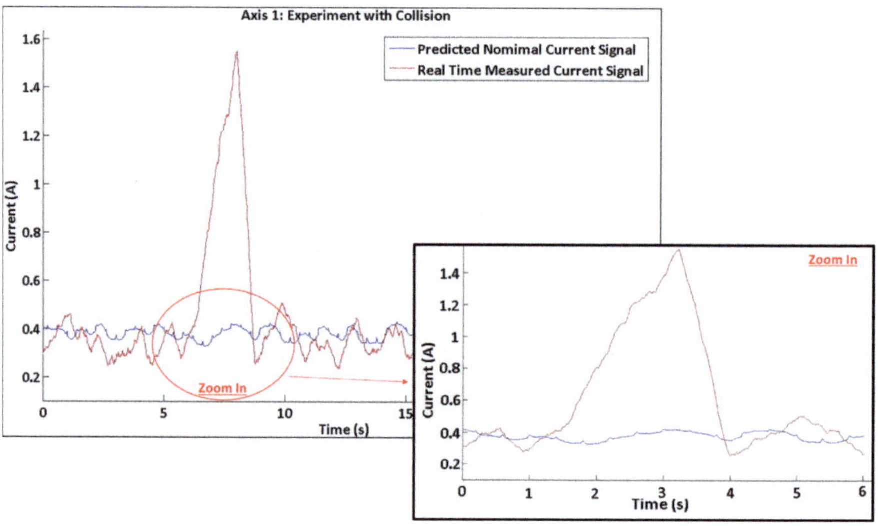

Fig. 21.16 Axis 1—Motion with collision

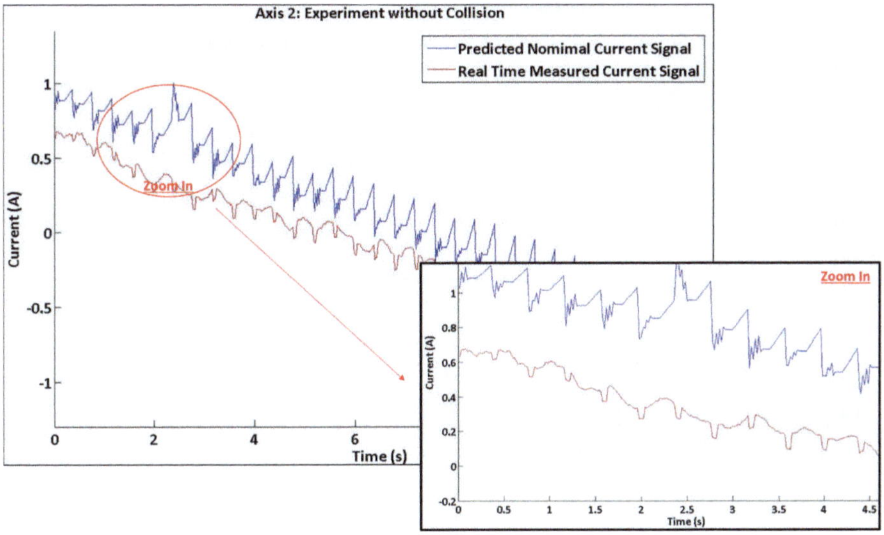

Fig. 21.17 Axis 2—Motion without collision

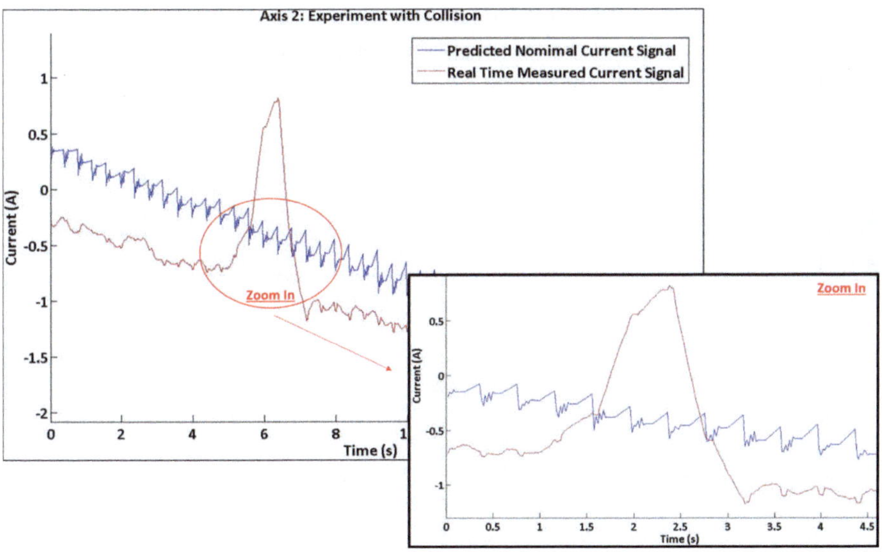

Fig. 21.18 Axis 2—Motion with collision

as the safe mode is activated. The frequency of this phenomenon is too low and it occurs usually in critical robot trajectories. As critical trajectories they mentioned to trajectories which contains singularities or motion in the limits of robot's envelope.

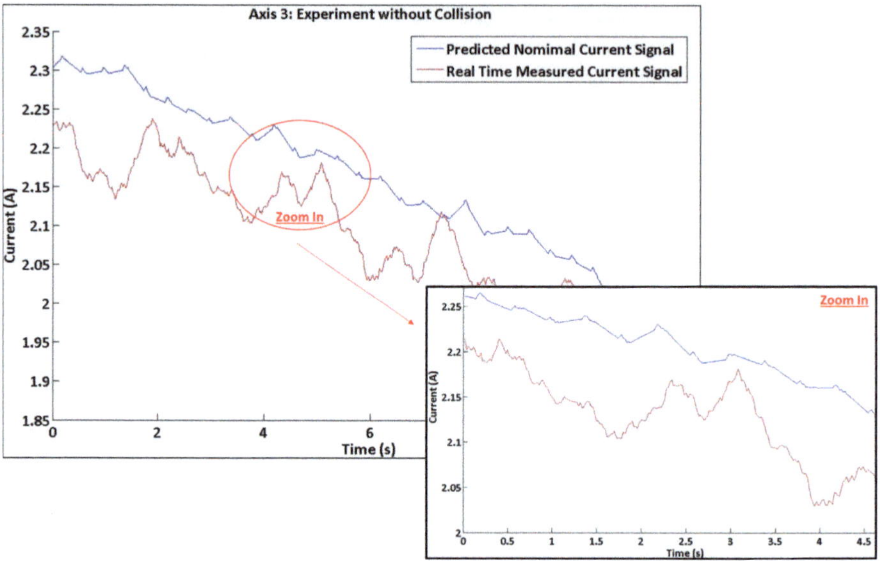

Fig. 21.19 Axis 3—Motion without collision

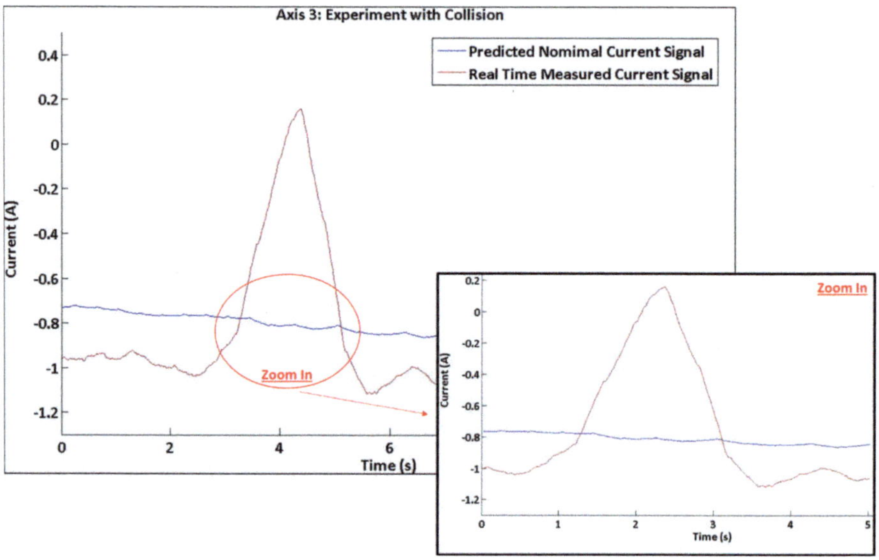

Fig. 21.20 Axis 3—Motion with collision

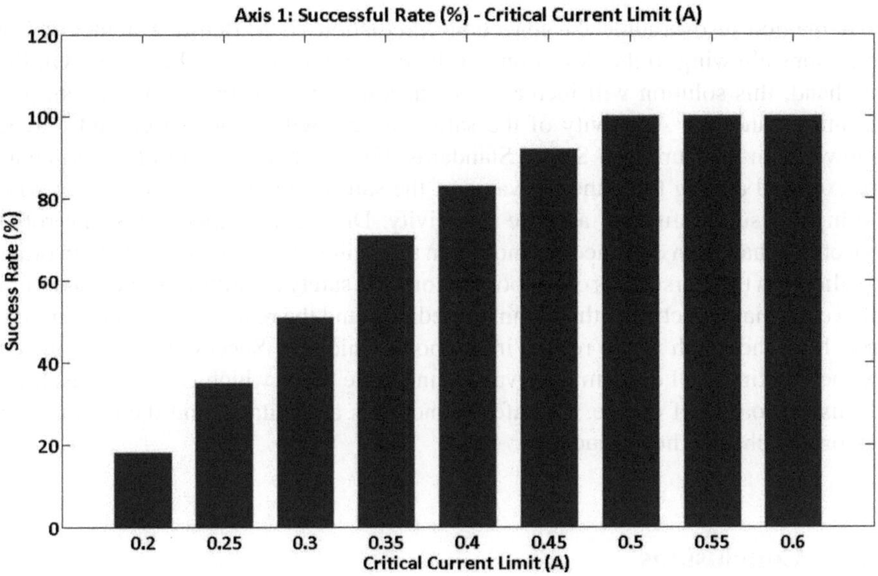

Fig. 21.21 Axis 1—Diagram of successful rate and critical current limits

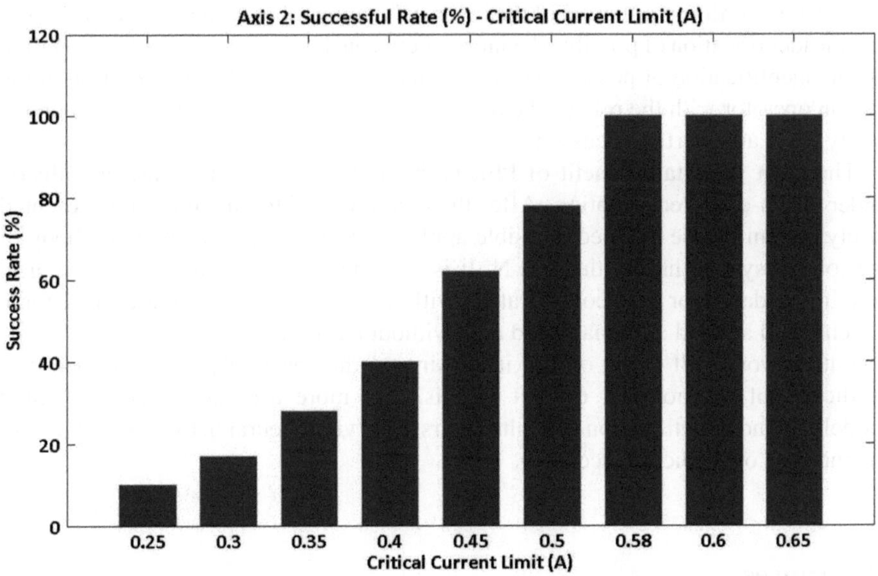

Fig. 21.22 Axis 2—Diagram of successful rate and critical current limits

A method to overcome the inaccurate predictions is to define a higher safety conditions allowing to the deviations of the compared signals to be higher. On the one hand, this solution will increase the successful rate of the safety system. On the other hand, the sensitivity of the safety system will be decreased and maybe is lower than the European Safety Standards. For this reason, a set of experiments are executed aiming to define the value of the safety condition of each motor with the highest successful rate and the sensitivity. During the experiments, the robot trajectories had been designed for motion in the limit of the robot's envelope in order to validate in the worst cases of a robot motion. The safety condition, as the maximum allowed deviation between the nominal predicted and the real time measured signal, arise from the graph of the results in the point which the Success Rate maximized for the first time. All of them have values under the 0.6 A which is our highest limit of this approach. Of course, the safety conditions are editable and the responsible person can change their values.

21.4 Conclusions

Power and Force Limiting software bases on the online monitoring of the real robot as well as the online simulation of the inverse dynamic robot model. The comparison and the identification of possible deviations between them leads to useful information for the identification of possible collisions and ensures the safe collaboration of the human operator with the robot. The direct benefits of the discussed approach on the safety HRC are shortly discussed.

The most important benefit of PFL is the high sensitivity to detect a collision under an un-expected situation. After the validation of the method, the presented safety system is able to detect possible applied external forces on the main body of the robotic system higher than 60 N. It is important to mention that according to certain standards for safe collaboration with the human, 60 N is the critical force which could applied at human head area without harm it.

Future work will focus on the increasing of the sensitivity via more accurate prediction of the nominal current signals. The more accurate inverse dynamic modelling and the utilization of multi-layers supervised neural networks will lead to the increase of prediction accuracy.

References

1. Michalos G, Makris S, Papakostas N, Mourtzis D, Chryssolouris G (2010) Automotive assembly technologies review: challenges and outlook for a flexible and adaptive approach. CIRP J Manuf Sci Technol 2:81–91
2. Chryssolouris G (2006) Manufacturing systems: theory and practice, 2nd edn. Springer-Verlag, New York

3. Paralikas J, Fysikopoulos A, Pandremenos J, Chryssolouris G (2011) Product modularity and assembly systems: an automotive case study. CIRP Ann Manuf Technol 60:165–168
4. Zhenyun S, Peijiang Y, Qishen W, Dongdong C, Tianmiao W (2016) New design of a compact aero-robotic drilling end effector: an experimental analysis. Chin J Aeronaut 29:1132–1141
5. Tsarouchi P, Makris S, Chryssolouris G (2016) Human-robot interaction-review and challenges on task planning and programming. Int J Comput Integr Manuf 29:916–931
6. Michalos G, Makris S, Spiliotopoulos J, Misios I, Tsarouchi P, Chryssolouris G (2014) ROBO-PARTNER: seamless human-robot cooperation for intelligent, flexible and safe operations in the assembly factories of the future. Procedia CIRP 23:71–76
7. Michalos G, Makris S, Tsarouchi P, Guasch T, Kontovrakis D, Chryssolouris G (2015) Design considerations for safe human-robot collaborative workplaces. Procedia 37:248–253
8. Michalos G, Karagiannis P, Makris S, Tokçalar Ö, Chryssolouris G (2016) Augmented reality (AR) applications for supporting human-robot interactive cooperation. Procedia CIRP 41:370–375
9. Chenjie G, Nikola K, Zhi L, Jie Y (2017) A spiking neural network model for obstacle avoidance in simulated prosthetic vision. Inf Sci 399:30–42
10. Alami R et al (2006) Safe and dependable physical human-robot interaction in anthropic domains: state of the art and challenges. In: IEEE/RSJ international conference on intelligent robots and systems, pp 1–16
11. Michalos G, Sipsas P, Makris S, Chryssolouris G (2016) Decision making logic for flexible assembly lines reconfiguration. Robot Comput-Integr Manuf 37:233–250
12. Michalos G, Makris S, Spiliotopoulos J, Misios I, Tsarouchi P, Chryssolouris G (2014) ROBO-PARTNER: seamless human-robot cooperation for intelligent. Flexible and safe operations in the assembly factories of the future (CATS 2014). In: 5th CIPR conference on assembly technologies and systems, 13–14 November, Dresden, Germany, pp 71–76

Part IV
Epilogue

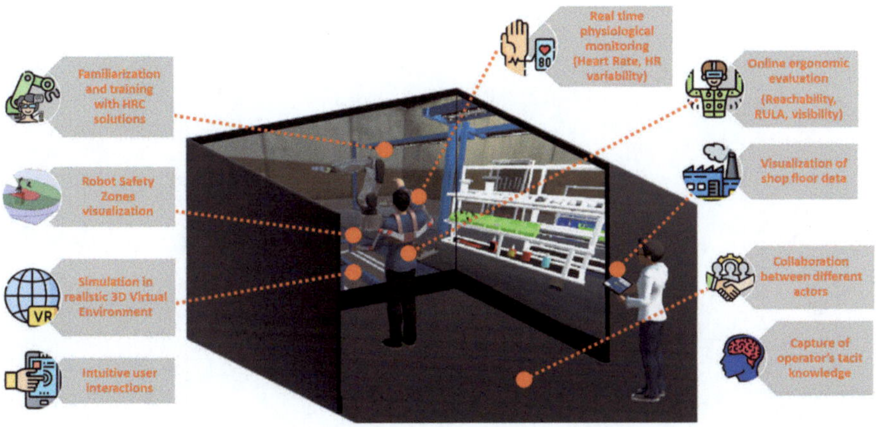

Chapter 22
Epilogue and Outlook

Over the last years both Academia and Industry have pursued the vision of flexible production by introducing technology [1] that allow robots cooperate autonomously [7], humans and robots to coexist and share production tasks [4, 6]. Several real-world examples have been discussed in this book and provided proof that robot can efficiently cooperate among themselves as well as with humans. Humans and robots' future lies in collaboration rather than competition.

It was revealed though that Human Robot Collaborative (HRC) applications present are offering a rather efficient way to advance production. The following drawbacks still need to be addressed but this does not limit the value of the overall concept.

- Diversified needs for robotics by the production tasks. Robot arms are not able to cover all applications scenarios (rigidity, adaptability, payload, working envelope etc.) and soft robotics may seem preferable.
- Low performance of collaborative operations. In the scenarios achieved so far, humans can perform tasks faster when working on their own as are robots when they function independently. This is due to:
  - Complexity of safety systems that create more separation: Resolution, response time and robustness in uncertain conditions cannot be handled by existing safety/sensing/perception and hardware/software.
  - Design aspects of the system as HRC hazards are not systematically evaluated in the design process.
  - Operator acceptance and efficient integration in the workflow: People are not used to work cooperatively with robots as they would with other humans. Workers need to develop familiarity and trust in robots through intuitive interaction and accurate anticipation of robot actions/intentions.
- Not well developed cognition capabilities for robot autonomy: the robot needs to adjust its behavior to shape its operation around the human behavior. There

S. Makris, *Cooperating Robots for Flexible Manufacturing*,
Springer Series in Advanced Manufacturing,
https://doi.org/10.1007/978-3-030-51591-1_22

are not yet well established structured methods to validate implementations of autonomous behavior.

- Programming of the robots still needs to be more user friendly, requiring programming skills knowledge and is detached from the method of interacting with it during execution.

Nevertheless, while Human-robot collaboration is in close reach, the real breakthrough is expected through the advance to robot to robot cooperation. The technology discussed in Part 2 of this book, demonstrated the potential as well as the technology involved. There are numerous forms of cooperating robots, ranging from Dual arms, mobile cooperating roots and so and so forth. Many aspects of synchronizing their actions, reasoning over their decisions, handling dexterously real world parts have been discussed. These topics are further being investigated today by the team in LMS and a few other groups globally. It is deep belief of the author that this technology will be the answer to plenty of challenges, helping to increase productivity, quality and flexibility cost efficiently.

22.1 Emerging Technology

Ongoing research is advancing towards exploiting emerging technology in a number of prioritized areas, such as high payload collaborative robots, dynamically reconfigurable safety monitoring systems and smart HR interfaces allowing the seamless integration of operators and robots in a common workflow. To advance the Human robot collaboration paradigm, Industrial Psychology, Ethical Aspects and Human Factors is being investigated for advancing human acceptance.

In addition, new robotic systems such as exoskeletons have been advancing over the past years offering additional capabilities to manufacturing. Furthermore, AI enabled cognition for autonomous HR collaborative applications is envisaged to enable the robotic resources to perceive their operating environment, reason over the perceived status and learn to adapt their behavior. For this purpose, artificial intelligence algorithms offer the means for of autonomy through multi-level perception for assessing the process and operating environment status by the robot. Unlike traditional robotics applications where the design focuses on performance and it is only sufficient to ensure separation of humans and robots, the new generation of collaborative applications requires higher consideration of the human factors. For this purpose, a series of safety developments can help to cover the whole chain from design to operation of such cells. Software modules able to apply the Risk Assessment and Hazard Identification routines foreseen by the international safety standards, on top of the existing CAx design and simulation tools are emerging. VR/AR tools for the validation of collaborative operations and the reduction of the certification time help for the control systems to digitally represent all safety/interaction functions and allow the users to validate the outcome of different interaction within a totally safe and immersive environment.

Further developments in computing will contribute to the advancement of manufacturing and cooperating robots. A new generation of computing resources such as quantum computers, biotechnology, new networks as well as the emergence of cloud infrastructures are expected to advance the way that cooperating robots are used in the factories to advance their operation [2, 3, 8, 9].

Despite the advances expected in technology, one thing seems to be for sure. Humans remain in the center and they are expected to still be the ones to drive the future of manufacturing with the help of robotics. Human ambition and creativity will be the driving force to exploit our abilities to bring future factories to existence.

22.2 Social and Ethical Aspects

With the widespread technology adoption in industry, human is confronted with fear for the future factory. A factory where robots and machines at large prevail. Machines involved encompasses physical smart robots and machines, intelligent algorithms, fast communication and data sharing. The human feels being small and incapable in such an environment. However, the reality is rather different.

Cooperating robots is the means that can offer additional freedom to the worker than the one available today; such a freedom offers the choice on which tasks to perform. It allows to the robots to take over heavy tasks and it offers the opportunity to perform value added tasks exploiting human creativity. Also, these advanced robotic systems, require for human intervention, programming and intuitiveness. The human is reclaiming his natural position in being the lead in the human-robot relation and mastering the shop floor operation rather than having the role of trying to behave similarly to a machine. Additionally, cooperating robots in factories accompanied by the technology discussed in this book, is the means to bring advanced robotics to the masses of factory workers. Workers have at this stage the means to perform better, and waive the glitches of the past.

When the potential of technology is great, there still needs to mention responsibility and ethical use of it. Workers, production planners, safety engineers, data analysts need to develop a new deal for running the future factory. Keeping the human at the center will ensure his highest performance and will result into building the products of the future.

Since we have the opportunity to apply such a powerful technology and apply it in many factories, with different cultures and maturity of its users, we should be doing so with *Φρόνησις (-Phronesis)* meaning *Prudence* [10–12], respecting Ethics and Human values.

Engineers will have to balance between traditional and new methods and tools. We should not miss the original al human need which is the core value of modern societies. Coordinated cooperation is critical. From massive effort we should aim to coordinated effort. Human being in the center will only enable to exploit any technology to its maximum.

This book intends to highlight the potential and act as an eye opener in some extent. The industry is running out of time. The global population from 700 million people in 1760 reached 3 billion people in 1960, 7 million people in 2011, 7.7 billion in 2020. It will soon reach 20 billion [13]. More people will be claiming the product initially thought to be offered at smaller quantities. Creating future factories, capable of coping with such level of demand, can only be achieved by advancing technology.

Manufacturing reside in the middle of technology and philosophy, with the term philosophy carrying its original meaning of being friend of knowledge and wisdom. It aims to balance advantages and disadvantages of technology and offers a unique framework where people cooperate and is the means to enable people to excel and thrive. And cooperating robots is a strong technology to physically augment the abilities of people.

22.3 The Need for Life-Long Education and the Teaching Factory

The technology discussed throughout the book represents the cutting edge of today's technology for controlling robots in manufacturing. Apparently, bringing this technology to the actual factory requires substantial amount of training. Given that technology development is realized in locations not always located nearby the actual production plants, transferring this knowledge to the industry is not always straightforward.

For this reason the concept of Teaching factory has been discussed and introduced as the means to facilitate this process of lining the factory with a large pool of creators, being product designers, production designers, production engineers, robot designers, control engineers, and so and so forth [1, 2]. The Teaching Factory concept is based on the knowledge triangle notion and its concept has its origins in the medical sciences discipline and specifically, in the paradigm of the teaching hospitals, namely the medical schools, operating in parallel with hospitals. It aims at incorporating the learning and working environment, which realistic and relevant learning experiences arise from.

References

1. Chryssolouris G (2006) Manufacturing systems: theory and practice, 2nd edn. Springer, New York
2. Dotoli M, Fay A, Miśkowicz M, Seatzu C (2017) Advanced control in factory automation: a survey. Int J Prod Res 55:1243–1259. https://doi.org/10.1080/00207543.2016.1173259
3. Krüger J, Wang L, Verl A, Bauernhansl T, Carpanzano E, Makris S, Fleischer J, Reinhart G, Franke J, Pellegrinelli S (2017) Innovative control of assembly systems and lines. CIRP Ann Manuf Technol 66:707–730. https://doi.org/10.1016/j.cirp.2017.05.010

4. Makris S, Karagiannis P, Koukas S, Matthaiakis A-S (2016) Augmented reality system for operator support in human–robot collaborative assembly. CIRP Ann 65:61–64. https://doi.org/10.1016/j.cirp.2016.04.038
5. Mavrikios D, Georgoulias K, Chryssolouris G (2018) The teaching factory paradigm: developments and outlook. Procedia Manuf 23:1–6. https://doi.org/10.1016/j.promfg.2018.04.029
6. Michalos G, Kousi N, Karagiannis P, Gkournelos C, Dimoulas K, Koukas S, Mparis K, Papavasileiou A, Makris S (2018) Seamless human robot collaborative assembly—an automotive case study. Mechatronics 55:194–211. https://doi.org/10.1016/j.mechatronics.2018.08.006
7. Michalos G, Makris S, Chryssolouris G (2015) The new assembly system paradigm. Int J Comput Integr Manuf 28:1252–1261. https://doi.org/10.1080/0951192X.2014.964323
8. Wang L, Gao R, Váncza J, Krüger J, Wang XV, Makris S, Chryssolouris G (2019) Symbiotic human-robot collaborative assembly. CIRP Ann 68:701–726. https://doi.org/10.1016/j.cirp.2019.05.002
9. Kurzweil R (2006) The singularity is near: when humans transcend biology. Penguin Books, New York
10. Φρόνηση. Βικιπαίδεια (2017)
11. Prudence. Wikipedia (2020)
12. Nicomachean ethics. Wikipedia (2020)
13. World Population by Year—Worldometer. https://www.worldometers.info/world-population/world-population-by-year/. Accessed 20 Apr 2020

MIX
Papier aus verantwortungsvollen Quellen
Paper from responsible sources
FSC® C105338

If you have any concerns about our products,
you can contact us on
ProductSafety@springernature.com

In case Publisher is established outside the EU,
the EU authorized representative is:
Springer Nature Customer Service Center GmbH
Europaplatz 3, 69115 Heidelberg, Germany

Printed by Libri Plureos GmbH
in Hamburg, Germany